ML
621.2 M289
Majumdar, S. R.
Oil hydraulic systems : principles and

WITHDRAWN

OIL HYDRAULIC SYSTEMS
Principles and Maintenance

S R Majumdar
Formerly Director
Central Staff Training and Research Institute
Kolkata

McGraw-Hill
New York   Chicago   San Francisco   Lisbon   London   Madrid
Mexico City   Milan   New Delhi   San Juan   Seoul
Singapore   Sydney   Toronto

The McGraw·Hill Companies

Cataloging-in-Publication Data is on file with the Library of Congress.

Copyright © 2003, 2001 by The McGraw-Hill Companies, Inc. All rights reserved. Printed in the United States of America. Except as permitted under the United States Copyright Act of 1976, no part of this publication may be reproduced or distributed in any form or by any means, or stored in a data base or retrieval system, without prior written permission of the publisher.

This book was first published by Tata McGraw-Hill in 2001.

2 3 4 5 6 7 8 9 0   BKM/BKM   0 9 8 7 6

ISBN 0-07-140669-7

The sponsoring editor for this book was Kenneth P. McCombs, the editing supervisor was David E. Fogarty, and the production supervisor was Pamela A. Pelton.

This book was printed on recycled, acid-free paper containing a minimum of 50% recycled, de-inked fiber.

McGraw-Hill books are available at special quantity discounts to use as premiums and sales promotions, or for use in corporate training programs. For more information, please write to the Director of Special Sales, Professional Publishing, McGraw-Hill, Two Penn Plaza, New York, NY 10121-2298. Or contact your local bookstore.

> Information contained in this work has been obtained by The McGraw-Hill Companies, Inc. ("McGraw-Hill") from sources believed to be reliable. However, neither McGraw-Hill nor its authors guarantee the accuracy or completeness of any information published herein, and neither McGraw-Hill nor its authors shall be responsible for any errors, omissions, or damages arising out of use of this information. This work is published with the understanding that McGraw-Hill and its authors are supplying information but are not attempting to render engineering or other professional services. If such services are required, the assistance of an appropriate professional should be sought.

*Dedicated
to the memories of
my parents
Narendra and Kiran*

Contents

Preface xi

1. **Basics of Industrial Hydraulics** 1
   1.1 History of Fluid Power *2*
   1.2 Basic System of Hydraulics—Major Advantages and Disadvantages *2*
   1.3 Physical Units for Fluid Power *4*
   1.4 Units of Measurement *6*
   1.5 Principles of Hydraulics: Fluid Power *10*
   1.6 Hydrostatics or Hydrokinetics *16*
   1.7 Fluid Energy in Motion *16*
   1.8 Flow Velocity *17*
   1.9 Laminar and Turbulent Flow *19*
   Review Questions *19*
   References *19*

2. **Hydraulic Oils and Fluid Properties** 20
   2.1 Types of Hydraulic Fluids and Selection Criteria *20*
   2.2 Properties of Hydraulic Fluids *21*
   2.3 Physical Characteristics *23*
   2.4 Reynolds Number *33*
   2.5 ISO Viscosity Grade *35*
   2.6 Influence of Temperature on Viscosity *37*
   2.7 Petroleum Oil as Hydraulic Fluid *39*
   2.8 High-water Based Fluid—History of Evolution *40*
   2.9 Fluid Preparation *42*
   2.10 Common Fire Resistant Fluids *45*
   2.11 Maintenance of Hydraulic Oils *47*
   2.12 Bio-degradable Oils *50*
   Review Questions *52*
   References *53*

3. **Filters and Filtration** 54
   3.1 Nature and Effect of Contamination *55*
   3.2 Sources of Contamination *56*
   3.3 Effect of Dirt on Individual Hydraulic Components *58*

vi Contents

- 3.4 System Failure *60*
- 3.5 Contamination Levels and Standardization *62*
- 3.6 Filter Rating *66*
- 3.7 Filter Terminology *68*
- 3.8 Filter Design *69*
- 3.9 Environmental Factor *72*
- 3.10 Types of Filter *73*
- 3.11 Filter Construction *77*
- 3.12 Location of Filter *81*
- 3.13 Physical Measurement of Contaminant Size and Fluid Analysis *82*
- 3.14 Magnetic Filter *85*
- 3.15 Optimum Filtration *85*
- 3.16 Automatic Particle Counter and its Performance Characteristics *86*
- 3.17 Conclusion *87*
  - *Review Questions* *87*
  - *References* *88*

4. Hydraulic Pumps— Construction, Sizing and Selection 89

- 4.1 Pump Classification *89*
- 4.2 Gear Pumps *93*
- 4.3 Internal Gear Pump *100*
- 4.4 Gerotor Pump *101*
- 4.5 Screw Pumps *102*
- 4.6 Vane Pumps *104*
- 4.7 Piston Pumps *113*
- 4.8 Selecting and Sizing of Hydraulic Pumps *127*
- 4.9 Oil-Compatibility *132*
- 4.10 Size *134*
- 4.11 Noise *135*
- 4.12 Pump Ripple *138*
- 4.13 Checklist *141*
  - *Review Questions* *141*
  - *References* *142*

5. Direction Control Valves 143

- 5.1 What to Control? *143*
- 5.2 Direction Control Valve *144*
- 5.3 Operating Methods and Parameters *146*
- 5.4 Control Technique *150*
- 5.5 Center Conditions of Spool Valves *154*
- 5.6 Balancing Groove on Spool Land *158*
- 5.7 Overlap in Sliding Spool Valve *159*

5.8 Spool Positioning, Back Pressure and Force Acting on Spool *164*
5.9 Few Points to Note About Valves *169*
5.10 DC Valve Specification *170*
  Review Questions *171*
  References *171*

6. Flow and Pressure Control — 172

6.1 Non-return Valves *172*
6.2 Pressure Control Valves *175*
6.3 Sequence Valve *179*
6.4 Counter Balance Valve *180*
6.5 Pressure Reducing Valve *182*
6.6 Unloading Valve *183*
6.7 Speed Control Device *183*
6.8 Principle of Flow Control *184*
6.9 Calculation of Flow Quantity through an Orifice *196*
6.10 Positioning of a Flow Control Valve *196*
6.11 Proportional Pressure and Flow Control Valves *197*
  Review Questions *198*
  References *199*

7. Hydraulic Servo Technique—Recent Trends — 200

7.1 Function of a Hydraulic System *201*
7.2 Mechanical Feedback and Application of Tracer Valve *204*
7.3 Feedback in the System *206*
7.4 Electro-hydraulic Servo System *207*
7.5 Torque Motor *221*
7.6 Types of Servo Valves *223*
7.7 Special Servo Valve Features *235*
7.8 Terms used in Servo Technology *238*
  Review Questions *241*
  References *241*

8. Linear Actuators — 242

8.1 Hydraulic Cylinders *243*
8.2 Construction of Cylinders *246*
8.3 Seals in Cylinder *251*
8.4 Cylinder Reliability *252*
8.5 Predicting Wear *255*
8.6 Cylinder Force, Acceleration and Losses *256*
8.7 Calculation of Cylinder Forces *259*
8.8 Flow Velocity *262*
8.9 Cylinder Efficiency *263*

- 8.10 Sizing of Cylinder Tubes 263
- 8.11 Piston Rod Design 266
- 8.12 Mounting Style of Cylinders 269
- 8.13 Cushioning of the Hydraulic Cylinder 275
- 8.14 Maintenance Tips on Cylinder Mounting 278
- 8.15 Hydraulic Cylinders and their Characteristic Application 279
- 8.16 Checklist for Cylinder Design 281
- 8.17 Some Common Cylinder Problems 283
    - *Review Questions* 284
    - *References* 284

9. Rotary Actuators and Hydrostatic Transmission   286

- 9.1 Hydraulic Motors 286
- 9.2 Vane Motors 288
- 9.3 Gear Motors 290
- 9.4 Piston Motors 291
- 9.5 Selection of Hydro-motors 297
- 9.6 Hydraulic or Electrical Motors 299
- 9.7 Hydraulic Motors in Circuits 299
- 9.8 Types of Hydraulic Transmission 301
- 9.9 Pump-motor Combination 302
- 9.10 Open Loop or Closed Loop System 305
- 9.11 Application of Hydrostatic Transmission 308
- 9.12 Hydrostatic Steering 311
- 9.13 Torque Converter 312
    - *Review Questions* 316
    - *References* 317

10. Heat Generation in Hydraulic System   318

- 10.1 Sources of Heat 318
- 10.2 Estimation of Heat Rise 321
- 10.3 Role of Hydraulic Oil Tank in Heat Dissipation 324
- 10.4 Use and Application of Heat Exchangers in Hydraulic Systems 325
- 10.5 Air Cooling of Hydraulic System 326
    - *Review Questions* 332
    - *References* 333

11. Hydraulic Reservoirs and Accumulators   334

- 11.1 Common Types of Reservoirs 335
- 11.2 Reservoir Mounting and Construction 335
- 11.3 Reservoir Shape and Size 338
- 11.4 Reservoir Accessories 339
- 11.5 About Piping 340
- 11.6 Maintenance 340

- 11.7 Integral Reservoirs *341*
- 11.8 Hydraulic Accumulators *341*
- 11.9 Safety Instructions *344*
- 11.10 Properties of Nitrogen *345*
- 11.11 Accumulators in a Circuit *346*
- 11.12 Important Points to Note Regarding Accumulator Selection *349*
- 11.13 Testing of Accumulators *349*
  - *Review Questions 350*
  - *References 350*

12. Design of Hydraulic Circuit — 351

- 12.1 Hydraulic Circuits *352*
- 12.2 Manual or Automatic Hydraulic Systems *352*
- 12.3 Regenerative Circuit *361*
- 12.4 Use of Check Valves in Hydraulic Circuits *364*
- 12.5 Selection of Pump *367*
- 12.6 Standards in Circuit Diagram Representation *368*
- 12.7 Speed Variation in Cylinder Motion *371*
- 12.8 Some Basic Circuits *373*
- 12.9 Functional Diagram *383*
- 12.10 Application of Functional Diagram *388*
- 12.11 Electrical Control of Hydraulic Systems *388*
  - *Review Questions 389*
  - *References 391*

13. Seals and Packings — 392

- 13.1 Classification of Hydraulic Seals *393*
- 13.2 Factors for Seal Selection *397*
- 13.3 Seal Form *410*
- 13.4 Compact Packing *418*
- 13.5 How to Fit an O-ring *419*
- 13.7 Dynamic Seals *422*
- 13.7 Failure of Seal *426*
- 13.8 Seals are Affected by Additives *432*
- 13.9 General Guidelines for Seal Selection and Installation *433*
- 13.10 Faulty Fitting of Seals in Cylinders *434*
- 13.11 Burn Test *434*
  - *Review Questions 436*
  - *References 437*

14. Hydraulic Pipes, Hoses and Fittings — 438

- 14.1 Iron Pipes and Steel Tubes *438*
- 14.2 Pipe Fittings *442*

x   *Contents*

14.3   Energy Loss   *451*
14.4   Estimation of Line Diameter   *455*
14.5   Synthetic Hydraulic Hoses   *457*
14.6   Compatibility of Fire Resistant Oils with the Hose Material   *463*
14.7   Installation of Hoses, Tubes and Pipes   *464*
14.8   Design of End Fittings   *472*
14.9   Quick Coupling   *475*
14.10   Right Size of Hose   *477*
14.11   Hose Selection Criteria   *478*
14.12   Reliability Test for Hoses   *479*
14.13   Guidelines for Pipe and Hose-Maintenance   *481*
14.14   Pipe/Tube Preparation for Installation   *483*
14.15   Use of Teflon Tape   *484*
14.16   Effect of Friction on Pipes/Tubes/Hoses   *484*
14.17   Standard Threads on Tubes   *485*
14.18   Easy Screw-Together Reusable Assembly   *485*
       Review Questions   *487*
       References   *487*

15. Hydraulic System Maintenance, Repair and Reconditioning   489

15.1   Common Faults in a Hydraulic System   *490*
15.2   Procedure for Repair   *492*
15.3   Contamination   *498*
15.4   Component Fittings and Failure due to Contaminants   *500*
15.5   Filter and Filter Maintenance   *504*
15.6   Pump Maintenance   *509*
15.7   Hydraulic System Maintenance   *513*
15.8   Maintenance and Performance Monitoring   *514*
15.9   Estimation of Seal Failure   *518*
15.10   Maintenance of Line-fittings   *520*
15.11   Noise   *523*
15.12   Fault Diagnosis of a Hydraulic System and its Components   *527*
15.13   General Safety Measures for Fluid Power System   *537*
15.14   Inspection Format for Hydraulic Systems   *539*
       Review Questions   *540*
       References   *541*

Index   **542**

Preface

A revolutionary change has taken place in the field of Fluid Power technology due to the integration of electronics as a control medium for hydraulic components and systems. Due to increased sophistication of hydraulics and allied fields of engineering, the hydraulically driven machines are now able to generate more power and higher accuracy in speed, force and position control. This book has been written with the sole objective of bringing the various selection and manufacturing techniques, control, procedure and application of hydraulic components from maintenance point of view to light so that the graduate and undergraduate level student pursuing a career in hydraulics may find it most useful.

This book is meant for the students who want to make a profession in hydraulics and fluid power, either as a system designer, a hydraulic construction mechanic and fabrication engineer or as a maintenance engineer. The various topics dealt in this book are concise and self contained with maximum possible pictorial illustration for easy understanding and clear conception. Efforts have also been made to include the latest trends in the field of hydraulics and allied control areas to keep the reader abreast of the ever changing state of the technology in oil hydraulics.

While writing the book, the author referred a good number of literature and titles written on the subject by renowned authors who have been acknowledged at the end of each chapter. However there might have been some inadvertent omission in the list of names, for which the author takes this opportunity to acknowledge the direct or indirect contribution of all the authors once again in shaping this book.

The author would like to thank Tata McGraw-Hill Publishing Company Limited, New Delhi for showing interest in the book. He is also grateful to Ms Vibha Mahajan, Mr Thomas Mathew Rajesh, for their spontaneous help and assistance in developing and publishing the book. His sincere thanks are also due to Ms Mini Narayanan and the production team of Tata McGraw-Hill Publishing Company Limited.

Last but not the least, I am thankful to my family members for their continued support without which this book would not have been possible. My friends and colleagues who encouraged me constantly to write this book are also acknowledged for their help and assistance.

S R Majumdar

1

Basics of Industrial Hydraulics

IN the recent past there has been a significant increase in the use of hydraulics in our industries. The use of oil hydraulic systems as a means of power transmission in modern machines evolved a few decades earlier in the western world. But its application in Indian industries is of comparatively recent choice and hence, there is a great deal of urgency and importance to master the art of its application and maintenance. Hydraulic systems are now extensively used in machine tools, material handling devices, transport and other mobile equipment, in aviation systems, etc. At the moment there exists a big gap between the availability and requirement of trained manpower in this vital field of modern engineering in India. To bridge the gap, it is essential that our design and application engineers and maintenance personnel from the lowest to the highest level are given extensive, on-the-job training so that the operational efficiency of machineries using a hydraulic system as the prime source of power transmission can be maintained at an optimum level. Apart from the fluid power system designer, a good maintenance and millwright mechanic should also have first-hand theoretical knowledge to enable him to tackle practical problems encountered during installation, operation and maintenance of the hydraulic equipment.

The basic principles that are associated with the science of oil hydraulics are to be explained in a manner so as to inculcate a sense of confidence needed to tackle problems without any ambiguity. The following chapters, it is hoped, will be able to provide the required knowledge and skill to the concerned personnel in this area of engineering with adequate analytical capability to tackle a problem most effectively.

1.1 HISTORY OF FLUID POWER

Since time immemorial, man has recognized and accepted fluids as a source of power. This is quite evident from the fact that in olden days simple machines like the Pelton wheel were developed to transmit irrigation water or water head was used to transmit power. In 1648, a French Physicist, Blaise Pascal, proved the phenomenon that water transmits pressure equally throughout a container. This principle was later on used in industry to generate fluid-force as in the Bramah's press.

However due to the advent of electricity around 1850, the use of water-power (water hydraulics) in industry declined until the outbreak of World War II. From 1920 onwards, oil-hydraulics started appearing in various machine tool controls in Europe, USA, Canada, etc. Engineers started using fluids for power transmission and basic elements like pumps, control-valves, cylinders, etc. were experimented with and perfected. Slowly oil-hydraulics assumed a place of importance in areas of power-transmission and replaced many mechanical elements like line-shafts, chains, gear boxes, electric drive motors, etc., in various mechanical systems.

1.2 BASIC SYSTEM OF HYDRAULICS–MAJOR ADVANTAGES AND DISADVANTAGES

The basic components of the above system are pump, strainer, oil reservoir, filter, pressure gauge, pressure relief valve, direction control valve, actuator (cylinder or motor), etc. These components are joined together or connected by means of pipes, tubes or hoses.

As has been mentioned, the industrial hydraulic system is a power transmission system using oil to carry the power. All systems require an input and an output.

The power may be transmitted directly into a load, or may be transmitted in the form of control. The greater and finer the signal, the more positive, reliable, accurate and responsive is the control. The inputs and outputs of any power and control system including the hydraulic system are mechanical such as a rotating shaft or a reciprocating plunger. An added advantage is that this system is easily adaptable to a variety of energy forms and the signals may be initiated by electrical, chemical, manual, optical, electronic/digital or accoustic means. Hand levers, plungers, springs, rollers and strikers, solenoids and torque motors are common examples of control inputs, while the output may be the movement of a piston rod or the turning of a shaft. Another very significant advantage of a hydraulic system is that there is a tremendous possibility of force amplification or in other words a high force may be generated from a small input signal as in the case of a servo system. Of course there are losses like friction losses, but the total gain is much higher than an ordinary mechanical system, this being the most important reason why one should go for hydraulic system. Hydraulic oil can be piped in all directions and can flow in several pipes at the same time, but all inputs and outputs will be balanced on the same pressure.

1.2.1 Advantages of a Hydraulic System

The basic advantages offered by a hydraulic system are as follows:

1. Hydraulic power is easy to produce, transmit, store, regulate and control, maintain and transform.
2. Weight to power ratio of a hydraulic system is comparatively less than that for an electro-mechanical system.
  (About 8.5 kg/kW for electrical motors and 0.85 kg/kW for a hydro-system).
3. It is possible to generate high gain in force and power amplification.
4. Hydraulic systems are uniform and smooth, generate stepless motion and variable speed and force to a greater accuracy.
5. Division and distribution of hydraulic power is simpler and easier than other forms of energy.
6. Limiting and balancing of hydraulic forces are easily performed.
7. Frictional resistance is much less in a hydraulic system as compared to a mechanical movement.
8. Hydraulic elements can be located at any place and controlled reversely.
9. The noise and vibration produced by hydraulic pumps is minimal.
10. Hydraulic systems are cheaper if one considers the high efficiency of power transmission.
11. Easy maintenance of hydraulic system is another advantage.
12. Hydraulics is mechanically safe, compact and is adaptable to other forms of power and can be easily controlled.
13. Hydraulic output can be both linear, rotational or angular. Use of flexible connection in hydraulic system permits generation of compound motion without gears etc.
14. Hydraulics is a better *over-load safe* power system. This can be easily achieved by using a pressure relief valve.
15. Absolutely accurate feed back of load, position, etc. can be achieved in a hydraulic system as in electrohydraulic and digital electronic servo system. Because of high power and accurate control possibility, in modern engineering language hydraulics is termed as the muscle of the system and electronics its *nerves*.

1.2.2 Disadvantages of a Hydraulic System

In spite of all the above advantages, hydraulic systems have some drawbacks which are mentioned below.

The disadvantages are:

1. Hydraulic elements have to be machined to a high degree of precision which increases the manufacturing cost of the system.

2. Certain hydraulic systems are exposed to unfriendly climate and dirty atmosphere as in the case of mobile hydraulics like dumpers, loaders, etc.
3. Leakage of hydraulic oil poses a problems to hydraulic users.
4. Hydraulic elements have to be specially treated to protect them against rust, corrosion, dirt, etc.
5. Hydraulic oil may pose problems if it disintegrates due to aging and chemical deterioration.
6. Petroleum based hydraulic oil may pose fire hazards thus limiting the upper level of working temperature. However, due to availability of synthetic fire resistant oils this problem is of academic interest nowadays. To combat environmental effects of petroleum and chemical based oils, efforts are on to use biodegradable oils now.

1.3 PHYSICAL UNITS FOR FLUID POWER

Before describing the physical principles of oils it may be necessary to know first the physical units used in fluid power measurement.

1.3.1 Philosophy of Mass/Weight/Force

Engineers throughout the world are now fully accustomed to the use of Systeme Internationale (SI) units and associated terminology. However, certain terms like weight, mass, etc. are still wrongly used by many. We know that weight is in reality a unit of force, i.e. $W = mg$, where m = mass and g = acceleration due to gravity, yet we persist in using it to designate a physical property of a body associated with its mass. To be consistent in the use of proper physical terminology as laid down by ISO, one should use the unit newton in SI system to denote weight. The discrepancy in the use of proper and appropriate terms may be made clear from the subsequent clarifications.

Pound (lb) was earlier used by the British system as a unit of mass or weight. Now, to convert 1 kg to 1b we state that 1 kg = 2.2 lb. But a layman may not appreciate such a statement. What the equality says is that 1 kg (mass, SI unit) is the same as a mass which weighs 2.2 lb in British units. But what this equality does not say is under which gravity condition this statement is true.

In fact this statement is true only where the acceleration due to gravity is 9.81 m/s^2 or 32.2 ft/s^2. If the above comparison is made at a reference where the acceleration due to gravity is 7.62 m/s^2 or 25 ft/sec^2, then 1 kg (mass) = 1.71 lb (weight).

There is nothing wrong in using these units as long as we recognize the fundamental difference. However, the inconsistency caused trouble until the FPS Units were phased out entirely. In the metric world there is still a lot of inconsistency in the use of mass/weight units. A troublesome practice is the one of kg (mass) and kg (force) or kilopound by some metric users.

There is an inconsistency of the same order as that we have seen in the preceding paragraph. The kilopound is not an SI unit and has been replaced by the

Newton. We have become so used to using kg (wt.) in the context of a fundamental physical property of a body that we forget it is really a derived term. The space program with its "weightless" environment provided some perspective on this situation but somehow we have not been able to carry over the concept of the variability of the force of gravity into a drafting room.

"Weight" means the force of gravity acting on a given mass; to the layman "weight" generally means mass—the amount of potatoes, oil or grease he buys at a store. Further complication arises by using "kilogram" for both mass and force in the metric technical (gravitational) system. Inconsistency increases when we say for instance, that potato is being sold by "weight" but is measured in newtons.

This confusion is only due to the practical method of measuring mass which is performed by gravitational methods, i.e. we measure the "force" by which that mass is attracted to the earth.

The spring balance is an example of this, but it is calibrated in mass unit, and should be adjusted for a given location by the use of known mass pieces. A double-pan balance needs no adjustment for location because the forces acting equally on the masses on both sides are effectively negated.

Because of the dual meaning of the term "weight" it is recommended that the technical community should not use the term weight and its derivatives. The SI system has separate and distinct units for mass and force. Instead of using weight, the term mass or force, whichever is appropriate and its respective SI unit, the kilogram or newton should be used.

When referring to the force of gravity, use newton. When referring to mass, use kilogram. The SI system is an absolute system in which mass is the base quantity and force is the derived quantity.

If we can religiously abstain from using the terms "weight", "to weigh" and "weights", the problem will resolve itself.

In some countries, this is done for the consumer as follows:

Old usage	New usage
1. weight	Mass
2. to weigh	to measure the mass
3. weights	mass pieces
4. Any device for weighing	mass meter

As a rule, engineers, when they work with forces, convert mass loads (kilograms) into force units (newtons) to solve a problem involving forces. Mass units multiplied by the acceleration of gravity for a specific location will give the force of gravity in newtons acting on that mass. ($30 \text{ kg} \times 8.8 \text{ m/s}^2 = 294 \text{ N}$).

A crane operator in a factory is interested only in the maximum load (mass units) that the crane structure can safely hold.

To stop the usage of the term "weight" is the ultimate solution to the confusion. Until that is achieved we have to live with this contradiction in physical terminology.

1.4 UNITS OF MEASUREMENT

1.4.1 Basic Units

Six basic SI units are given below:

Physical quantity	Name of SI unit	Symbol
1. Length	meter	m
2. Mass	kilogram	kg
3. Time	second	s
4. Electric current	ampere	A
5. Thermodynamic temperature	Kelvin	K
6. Luminous intensity	candela	cd

Symbols for units do not take a plural form.

Supplementary units

	SI	Symbol
1. Plane angle	radian	rad
2. Solid angle	steradian	sr

1.4.2 Definitions

1. Length The unit of length, called the meter is 1650763.73 wave lengths in vacuum of the radiation corresponding to the transition between the energy levels 2 P_{10} and 5 d_5 of the krypton-86 atom.

2. Mass The unit of mass, called the kilogram, is the mass of the international prototype of a platinum-iridium cylinder kept at Sèvres in Paris.

(The kilogram was originally intended to be the weight in vacuum of one cubic decimeter of distilled water at 4°C.)

3. Time The unit of time called second is the duration of 9192631770 periods of the radiation corresponding to the transition between the two hyperfine levels of the ground state of the caesium-133 atom.

$$\left[\text{old definition—second} = \frac{1}{86400} \text{ of a mean solar day.} \right]$$

4. Temperature Kelvin is $\frac{1}{273.16}$ of the thermodynamic temperature of water.

1.4.3 Common factors and multiples of Metric and SI units

Multiplication factor	Prefix	Symbol
10^{12}	tera	T
10^9	giga	G
10^6	mega	M
10^3	kilo	k
10^2	hecto	h

10^1	deca	da
10^{-1}	deci	d
10^{-2}	centi	c
10^{-3}	milli	m
10^{-6}	micro	μ
10^{-9}	nano	n
10^{-12}	pico	p
10^{-15}	femto	f
10^{-18}	atto	a

Example 1.1

(i) 14000 N = 14 kN
(ii) 0.00846 m = 8.46 mm
(iii) 16245 N/m² = 16.25 kN/m²
(iv) 220,00,0000 watt = 220 × 10⁶ W = 220 mega watt
 = 220 MW
(v) 15 giga bytes = 15 × 10⁹ bytes
 = 15 G bytes

1.4.4 Force and Mass in SI System

Mass is the fundamental property of a quantity of a particular matter. Conceptually it can be considered as the quantity's resistance to change in motion (inertia) in Newtonian systems. For example, the mass of a certain volume of iron is different from the mass of the same volume of wood, oil, cotton, etc.

Force is a derived function rather than a fundamental one. It is not the property of the quantity of material, but results from what is happening to it. It relates to energy being expended in moving a mass (kinetic energy), or being stored (potential energy) relative to the mass in non-relativistic systems.

Weight is considered to be an inherent property, like mass. Actually it is a special kind of force which is not constant. It depends on the acceleration of the mass, which because of the force of gravity equals 9.807 m/s² in the SI metric system.

1.4.5 Newton's Second Law of Motion

This law states that force is proportional to the rate of change of momentum M.

$$F \propto \frac{dM}{dt} \quad \text{or} \quad F = K \cdot \frac{dM}{dt}$$

where $\frac{dM}{dt}$ is the rate of change of momentum and K is a constant. Normal engineering tolerances can be held if the mass is assumed to be constant. Making this assumption,

$$F = K\left(d\left(\frac{mv}{dt}\right)\right) = Km\frac{dv}{dt} = Kma.$$

This is the most commonly used form of Newton's second law.

Let us consider three different ways in which Newton's second law of motion can be expressed.

They differ in the use of defined or derived units for F, K, m and a. These expressions do not use SI metric units.

Absolute Version This is based on fundamental Physics which is used to define force. Here the assumption is that mass, distance, and time units have been defined so that $K = 1$. The definition is $F = Kma$ with F in dynes, K in $(dynes \cdot sec^2)/(g(m) \cdot cm))$, m in $g(m)$ and a in cm/sec^2.

Practical Version Force, mass and acceleration have been defined previously in this version. K assumes a value to make the equation true. Again, the definition is $F = Kma$ where F is in $g(f)$, K is $(sec^2 g(f))/((980.7) cm \cdot g(m))$, m is in kg (m), and a is in cm/sec^2. Note that $K = 1/g$. This version of the second law can be written as $F = \dfrac{(W)}{(g)} a$.

One of the problems usually encountered when switching to the SI metric system is the use of gram-mass, $g(m)$ and gram-force $g(f)$. These terms were used in the CGS metric system which is now outdated. The SI metric system avoids ambiguity by using newton, N, as the unit of force. The kilogram and its multiples are used only as the unit of mass. The CGS metric unit $kg(f) = 9.807$ N. Unit of N is $kg \cdot m/sec^2$. Then for conversion purposes $kg(f) = 9.807 \ \dfrac{kg \cdot m}{s^2}$. Since acceleration due to gravity, g, in the SI metric system is 9.807 m/s², $kg(f) = g \cdot kg$.

This relationship is more easily recognised using the units: $W = mg$.

Fortunately, the absolute and engineering versions of Newton's second law are not commonly used in fluid power engineering.

The more familiar $F = ma = \dfrac{(W)}{(g)} a$ is used as it is more suitable for fluid power.

Pressure:

$$Pressure = \dfrac{Force}{Area}$$

The SI unit of pressure is $\dfrac{N}{m^2} \dfrac{(newton)}{(meter)^2}$

A more convenient unit is bar.

$$1 \text{ bar} = 10^5 \text{ N/m}^2$$

$$1 \text{ kgf/cm}^2 = 0.981 \text{ bar}.$$

If 1 bar is taken as 1 kgf/cm², there is an error of 2% approximately.

Force: The unit of force in SI units is newton (N).

$$1N = 1 \text{ kg} \cdot m/s^2$$

$$= 1000 \text{ gm} \cdot 100 \text{ cm/s}^2 = 10^5 \text{ gm cm/s}^2 = 10^5 \text{ dynes}.$$

1 kilogram force = 1 kg × 9.81 cm/s² = 9.81 kg m/s² = 9.81 N.

Work: Work = $F \times D$ (force × distance). The unit is joule.

1 Joule = 1N × 1 meter = 1 Newton meter = 1 Nm.

Power: Work divided by time.

1 newton = meter or joule per second is unit of power = 1 watt.

$$1 \text{ joule} = 1\text{N} \times 1 \text{ m}$$
$$= 10^5 \text{ dyne} \times 100 \text{ cm} = 10^7 \text{ erg}.$$
$$1 \text{ watt} = 1 \text{ joule/s} = 10^7 \text{ ergs/sec}.$$

1.4.6 Some Derived SI Units

Area	$= m^2$
Volume	$= m^3$
Frequency	= hertz (cycles per second) = Hz (1/s)
Density	$= kg/m^3$
Speed	= m/s; angular velocity = rad/s
Acceleration	$= m/s^2$; angular acceleration = rad/s^2
Force	= N
Pressure	$= N/m^2$ (bar is very often used)
Viscosity (dynamic)	$= \dfrac{Ns}{m^2}$
Viscosity (kinetic)	$= \dfrac{m^2}{s}$
Work, heat	= joule = J (Nm)
Power	= watt = W(J/s)
Customary temperature	= deg. celsius °C
Specific heat	= joule per kg kelvin = J/kg K.

1.4.7 SI Units vs Metric Technical Units

International System of Units (SI units)	Metric Technical Units (Technical Measuring System)
The basic units are: Length (m), time (s), mass (kg). force = product of mass and acceleration resulting in the unit kg · m/s² defined by 1 newton = 1 kg · m/s²	The basic units are: length (m), time (s), force (kgf) Mass = force divided by acceleration hence unit kg · s²/m We cannot deduce the unit kg from the

Pressure	pressure = force divided by area resulting in the unit N/m². More convenient figures result from using the unit bar 1 bar = 10⁵ N/m²	technical measuring system. The conversion to SI units comes from 1 kgf·s²m⁻¹ = 9.81 kg because 1 kgf is defined as the force which gives a mass of 1 kg an acceleration of 9.81 m/s². pressure = force divided by area hence, the unit kgf/cm² introduces the relationship. 1 kgf/cm² = 0.981 bar. If we take 1 bar = 1 kgf/cm², we have an error of 2%.
Work (energy)	work = product of force and distance hence the definition 1 joule = 1 N·m = 1 W·S	Work = product of force and distance hence the unit kgf·m
Power	power = work divided by time. Power has the definition 1 W = 1 J/s = 1 N·m/s.	power = work divided by time. Power has the unit kgf·m/s.
Density—specific weight	density = mass divided by volume with the unit kg/m³. The term specific weight is no longer used in the SI unit system. In order to be exact, we always take in physical formulae density = specific weight divided by acceleration of free fall $p = \gamma/g$.	specific weight = weight divided by volume. The unit is kgf/m³. As a mass of 1 kg, by an acceleration of free fall of 981 m/s² imparts a force of $F = 1$ kgf, the values of density and specific weight are the same.

1.5 PRINCIPLES OF HYDRAULICS: FLUID POWER

In the language of physical science anything that can flow is a fluid. As per the definition of fluid, air, oil and water are nothing but fluids since all of them can flow. In our discussion here, we will concentrate only on oil as the present day hydraulic system uses oil. The operation of a fluid power system, i.e. a hydraulic system using oil is governed by the basic physical laws of fluid flow as developed by the great scientist Blaise Pascal (1648). This law is known as "Pascal's law".

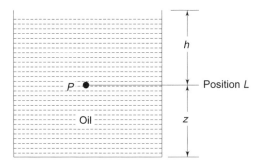

Fig. 1.1 *Potential head*

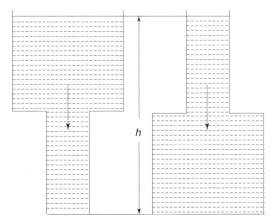

Fig. 1.2(a) *Potential head is independent of shape and size*

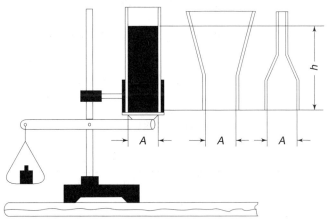

Fig. 1.2(b) *Potential head is independent of container shape*

1.5.1 Law of Hydrostatics

The law of hydrostatics states that the pressure p of a fluid at rest increases on increasing the depth. This means that,

p = Pressure = wh where,

w = specific weight of the liquid

h = depth or 'head' of the fluid.

It refers to a specific datum (Fig. 1.1)—the head defined by position L is called the *potential head*, h and the total head represented by $(h + z)$ is the *piezometric head*.

∴ $$h + Z = \frac{p}{W} + Z$$

From the above we understand that every liquid exerts a pressure on its base area by its own weight.

The pressure is dependent on the height of the liquid column and its density irrespective of the shape or geometry of the container (Figs 1.2a and 1.2b).

Mathematically Hydrostatic Pressure is equal to

$$p = h \times \rho \times g$$

h = height of fluid column

ρ = density of liquid

g = acceleration of free fall.

Example 1.2 *Calculate the hydrostatic pressure at the bottom of a hydraulic oil container filled with oil ($\rho = 0.8$ kg/dm³) up to a height of 800 mm.*

Solution $p = h \times l \times g$

$$= 800 \text{ mm} \times 0.8 \, \frac{\text{kg}}{\text{dm}^3} \times 9.81 \, \frac{\text{m}}{\text{s}^2}$$

$$= 8 \text{ d} \times 0.8 \, \frac{\text{kg}}{\text{dm}^2} \times 9.81 \, \frac{\text{m}}{\text{s}^2}$$

$$= 8 \times 0.8 \, \frac{\text{kg}}{\text{dm}^2} \times 9.81 \, \frac{\text{m}}{\text{s}^2}$$

$$= 8 \times 0.8 \times 9.81 \, \frac{\text{N}}{\text{dm}^2}$$

$$= 8 \times 0.8 \times 9.81 \, \frac{\text{N} \times 100}{\text{dm}^2 \times 100}$$

$$= 8 \times 0.8 \times 9.81 \, \frac{100 \cdot \text{N}}{\text{m}^2}$$

$$= 6278 \, \frac{\text{N}}{\text{m}^2} = 0.063 \text{ bar}$$

The hydrostatic pressure is mostly negligible in common workshop machine tools using oil hydraulics which however operate at a significantly higher pressure rating.

1.5.2 Head of Oil

Pascal's law: Pascal's law states that the pressure generated by exerting a force on a confined mass of liquid at rest acts undiminished in equal magnitude and in all directions normal to the inside wall of the fluid container. If we elaborate Pascal's law, we find in Fig. 1.3 that if a force F is exerted on the confined liquid in the jar, a pressure p is generated and this pressure p will act on the inside wall of the container undiminished, i.e. in equal magnitude and at right angles to the point where pressure acts on the surface.

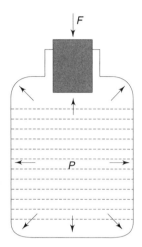

Fig. 1.3 *Pressure exerted in a confined mass of liquid is equal in all directions*

1.5.3 Bramah's Press Principle

The principle of Pascal's law was successfully applied by an English Engineer, Mr. Joseph Bramah (1795), to develop a hydraulic press in which by applying a small input force a large output force was generated. The Bramah's press principle may be understood by explaining the principle of Pascal's law.

Let us assume two oil containers both cylindrical in form and connected together contain some oil as shown in Figs 1.4 a and b. The two cylinders are of different diameter say D_1 and D_2 respectively, where D_1 is smaller than D_2. Both the cylinders have a piston with piston rod as shown. If a force F_1 is exerted on the smaller piston, then according to Pascal's law, pressure p will be generated in the oil and this will be constant and act equally in all directions.

Mathematically, we can determine this pressure p by dividing the force by the area of the piston, or, $P = \dfrac{F_1}{A_1}$

where A_1 = area of the smaller piston = $\dfrac{\pi}{4} D_1^2$

As this pressure will also act at the bottom surface of the bigger piston, we can calculate force F_2 which will be generated by the bigger piston.

$F_2 = P \cdot A_2$ = where A_2 = area of the bigger piston.

$= P \cdot \dfrac{\pi}{4} D_2^2$ or $P = \dfrac{F_2}{A_2}$

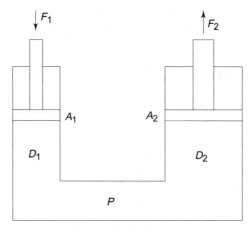

Fig. 1.4(a) *Principle of Bramah's press*

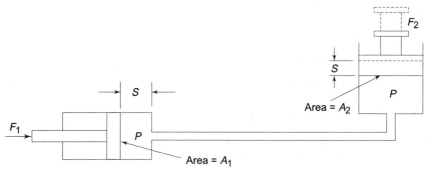

Fig. 1.4(b) *A small effort F_1 produces a higher force F_2*

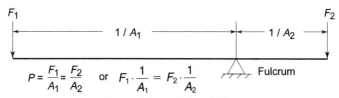

Fig. 1.4(c) *A hydraulic lever*

Similarly $$P = \frac{F_1}{A_1}$$

∴ $$P = \frac{F_1}{A_1} = \frac{F_2}{A_2} \qquad (1.1)$$

or $$F_2 = F_1 \times \frac{A_2}{A_1}$$

But as $A_2 > A_1$, $\frac{A_2}{A_1}$ is > 1.

or F_2 is higher than F_1.

It can therefore be concluded that by applying a smaller force F_1 on the smaller piston, a bigger force F_2 can be generated in the bigger piston. From equation (1.1) we can also write:

$$P = \frac{F_1}{A_1} = \frac{F_2}{A_2} \text{ or } \frac{F_2}{F_1} = \frac{A_2}{A_1} \text{ or } \frac{F_1}{F_2} = \frac{A_1}{A_2}$$

or $\quad F_2 : F_1 = A_2 : A_1$

or $\quad F_2 : F_1 = \dfrac{D_2}{D_1}$

$$F_2 : F_1 = D_2^2 : D_1^2$$

By applying the equation for work done, we can say work done by the smaller piston = $F_1 \cdot S_1$ and work done by larger piston = $F_2 \cdot S_2$.
where S_1 and S_2 are the distances moved by the smaller and bigger pistons respectively. It is obvious that S_1 is greater than S_2.

Equating the two we can write,

$$F_1 \cdot S_1 = F_2 \cdot S_2 \text{ which also gives } \frac{F_1}{F_2} = \frac{S_2}{S_1}$$

or $\quad F_1 \cdot S_1 = P \cdot A_2 \cdot S_2$

or $\quad F_1 \cdot S_1 = \dfrac{F_1}{A_1} \cdot A_2 \cdot S_2 = F_1 \cdot \dfrac{A_2}{A_1} \cdot S_2 = \left[F_1 \cdot \dfrac{A_2}{A_1} \right] \left[\dfrac{A_1}{A_2} \cdot S_1 \right]$

or $\quad \dfrac{S_1}{S_2} = \dfrac{A_2}{A_1} = \dfrac{D_2^2}{D_1^2}$

Hence, $\quad \dfrac{F_2}{F_1} = \dfrac{A_2}{A_1} = \dfrac{S_1}{S_2} = i$, where i = conversion factor.

The equation $\dfrac{F_1}{A_1} = \dfrac{F_2}{A_2}$ can also be written as

$$F_1 \cdot \frac{1}{A_1} = F_2 \cdot \frac{1}{A_2}$$

Therefore mathematically we have shown that the arrangement as in Fig. 1.4b is equivalent to a mechanical lever as shown in Fig. 1.4c with the length of the lever arms being inversely proportional to the piston areas. The system is therefore a force amplifier, the only limitation being the ability of the cylinders containing liquid to withstand the pressure developed.

Due to the presence of friction between the cylinder, piston and body, oil-friction and losses of pressure oil, the conversion factor i is always less than the theoretical value $\dfrac{A_2}{A_1}$ and therefore should be multiplied by the hydraulic efficiency η which in all practical probability may lie between 0.8 and 0.9.

$$\therefore \quad F_2 = \left(\frac{D_1}{D_2} \right)^2 \cdot F_1 \cdot \eta$$

1.6 HYDROSTATICS OR HYDROKINETICS

Pascal's law holds good when the fluid be it oil, water or any other liquid, is in rest or in the static state. Hence we may bracket the above theories as *hydrostatics*. One must differentiate this from the hydraulic machines where the oil is in motion. Though up to a tolerable limit the hydrostatic pressure as calculated from the previous discussion may be same as in a moving liquid, there may be a mistake in our calculation if the *hydrokinetic* effect is not taken into consideration in cases of industrial hydraulic systems. *Hydrokinetic* theories are not like those of *hydrostatics*. Due to the motion of oil or liquids inside the pipes and other elements, the flow of fluid will introduce friction losses and effects due to inertia.

1.7 FLUID ENERGY IN MOTION

The sum total of energy in a liquid in motion may be classified as
  (i) Kinetic energy
  (ii) Potential energy
  (iii) Pressure energy.

By definition, energy is work done. More appropriately we can define it as the ability of a body to do work. The total hydraulic energy consists of three different energies.

Potential energy: Energy by virtue of the position of a body is termed potential energy. It is also called as stored energy. The potential energy in a hydraulic system is the equivalent of the height of liquid column in the system (h) · $W_{pot} = V \times h \times \rho \times g$ where V = volume of liquid, ρ = density of liquid and g = acceleration due to gravity.

Kinetic energy: By definition, kinetic energy is the energy by virtue of motion of a body. Hence kinetic energy is closely associated with the velocity of a moving body. For a freely falling body the velocity (v) attained at a time (t) is $v = gt$, where g = acceleration due to gravity (9.81 m/s^2). The distance (h) travelled by the body may be expressed as $h = gt/2 \cdot t$, where, $gt/2$ is the average velocity.

$$\therefore \quad h = \frac{gt^2}{2} = \frac{g \cdot v^2}{2g^2}; \quad \text{as } tg = v$$

$$\therefore \quad t = \frac{v}{g}$$

and,
$$h = \frac{v^2}{2g}$$

For a body of mass m, the kinetic energy $h = \dfrac{mv^2}{2g}$

where mass m = volume of liquid × density = $V \times \rho$

$\therefore W_{kin} = V \times \rho \times \dfrac{v^2}{2}$. This is also called *velocity pressure*.

Basics of Industrial Hydraulics 17

Pressure energy: Pressure energy is the energy of a body by virtue of its condition. The pressure energy of a liquid column is expressed by dividing pressure by the density of the liquid.

$$\therefore \quad h = \text{pressure energy} = \frac{P}{\rho}$$

where P = oil pressure

ρ = density of oil.

W_{pr} = volume displaced × pressure = $V \times P$.

According to Bernoulli's principle, the total energy of a liquid is constant.

Neglecting the losses $W_{pot} + W_{pr} + W_{kin}$ = constant

As all the three forms of energy can exist in a hydraulic system, we can represent the sum total of energy at a time in an oil system as

$$E = h + \frac{v^2}{2g} + \frac{P}{\rho}$$

From the first law of thermodynamics, we know that energy cannot be created or destroyed even though the energy can be transformed to other forms. Chemical energy can be converted to heat energy or heat energy could be converted to electrical or mechanical energy, etc. Now applying this analogy to fluid flow, we understand that the total energy (E_1) contained in the liquid at a point denoted by subscript 1 may not change from the energy (E_2) at another point denoted by subscript 2 or in other words, $E_1 = E_2$. (Fig. 1.5)

But $\quad E_1 = h_1 + \dfrac{v_1^2}{2g} + \dfrac{P_1}{\rho}$

and $\quad E_2 = h_2 + \dfrac{v_2^2}{2g} + \dfrac{P_2}{\rho}$

$\therefore \quad h_1 + \dfrac{v_1^2}{2g} + \dfrac{P_1}{\rho} = h_2 + \dfrac{v_2^2}{2g} + \dfrac{P_2}{\rho}$

This is called Bernoulli's theorem.

1.8 FLOW VELOCITY

Hydrodynamics (motion of a moving liquid).

Constant Flow: The velocity of a liquid flowing through a pipe of variable cross-section changes to maintain continuity of flow as may be seen from Fig. 1.5.

$$\text{Flow rate} = \frac{\text{volume}}{\text{time}}$$

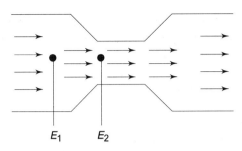

Fig. 1.5 *Continuity of flow*

∴ $Q = \dfrac{V}{t}$; where Q = flow rate in l/min and t = time in min.

or, $Q = A\dfrac{S}{t}$

or, $Q = A \cdot v$

But, $Q_1 = Q_2$

∴ $Q = A_1 \times V_1 = A_2 \times V_2$

$Q = 0.1\, A \cdot v$ where A is in cm², v is in m/min. and Q in l/min.

Example 1.3 *Through a hydraulic pipe of 15 mm dia flows oil at a flow rate of 12 l/min. Find out the flow velocity.*

Solution We know,

$$\text{Flow velocity }(v) = \dfrac{\text{Flow rate }(Q)}{\text{Area of pipe cross section }(A)}$$

or $V = \dfrac{Q}{A}$

Here $Q = 12\, l/\text{min} = \dfrac{12 \times 1000 \text{ cm}^3}{60\, s} = 200\, \dfrac{\text{cm}^3}{s}$

Diameter of pipe = 15 mm = 1.5 cm.

∴ Area of pipe cross section = $\dfrac{\pi}{4}(1.5 \text{ cm})^2$

= 1.76 cm²

∴ $V = \dfrac{Q}{A} = \dfrac{200 \text{ cm}^3/s}{1.76 \text{ cm}^2} = 113\, \dfrac{\text{cm}}{s}$

$= \dfrac{113}{100}$ m/s. = 1.13 m/s.

In flow lines which are located with only a small difference in level, the potential pressure can be neglected and hence the principle of conservation of energy is simplified.

The static pressure falls with increasing velocity pressure and vice versa, i.e. with increasing flow velocity the static pressure diminishes.

1.8.1 Friction

As soon as liquid flows in a system heat is produced due to friction so that a part of the energy is lost in the form of heat energy.

Although friction is never quite eliminated, it can nevertheless be limited. Causes of unnecessarily large friction losses in hydraulic flow lines are:
(a) Flow lines too long
(b) Too many bends
(c) Excessive velocities due to fittings.

Influence of friction on pressure: As stated earlier, friction generates heat and heat can reduce the oil viscosity. This may result in undesirable leakage, thereby reducing the pressure.

First of all, due to the weight of the liquid, potential energy is permanently consumed by the continuous motion of liquid particles. Complying with the energy conservation principle, it is not destroyed but is converted into heat energy due to friction on the pipe walls.

1.9 LAMINAR AND TURBULENT FLOW

Laminar flow exists when no irregularities occur in the magnitude and direction of velocity. Turbulent (whirling) flow is when the main flow stream exhibits unstable and random motion. The variations in velocity act on all sides.

The nature of the flow is determined by the *Reynolds Number, Re*:

$Re \leq 2320$ laminar flow
$Re \geq 2320$ turbulent flow.

A direct relationship exists between the magnitude of velocity, viscosity of the flow medium, internal and external friction and the nature of the flow.

Review Questions

1. State the basic advantages of a hydraulic system over mechanical system.
2. Differentiate between mass, weight and force. What type of confusion occurs while using the term 'weight'?
3. What is hydrostatics? Does it differ from hydrokinematics?
4. Define pressure. Does friction influence pressure in a hydraulic system?

References

1.1 Herion Taschenbuch, M/s Herion—Werke K G, Stuttgart, Germany, p. 373.
1.2 S.I. Units-BI S No. 10005 – 1985, SI Units & Recommendations, Bureau of Indian Standards Publications, New Delhi.
1.3 *Fluid Power Directory Hand Book '89*, Fluid Power Society of India, Bangalore, India.
1.4 *Westermann Tables*, Westermann Publications, Germany, p. 71.
1.5 Mcliele, L S, *Simplified Hydraulics*, McGraw Hill Co. Ltd., New York.
1.6 Krist Thomas Dr. Ing., *Hydraulik*, Vogel Verlag, Wuerzburg, Germany, pp. 37-38.

2

Hydraulic Oils and Fluid Properties

A HYDRAULIC fluid power system may be defined as a means of power transmission in which a relatively incompressible fluid is used as the power transmitting media. The primary purpose of a hydraulic system is the transfer of energy from one location to another and the conversion of this energy to useful work.

Hydraulic power is usually generated by pumps and the energy generated is converted to useful work by hydraulic cylinders or other actuators (linear or rotary).

The transmission of this energy is accomplished by movement of the hydraulic fluid through metal tubes or elastomeric hoses, while the control of the power is achieved by means of values. As no hydraulic system can perform the assigned task without the hydraulic fluid, this fluid is of utmost importance in a hydraulic system.

The broad tasks of hydraulic oil can be classified broadly as follows:

1. to transfer hydraulic energy
2. to lubricate all parts
3. to avoid corrosion
4. to remove impurities and abrasion
5. to dissipate heat

2.1 TYPES OF HYDRAULIC FLUIDS AND SELECTION CRITERIA

There are innumerable types of materials in use as hydraulic fluids. These range from water to inorganic salt solutions to water oil emulsions, synthetic and naturally occurring organic materials.

Though water was the first hydraulic fluid and was used during the early stages of the Industrial Revolution, petroleum based hydrocarbon type fluids are widely used today. Nevertheless, a specific requirement of the hydraulic fluid is determined by the design of the system and by the function the system is designed to perform. Certain characteristics are considered desirable in a good hydraulic fluid as explained in Section 2.2.

2.2 PROPERTIES OF HYDRAULIC FLUIDS

2.2.1 Good Lubricity

The components in a hydraulic system contain many surfaces which are in close contact and which move in relation to each other. The hydraulic fluid must separate and lubricate such surfaces. Protection against wear is a principal reason for selecting a fluid having good lubricating characteristics as a hydraulic medium.

2.2.2 Stable Viscosity Characteristics

Viscosity is a very important fluid property from the point of view of actual use. Viscosity may be considered as the resistance of the fluid to flow or as a measure of internal friction. Viscosity varies with temperature and pressure.

Fluids having large changes of viscosity with temperature are commonly referred as *low viscosity index fluids* and those having small changes of viscosity with temperature are known as *high viscosity index fluids*. Viscosity is also important with regard to the ability of fluid to lubricate.

2.2.3 Stable Chemically and Physically

Fluid characteristics should remain unchanged during an extended useful life and during storage. The fluid in a working hydraulic system is subjected to violent usage—large pressure fluctuations, shock, turbulence, aeration, cavitation, water and particulate contamination, high shear rates, and large temperature variations. Since many aspects of stability are chemical in nature, the temperatures to which the fluid will be exposed is an important criterion in the selection of a hydraulic fluid.

2.2.4 System Compatibility

From the design point of view, it is expected that the hydraulic fluid should be inert to those materials used in or near the hydraulic equipment. If the fluid in anyway attacks, destroys, dissolves or changes parts of the hydraulic system, the system may lose its functional efficiency and may start malfunctioning. Similarly, changes in the hydraulic fluid itself caused by interaction with the system material can also cause system malfunction. Replacement of one hydraulic fluid with another involves the consideration of compatibility.

2.2.5 Good Heat Dissipation

An important requirement of the fluid is to carry heat away from the working parts. Pressure drops, mechanical friction, fluid friction, leakages, all generate heat. The fluid must carry the generated heat away and readily dissipate it to the atmosphere or coolers. Therefore high thermal conductivity and high specific heat values are desirable in the fluid chosen.

2.2.6 High Bulk Modulus

In general, oil is taken as incompressible. However, in practice, all materials are compressible and so is oil. The bulk modulus is a measure of the degree of compressibility of the fluid and is the reciprocal of compressibility. The higher the bulk modulus, the less the material will be compressed with increasing pressure. Bulk modulus is an important characteristic of a hydraulic fluid because of control problems, especially in servo hydraulics.

2.2.7 Adequate Low-temperature Properties

This is an important consideration for hydraulic systems which must operate outdoors, in low temperature environments or at high altitudes. Low-temperature properties may be described by the *pour point* or viscosity-temperature characteristics of the fluid.

2.2.8 Flash Point

The flash point of a hydraulic oil is defined as the temperature at which flashes will be generated when the oil is brought into contact with any heated matter, e.g., a heated stick. The fire point is actually the ignition point of the oil.

2.2.9 Low Foaming Tendency

A liquid has a property to absorb a portion of gas or air with which it comes in contact. Though the accumulation of air is not detrimental when it is within a certain limit, it may create acute problems in proper functioning of the system if the limit is crossed.

The ability of a fluid to release air or other gases without the formation of foam is an important characteristic of a hydraulic fluid. Excessive foaming results in loss of fluid if the volume of the hydraulic system is exceeded. Compression of air-oil mixture by pump or actuators will increase its temperature which in turn may cause fluid deterioration by thermal breakdown or oxidation.

2.2.10 Fire Resistant

Fire resistance is one of the properties that is optional in a good usable hydraulic fluid. The commonly used hydraulic liquids are petroleum derivatives, and consequently they burn vigorously once they pass the fire point. For critical applications, artificial or synthetic hydraulic fluids are used which have high fire resistances. Various grades of fluids with high water content are also available nowadays for oil hydraulic systems.

2.2.11 Prevent Rust Formation

Moisture may be present to some extent in hydraulic systems. Moisture and oxygen cause rusting of iron parts in the system. Rust particles can cause abrasive wear of system components and also act as a catalyst to increase the rate of oxidation of the fluid. Fluids with rust inhibitors help to minimize rust formation in the system.

2.2.12 Low in Volatility

The fluid should have a low volatility, i.e., low vapor pressure or high boiling point characteristic. High vapor pressure may cause high back pressures or vapor-lock resulting in lack of adequate flow. The vapor pressure of a fluid varies with temperature and hence the operating temperature range of the system is important in determining the suitability of the fluid.

2.2.13 Good Demulsibility

Moisture or water may enter a hydraulic system through contamination or condensation. This water may either dissolve in the fluid or form two layers. Dissolved water may produce corrosion, rusting or sludge in the fluid. Fluids with emulsifiers easily separate the water from its main body. Generally used or contaminated fluids are more likely to emulsify with water than new fluids.

2.2.14 Low Coefficient of Expansion

A low coefficient of expansion is usually desirable in a hydraulic fluid to minimize the total volume of the system required at the operating temperature.

2.2.15 Low Specific Gravity

Specific gravity of fluid is of importance only in those cases where the overall system weight must be kept to a minimum. High specific weight means more weight for a given volume of fluid. Heavy fluids can also cause pump cavitation and malfunction. This aspect is important especially in the aircraft industry.

2.2.16 Non-toxic, Easy to Handle and Available

These characteristics refer to the interaction of the fluid with people who repair, handle, use or pay for the hydraulic system or hydraulic fluid. Obviously, it is desirable that the fluid be as simple to handle and as available and cheap as possible.

2.3 PHYSICAL CHARACTERISTICS

It will be appropriate here to describe in brief the various physical characteristics of the hydraulic oil in terms of mathematical formulae.

2.3.1 Density

Density may be defined as the mass of oil per unit volume. The unit will be kg/cm³ or kg/m³

$$\therefore \quad \rho \text{ (density)} = \frac{\text{mass}}{\text{volume}}$$

Density of any liquid is generally measured by an instrument called hydrometer. If one dips the instrument into the liquid or oil, the density can be directly read. Hydraulic oils which are used in industrial hydraulic systems may have a density of 0.8 to 0.9 gm/cm³.

2.3.2 Specific Gravity

Specific gravity of oil is defined as the ratio of densities of oil and water. Specific gravity of a fluid is important in those cases where the overall system weight must be kept minimum.

$$\therefore \quad \text{specific gravity} = \frac{\text{density of oil}}{\text{density of water}}$$

As per standards, density of water is accepted as 1. Hence if we say that the specific gravity of oil is 0.80, then the logical inference is that the oil in question has a density of 0.8 gm/cm³. Specific gravity and density may seem to be the same; it is necessary that we understand the mathematical relationship so that while solving problems we are not confused. A heavy fluid can cause pump cavitation and resultant malfunction of the system.

2.3.3 Specific Weight

Specific weight of hydraulic oil is calculated by multiplying density of oil by acceleration due to gravity.

2.3.4 Viscosity

Viscosity is a very important property of oil. It is the measure of the ability of a liquid to flow. It can actually be defined as the resistance to flow. To understand the physical concept of viscosity, let us take few simple examples. When one swims in a pool of water, one experiences a resistance to the motion. One might have noticed that when a liquid kept in a container is stirred and left to itself, the motion will disappear after sometime. This indicates that there is some kind of frictional force in all types of fluids. This force is called the *viscous force*.

Viscous flow Let us assume $ABCD$ is a layer of oil (Shown in Fig. 2.1) where AB is in contact with a stationary or fixed surface and CD is free. If a force is applied to the surface CD, CD will change its position to $C'D'$ relative to its lower surface AB; position of AB remaining unchanged as shown in Fig. 2.1. This means that there is a velocity gradient between the two surfaces AB and CD.

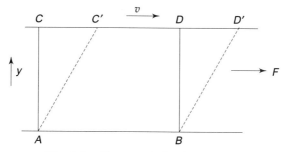

Fig. 2.1 *Shearing of oil in layer ABCD*

To understand it more clearly, let us imagine that a fluid is in contact with a fixed base plate B at the bottom as shown in Fig. 2.2(a). If a plate A is placed on its surface and moved in the direction as shown in the figure, the liquid begins to flow as the liquid can not withstand the shearing strain caused by the moving plate, A. The film of liquid in contact with plate A will move with the velocity, say V, of plate A. The film of liquid in contact with the stationary plate B however, will be at rest and velocity will be zero. In other words, the velocity of flow is zero at the stationary plate B and increases continuously as one moves up into the liquid at right angles to the direction of flow. The magnitude of velocity of various layers is represented in Fig. 2.2(b) by the arrow, as shown, which is the velocity profile of the flowing liquid. However, it is to be noted here that when a liquid such as a hydraulic oil flows through a pipe, the layer of liquid in contact with the internal wall of the pipe will be stationary, i.e. velocity of flow will be zero whereas the central layer of oil at the center of the pipe will move with the maximum velocity. The velocity profile will be parabolic as shown in Fig. 2.2(c).

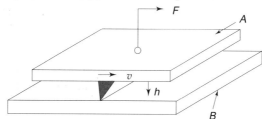

Fig. 2.2(a) *Plate A slides over Plate B to show principle of shear*

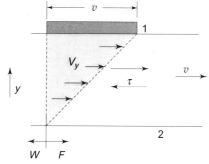

Fig. 2.2(b) *The gradient of oil velocity*

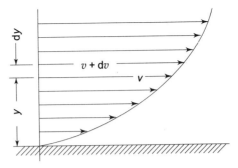

Fig. 2.2(c) *Parabolic velocity profile*

Viscous force and viscosity We have already noticed from Fig. 2.2(b) that the uppermost layer has the maximum velocity; the velocity decreasing gradually below the uppermost layer. Applying this analogy to a liquid flowing through a pipe, we can assume a similar velocity gradient with a parabolic velocity profile as shown in Fig. 2.3(a). At this point let us consider two adjacent layers of liquid at a certain distance y and $y + dy$ having velocities v and $v + dv$ respectively, where dv is a small increase in velocity at the upper layer. The flow-force on the upper layer tends to accelerate the motion of the lower layer whereas the force acting on the lower layer tends to retard the motion of the upper layer. Thus there acts on the liquid two opposite forces subjecting the layers to a shearing stress (τ). But a shearing stress cannot produce shearing strain in a liquid and instead causes the liquid to move. One can thus conclude here that the force which is nothing but viscous force acting on the adjacent layers, tends to destroy the relative motion. The property which gives rise to this viscous force is called *viscosity* which is nothing but internal friction between the layers. It may be mentioned that if the relative motion between the layers is to be maintained, an external force should be provided.

Absolute Viscosity According to Newton's hypothesis of laminar viscous flow (formulated in 1686), the shear strain on a certain layer of a liquid is dependent on:
- Fluid
- Type of flow

In laminar flow it is directly proportional to the velocity gradient at right angles to the flow. Mathematically it is written as

$$\tau \propto (-) \frac{dv}{dy}$$

The negative sign indicates that the viscous force acts in a direction opposite to the direction of fluid flow.

$$\therefore \quad \tau = -\mu \frac{dv}{dy} \quad \frac{kg}{m^2}$$

Where μ = Absolute viscosity of fluid, dv = velocity increment corresponding to the increment dy (Fig. 2.2c), τ = shear stress. μ is also termed dynamic viscosity. In the case when the shear stress is uniform over an area of oil layer, the total shearing strain or the friction force acting over the area is $F = \tau \times A$
where τ = shear stress
F = friction force
A = area of shear strain.

Let us consider here a laminar flow between two parallel flat walls 1 and 2 as shown in Fig. 2.3(a). Let us assume the origin of the coordinate system halfway between the two walls with x-axis being the direction of flow and the y-axis at right angles to the walls. Let us assume here a rectangular volume of oil-film (Fig. 2.3(b)) between two cross sections A and B normal to the flow. Let the thickness of the oil-films be zy and the width b and length l.

$\therefore$ Force $F = \tau$. (area of shear strain)
$\therefore$ We can write that friction force

$$F = -\mu \cdot \frac{dv}{dy} \text{ (Area of shear strain)},$$

But area of shear strain = $A' = 2bl$

$\therefore \qquad F = -\mu \cdot \frac{dv}{dy} \cdot 2bl$

But F is also equal to the force of flow, i.e.
F = Pressure · area
= $\Delta P \cdot 2y \cdot b$, where ΔP = oil pressure differential between A and B
y = Thickness of oil layer
b = width of oil layer

Now $\qquad F = -\mu \cdot \frac{dv}{dy} \cdot \text{Area}$

or, $\qquad dv = -\frac{F}{A'\mu} \cdot dy = \frac{\Delta P \cdot 2y \cdot b}{2b \cdot l \cdot \mu} dy$

(Substituting for F and A' from the above equation)

or, $\qquad dv = -\frac{\Delta P}{\mu \cdot l} \cdot y \cdot dy$

Integrating both sides we get,

$$\int dv = -\int \frac{\Delta P \cdot y}{\mu \cdot l} \cdot dy$$

$\therefore \qquad v = -\frac{\Delta P}{\mu \cdot l} \cdot \frac{y^2}{2} + c$, where c is the constant of integration

Substituting the limiting value of $y = 1/2 \cdot h$ when $v = 0$,

$$V_x = -\frac{\Delta P}{l \cdot \mu} \cdot \frac{1}{2} \left(\frac{1}{2}h\right)^2 + C = 0$$

∴ $$C = \frac{\Delta P}{8\mu l} \cdot h^2$$

∴ $$V_x = \frac{\Delta P}{2\mu l}\left(\frac{h^2}{4} - y^2\right)$$

The maximum velocity V_{max} of the oil is at the centre of the pipe where $y = 0$.

This will give, $V_{max} = \dfrac{\Delta P \cdot h^2}{8 \mu \cdot l}$

To determine the rate of flow Q, let us take two elementary areas located symmetrically relative to the Z-axis.

∴ $$dQ = v \cdot dA'$$

where dA' = increment in area due to increment of v and y.

∴ $$dQ = \frac{\Delta P}{2\mu l}\left(\frac{h^2}{4} - y^2\right) \cdot 2\, dy.$$

∴ $$Q = \frac{\Delta P}{\mu l} \int_0^{h/2} \left(\frac{h^2}{4} - y^2\right) dy$$

$$= \frac{\Delta \rho h^3}{12 \mu l}$$

Laminar flow in circular pipe Let us assume a flow of fluid is a section of a straight horizontal pipe of diameter $d = 2\,r$ between two cross sections at a distance of l apart.

Let us now consider a coaxial cylindrical control volume of radius r inside the stream as shown in Fig. 2.3(c).

Now, equating the force of flow and friction force acting on the volume of oil, one can write:

$$\Delta P \cdot \pi r^2 - 2\pi r\, \tau = 0 \quad \text{where } \tau = \frac{\Delta P r}{2l}$$

But $$\tau = -\mu \frac{dv}{dr}$$

∴ $$\frac{\Delta P \cdot r}{2l} = -\mu \frac{dv}{dr}$$

or, $$dv = -\frac{\Delta P}{2\mu l} r\, dr$$

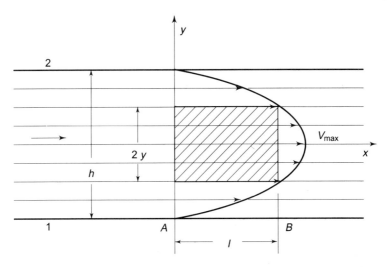

Fig. 2.3(a) *Section of an oil-film between two parallel Plates 1 and 2 showing position of maximum velocity and rectangular cross section of an oil-film between parallel plates*

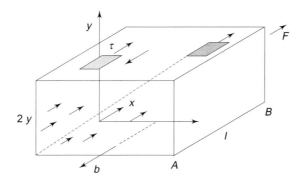

Fig. 2.3(b) *Cross section of an oil-film*

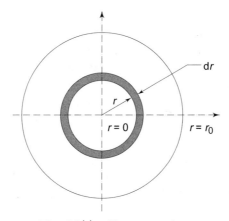

Fig. 2.3(c) *Pipe cross section*

Integrating we have,

$$V = -\frac{\Delta P}{2\mu l} \cdot \frac{r^2}{2} + C, \quad \text{where } C = \text{constant of integration.}$$

Now substituting the boundary condition where $r_0 = r$, $v = 0$,

when
$$C = \frac{\Delta P}{4\mu l} \cdot r_0^2$$

$$\therefore \quad V = \frac{\Delta P}{4\mu l}(r_0^2 - r^2)$$

The maximum velocity is at the center of the cross section where $r = 0$ and

$$V_{max} = \frac{\Delta P}{4\mu l} \cdot r_0^2$$

To calculate Q (flowrate), let us express the elementary discharge across a differential area dA' which is equal to

$$dQ = v \, dA'$$

$$\therefore \quad dQ = \frac{\Delta P}{4\mu l}(r_0^2 - r^2) \, 2/\pi \, r \, dr$$

$$\therefore \quad \int dQ = \frac{2\Delta P \pi}{4\mu l} \int_{r=0}^{r=r_0} (r_0^2 - r^2) \, r \, dr$$

or
$$Q = \frac{\pi \Delta P}{2\mu l} \cdot \frac{r_0^4}{4}$$

$$= \frac{\pi \Delta P}{8\mu l} \cdot r_0^4$$

The mean velocity, $v_m = \frac{1}{2} V_{max}$

Measurement of viscosity: Dynamic viscosity is measured by an instrument called viscometer. It is expressed in the metric system unit as Ns/m^2. However in most countries people use a number of practical units to measure viscosity. The most common practical units are:

°E – Degree Engler in Germany and continental Europe
RS – Redwood seconds in UK
SUS – Saybold Universal Seconds in USA.

In each case the viscosity is measured by a specially designed viscometer. For example, for °E, an Engler viscometer is used. In an Engler viscometer the drainage time of 200 gm of oil at its working temperature is measured. The oil is allowed to pass through an aperture of 4 mm as shown in Fig. 2.4. Similarly an equal amount of water is allowed to pass through the aperture and the time noted. The ratio of times measured is °E. The water should be distilled water and is at normal room temperature, i.e. at 20°C.

$$°E = \frac{\text{time to pass 200 g of oil at its operating temperature}}{\text{time to pass 200 g of distilled water at } 20°C}.$$

Mostly, the liquid flows out in the viscometer under its own weight and hence the density of the liquid is to be considered. When the dynamic viscosity is divided by the density of oil we get kinematic viscosity (v)

$$v = \frac{\mu}{\rho}.$$

The unit of kinematic viscosity is m²/s.

Stoke (St) and centistoke (cSt) are also used as kinematic viscosity units.

$$1\,\text{cSt} = 10^{-6}\,\text{m/s}.$$
$$1\,\text{stoke} = 100\,\text{cSt}$$

Unit of viscosity: Unit of viscosity is expressed in various ways. In physics one defines viscosity either as
  (a) dynamic or absolute viscosity
  (b) kinematic viscosity

Dynamic viscosity is the ratio between shear stress and rate of shear.

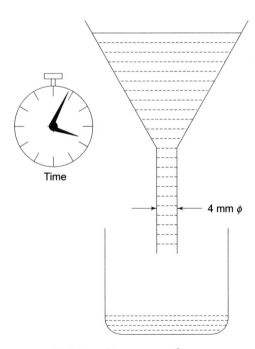

Fig. 2.4 *Measurement of viscosity*

Dynamic viscosity = $\dfrac{\text{shear stress}}{\text{rate of shear}}$

Its unit is poise (P).

1 poise = 100 centipoise (cp) = 0.1 newton second per square meter (Ns/m²)

When one considers the head of a liquid one uses kinematic viscosity.

$$\text{Kinematic viscosity } \nu = \dfrac{\text{dynamic viscosity}}{\text{density of oil}} = \dfrac{\mu}{\rho}$$

The unit of kinematic viscosity is stoke.

$$1 \text{ stoke} = 100 \text{ centistoke} = (\text{cSt}), 1 \text{ c St} = 10^{-6} \text{ m}^2/\text{s}$$

To convert °E to m²/s one can use the following formula:

$$\nu = (7.32 \text{ °E} - 6.31/\text{°E}) \cdot 10^{-6} \text{ m}^2/\text{s}.$$

However, people prefer to use various types of Nomograms for the conversion of viscosity units. One such Nomogram is shown in Fig. 2.5. For example, when

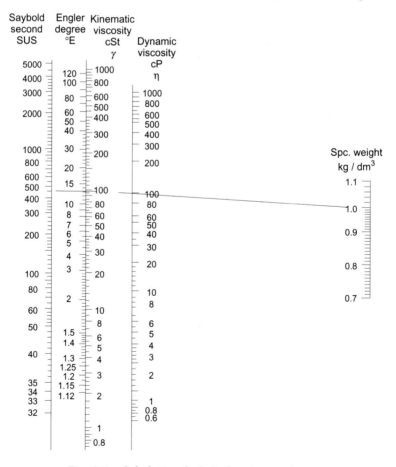

Fig. 2.5 *Calculation of velocity by using nomograms*

the kinematic viscosity of an oil (ν) is 100 cSt and specific weight is 1 kg/dm³, joining the points in the scale we get, ν = 13 °E or 500 SUS. Another conversion scale is shown in Fig. 2.6 for °E, SUS and RS if ν is known.

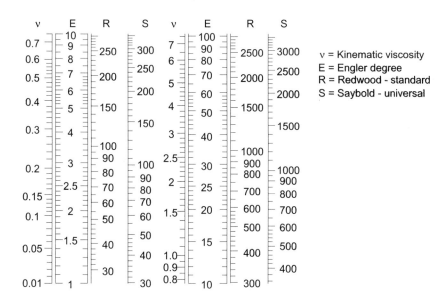

Fig. 2.6 *Viscosity conversion scale from cSt to °E, SUS and RS*

2.4 REYNOLDS NUMBER

The nature of flow of a fluid through a pipe can be either smooth or rough depending on flow conditions. Various factors determine the nature of flow. We can divide the flow of oil as either laminar flow, i.e. steady, smooth flow as in Fig. 2.7 (a) or turbulent flow, i.e. flow is disturbed, as in Fig. 2.7 (b).

From a practical point of view, the Reynolds number indicates if a flow is laminar or turbulent. However one should keep in mind that other physical factors like flow path, working conditions, etc. also determine if the flow will be laminar or turbulent. Viscosity is also an important factor. The relationship between viscosity and Reynolds number can be determined using the following formula:

$$\text{Re} = \frac{v \cdot d}{\nu},$$

where Re = Reynolds No.
  v = flow velocity,
  d = pipe diameter
  ν = kinematic viscosity.

The value of Re thus calculated may be used to determine whether the flow will be laminar or turbulent.

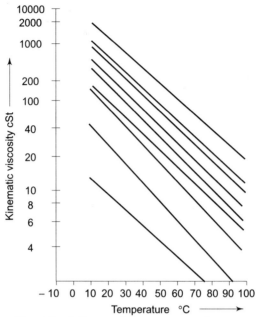

Fig. 2.7 *Influence of temperature on oil viscosity*

In practice, if the value of Re is below 2320, then the flow is laminar. This means, *Re = 2320 is the critical Reynolds number value* for a circular tube as shown in Fig. 2.7 (c).

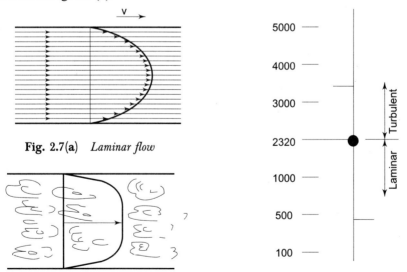

Fig. 2.7(a) *Laminar flow*

Fig. 2.7(b) *Turbulent flow*

Fig. 2.7(c) *Laminar to turbulent flow*

The critical value of Reynolds number for various pipes and tubes are given in Table 2.1.

Table 2.1 Reynolds Number for Various Pipes and Tubes

Type	Value (Re)
1. Round polished tube	2200 to 2320
2. Concentric polished opening	1100 to 1200
3. Eccentric polished opening	1000 to 1050
4. Concentric opening with recess	700
5. Rotary spool	550 to 750
6. Eccentric opening with recess	400
7. Valve with ball seat	25 to 100

2.5 ISO VISCOSITY GRADE

Viscosity of a lubricating oil being the most critical property in lubrication, the Society of Automotive Engineers, USA, standardized automotive engine oils and gear oils on the basis of viscosity only and identified such oils by SAE numbers. In the field of industrial lubricants, the International Organization for Standardization played a major role in establishing international standards for viscosity which led to the I.S.O. 3448 for industrial oils. This classification defines 18 viscosity grades in the range of 2 cSt to 1500 cSt @ 40°C and implies no quality evaluation. ISO 3448 also provides that to meet specific market needs there is nothing to prevent the continued use of such products not falling within the ISO VG range. However, the kinematic viscosities of such out-of-classification oils should be reported at 40°C.

The ISO VG system is to establish a series of definite kinematic viscosity levels so that lubricant suppliers, lubricant users and equipment designers will have a uniform and common basis for designating or selecting industrial lubricants. The standard of industrial oils under the new ISO VG standard is being adopted by most of the International oil companies and the oil industry in India has also decided to switch over to the ISO VG standard.

Various SAE grades for industrial oils are given in Table 2.2.

Table 2.2 SAE Viscosity Grades for Engine Oils

SAE Viscosity grade	Centipoises (cP) at 18°C (ASTM D-2602)	Centistokes (cSt) at 100°C (ASTM D-445)	
	Max.	Min.	Max.
5W	1250	3.8	—
10W	2500	4.1	—
20W [a]	1000	5.6	—
20	—	5.6	Less than 9.3
30	—	9.3	Less than 12.5
40	—	12.5	Less than 16.3
50	—	16.3	Less than 21.9

[a] To differentiate between SAE 20 and SAE 20 W, SAE 15W may be used to identify SAE 20W oils which have maximum viscosity at minus 18°C of 5000 cP.

Note: W Stands for winter.

The ISO VG classification of oils and their kinematic viscosities are given in Table 2.3.

Table 2.3 ISO VG Classification

ISO viscosity grade	Kinematic viscosity centistokes @ 40°C	
ISO VG	Min.	Max.
2	1.98	2.42
3	2.88	3.52
5	4.14	5.06
7	6.12	7.48
10	9.0	11.0
15	13.5	16.5
22	19.8	24.2
32	28.8	35.2
46	41.4	50.6
68	61.2	74.8
100	90	110
150	135	165
220	198	242
320	288	352
460	414	506
680	612	748
1000	900	1100
1500	1350	1650

Before ISO classification of oil in India, IOC (Indian Oil Corporation) was using oil grades as per their own standard. Thus hydraulic oils were being designated as servo System (SS) 311, SS314, etc. In Table 2.4 we have provided the new VG No. for such old gradation by IOC.

Table 2.4 Hydraulic and Circulation System Oils

Old No.	New VG No.
Servosystem 311	Servosystem 32
Servosystem 314	Servosystem 46
Servosystem 317	Servosystem 57*
	Servosystem 68
Servosystem 321	Servosystem 81*
—	Servosystem 100
Servosystem 526	Servosystem 121
—	Servosystem 150
Servosystem 533	Servosystem 176

Note: *indicates non ISO VG product.

Old No.	New VG No.
—	Servosystem 220
Servosystem 563	Servosystem 320
Servosystem 563	Servosystem 460
Servosystem 711	Servosystem A 32
Servosystem 733	Servosystem A 176*
Servocirol 11	Servocirol 32
Servocirol 14	Servocirol 46
Servocirol 17	Servocirol 57*
—	Servocirol 68
Servocirol 21	Servocirol 81*
—	Servocirol 100
Servocirol 26	Servocirol 121*
—	Servocirol 150
Servohydrex 14	servohydrex 32
Servohydrex 21	Servohydrex 57*

* indicates non-ISO VG product.

2.6 INFLUENCE OF TEMPERATURE ON VISCOSITY

Temperature has an adverse effect on the viscosity of hydraulic oil. Hence a maintenance person should see to it that the operating temperature of a hydraulic machine should be kept at a reasonably constant level. Otherwise there will be tremendous losses in the system which may prove detrimental to the overall working efficiency of the system.

A thick fluid is said to have a higher viscosity and its flow becomes sluggish. On the other hand a thinner fluid is said to have a low viscosity and flows more easily. Higher viscosity generates more friction and heat whereas low viscosity causes leakage. Both these conditions lead to a drop in hydraulic pressure which may be detrimental to the overall functioning of the system. The thickness or thinness of the oil is also dependent on the temperature of the system. Hence it is always better to compromise between oil viscosity and working temperature for an optimum working condition.

In order to have an optimum viscosity under varying temperatures, it is better to opt for a higher viscosity Index No. (VI No.). Hydraulic oils are available with VI Nos. as low as zero to as high as 100 and much above this today. Generally, VI No. 100 or above means a better oil, having minimum change or practically no effect of temperature on viscosity. However for general purpose use, a VI No. of 90 or so may be adequate.

2.6.1 Viscosity Index

The viscosity of petroleum-based lubricating and hydraulic oils changes with temperature. The degree of change varies with different oils. This is illustrated in Fig. 2.8 where different types of oils and their change in viscosity with temperature is shown. This temperature-dependent characteristic of oils is expressed in Viscosity Index No. In short this is written as "VI" and is expressed in some arbitrary numerical values. The viscosity of high viscosity index oils is less sensitive to changes in temperature than the viscosity of low viscosity index oils. That means the rate of change in viscosity with changes in temperature is relatively less with high VI oils than with low VI oils.

Viscosity index expresses this change as a single number on a still usable, but outdated scale devised many years ago by two scientists Dean and Davis. The scale assumed zero as the lowest index of naphthenic behaviour and 100 as the highest index of paraffinic behavior. While all other

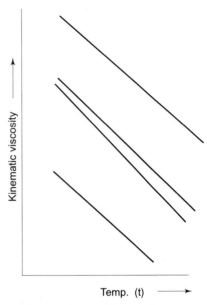

Fig. 2.8 *Influence of heat on the kinematic viscosity of oil*

oils were expected to fall between these values, the advent of solvent refining and use of chemical additives and synthetic fire resistant fluids have produced oils far outside the scale in both directions. A proposal by ASTM D 2270 modified the earlier system and is in use to indicate viscosity-temperature properties to determine VI above 100 and also from 0 to 100. The VI of an oil is not a criterion of the quality of an oil but reflects rather its crude source and refining process. As a thumb rule, oils with VI below 50 are considered naphthenic in nature, while these between 50 and 80 or medium VI oils have characteristics of both naphthenes and paraffines and those above 80 are classified as paraffinic. Paraffinic oils are considered to be more oxidation resistant and show less change in viscosity with changes in temperature and hence are high VI oils. However, modern additives control oxidation to satisfactory levels for a wide variety of oils regardless of their viscosity indices. Hence VI indicates mainly the viscosity behavior over a wide range of temperatures.

[Source: Caltex Lubrication, Vol. 29, No. 1, 1974 Jan–March, K.D. Rel Yea.]

As has been explained earlier, due to change in viscosity, a hydraulic system undergoes a tremendous amount of power loss as shown in Fig. 2.9.

If the oil viscosity is high, there are losses due to leakage which is also detrimental to the overall efficiency of the system. With higher working temperatures there will be a proportionate increase in loss of pressure.

However with correct selection of oil compatible with working temperature, powerloss, etc., one can optimize the overall efficiency as may be seen in Fig. 2.10.

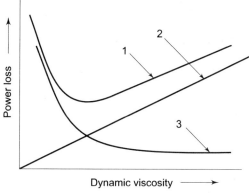

1 – Loss of power,
2 – Loss due to friction,
3 – Loss due to leakage

Fig. 2.9 *Powerloss and viscosity*

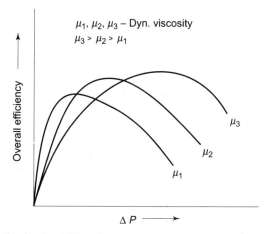

Fig. 2.10 *Effect of viscosity on presssure loss and efficiency*

2.7 PETROLEUM OIL AS HYDRAULIC FLUID

Even today, the most highly used base for hydraulic oil is petroleum. The characteristics of a petroleum oil depend on:
(a) The type of crude oil used
(b) The degree and method of refining
(c) The chemical treatment given to it.

Some of the advantages of petroleum oils are:
(i) Protects against rust
(ii) Has better sealing property
(iii) Has a better heat-dissipating capacity
(iv) Is easy to keep clean.

The other desirable properties could be imparted by suitable refining and chemical treatment. But a principal disadvantage is that it will burn. Hence in hazardous places, petroleum oil may not be the first choice.

2.7.1 Comparison of Water with Petroleum Oil

In comparison to oil, water has a number of specific advantages as stated below.
(i) The viscosity of water is very low and does not change much. The kinematic viscosity of water is 1.5 cSt at (40°F) and 0.3 centistoke at (200°F).
(ii) Water does not cost much due to its abundant availability.
(iii) It does not catch fire.
(iv) Water is found to be chemically stable for a longer duration of time.
(v) Leakage of water may not make the shop floor slippery as in the case of oil leaks.

But the disadvantages weigh much more in the case of water.
(i) Water can be used only over a limited range of temperature.
(ii) Water does not possess lubricating property.
(iii) It is highly corrosive and forms rust.

Hence it is not used as a hydraulic fluid in spite of the fact that the first hydraulic machine *was operated using water only*.

2.8 HIGH-WATER BASED FLUID—HISTORY OF EVOLUTION

Petroleum users all over the world faced a crisis in 1973 due to the sudden increase in the price of all petroleum products by OPEC. Though synthetic fire-resistant fluids were already in use, this increase forced engineers to devise and develop new oil blends in the early '70s. Some chemical and automobile firms in USA started experimenting with water as a fluid and eventually developed the high-water based hydraulic fluid with better wear protection than was available from the petroleum oil of the time. This fluid is usually referred to as the high-water based fluid but American National Standards Institute (ANSI) made use of the term High-Water Content Fluid. It may be mentioned here that water was used as the first hydraulic fluid and even today certain industries operate their system using pure water as the transmission fluid with machine components specially designed to suit the water. A very common example of such high-water based fluid is the solvable oil cutting fluid. This is the first generation of such oils, synthetic oil being the second generation since the high-water content oil. The next in the order was the microemulsion. The latest in the line is a thickened microemulsion introduced as late as 1981. These high water content fluids are referred as 95/5 or 95/10 fluids, i.e. 95% or 90% water and 5 to 10% additives are the principal ingredients of these fluids. They are very low viscosity oils, for all practical purposes the same as water and hence leakage prone. They are also rust prone and components in a system using 95/5 oils should be well protected from air.

It is told that these fluids will be main fluids in many engineering fields including hydraulic transmission in the future.

These HWCF fluids will have a major advantage over the others because of their lower cost. The following oils have been arranged with descending cost per liter.

1. Phosphate esters
2. Water-glycols
3. Transmission fluids
4. Common petroleum based hydraulic oils
5. Invert emulsions
6. HWCF microemulsions
7. HWCF synthetic solutions
8. HWCF soluble oils

2.8.1 Advantages of HWCF

- *Flash point* The flash point of HWCF is nearly 150°C. In case of fire, the water gets evaporated and the residue of additives will not ignite providing higher safety against fire.
- *Cooling characteristics* Hydraulic systems operating with HWCF cools the system down by at least 10°C more than if the same systems were operated by petroleum oil.
- *Biodegradability* Most chemical additives are biodegradable and are much sought after to meet the environmental protection regulations.
- *Cleanliness* Synthetic HWCFS are easy to handle and can be flushed using ordinary sewage disposal system.
- *Viscosity* As viscosity of water remains virtually constant operating speeds of hydraulic machines and oil flow through pipe under a given parameter of conditions remains unaffected.
- *Cost* Costwise, HWCFS are the cheapest oils produced so far. The cost of other synthetic oils like phosphate esters and water-glycols are much higher.
- *Transport* As most of the users may have to procure only the concentrate (i.e. the 5% to 10% chemical concentrate and additives) the cost of transportation will be much cheaper.
- *Storage* Only the concentrate has to be stored and hence a lot of space will be saved.

2.8.2 Some Disadvantages of HWCF

Contamination The contamination problem with high water content fluids is greater than with oil due to the higher density of the fluid. Contaminants are more likely to settle out causing difficulties for proper filtration.

- *Corrosion* Major part of the fluid being water, special care needs to be taken to reduce oxidation. Dissimilar and incompatible metals may induce galvanic action while used with such fluids necessitating proper and appropriate care.

- *Aeration*   This fluid has more tendency to foam than a normal petroleum-based oil. This can be minimized by proper design of pipes or bends and their sizes, which will reduce the possibility of turbulent flow. Laminar flow will reduce the foaming tendency of the fluid.
- *Evaporation*   Though evaporation is certainly a problem with this fluid, it can be controlled if the operating temperature is maintained within 5 to 50°C.
- *Maintenance*   To enhance the life cycle of such fluids, the water to concentrate ratio should be maintained within 25% of the water content. It is prescribed to keep the fluid mix alkaline with a pH factor of 7–5 to 9.0. It is found that a pH factor below 7.0 may allow bacteria to flourish creating fina-filtration problem.
- *Paint compatibility*   These fluids are not very compatible with the available commercial paints and hence need proper care before using the paint in the reservoir.

2.9 FLUID PREPARATION

It is possible that if a firm purchases the concentrate, the fluid can be prepared in the plant itself by mixing plain pure water to the concentrate in the correct proportion. This is done by pouring water in a premixing container and adding the proper amount of concentrate while stirring the mixture vigorously. Care must be taken that small undisposed or undissolved lumps of concentrate do not remain in the mixture. After conducting various tests to ensure the quality of the fluid, the same can now be added to the hydraulic tank or system reservoir. Utmost care needs to be taken to maintain the ratio of water to concentrate through various checks and if found diluted or concentrated, additional premixed fluid has to be added to maintain the correct ratio.

2.9.1 Special Care for HWCF

1. Biocide recommendation from supplier must be taken care of to prevent growth of bacteria.
2. Potential environmental problems must be taken care of for waste treatment.
3. It will be better to use closed reservoirs to minimize water evaporation.
4. The maintenance crew of the plant must take adequate precaution to maintain the fluid level and proper water concentrate ratio.

2.9.2 Few Important Points to Note

- Most HWCFs have a greater dispersant action than petroleum-based oils and are of higher density. This tends to pick up and hold particles in suspension longer than petroleum-based fluids.
- HWCF experiences higher external leakage and hence more make-up fluid is used. This necessitates proper filtration of the fluid to the level of as fine as the system filter used by the manufacturer of the machine tool.

- Probably the hydraulic elements and seals used with normal hydraulic system using petroleum based oil may have some problems of leakage, filtration and compatibility. Though manufacturers of such fluids take all precaution to optimize the use of such high water content fluids with the available components, system designers have to specify the types of seals, sealants, geometrical clearance between component mating parts, paints to be used, filtration ratings, so that the oil life, cavitation losses, leakage losses, prevention of seal failures could be maintained at the minimum level.

2.9.3 Important Additives Used with HWCF to Provide Required Physical Characteristics to HWCF

A high-water content fluid should have all the important characteristics that a petroleum-based oil generally possesses. Various additives are added to the fluid to sustain the important characteristics. Few such additives are mentioned here.

1. *Anti-foaming* Antifoam additives are added to reduce foaming of fluid.
2. *Anti-Wear* Wear resistant chemicals are added to the fluid to protect critical hydraulic components from wear.
3. *Corrosion inhibitor* Chemicals are added to protect surfaces from chemical attack by water.
4. *Biocide* Emulsifying chemicals are added to the fluid to inhibit growth of water-borne bacteria.
5. *Emulsifier* Emulsifying chemicals to facilitate formation and stabilization of an emulsion are added.
6. *Lubrication* Oiliness agents, EP agents are added to the fluid to enhance lubrication characteristics for effective full film boundary lubrication between the mating parts.
7. *Extreme Pressure* EP Chemicals are added to the fluid to improve anti-friction property, wear resistance or scoring. The additives chemically combine with the metal to form a film surface at high local temperature.
8. *Bacterial prevention* The fluid should be such as to permit the substance to be broken down by micro-organisms and should be bio-degradable.
9. *Flocculants* Chemicals added to dispersion of solids in a liquid to combine fine particles to form floe or small solid masses in the fluid.
10. *Emulsion* The oil should be fully dispersed in water to form a stable micro-emulsion.
11. *Deionisation* Elements which provide hardness like calcium, manganese, iron and aluminium salts are removed through deionization of the water.
12. *Oxidation inhibitor* Antioxidation additives are added to provide anti-oxidation characteristics. Oxidation changes the chemical characteristics of the fluid.
13. *Vapour phase inhibitor* Prevention of oxidation or corrosion of metals in contact with the vapor phase of the fluid is ensured by addition of appropriate chemicals.

2.9.4 Fire Resistant Fluids—Classification

Four types of fire resistant fluids are available. They are denoted by letters like HF-A, HF-B, HF-C, HF-D. A brief description of each is given here. This includes synthetic fire resistant oils also.

HF-A: Contains more than 80% water. One of the common variety is oil-in-water and solutions which are blends of selected additives in water.

HF-B: This is a water-in-oil emulsion containing petroleum oil, water emulsifiers and selected additives.

HF-C: Water-glycols are a solution of water, glycols, thickeners, and additives to provide viscosity as per need.

HF-D: Synthetic fluids are non-water type e.g. phosphate or blends of phosphate with petroleum oils.

2.9.5 Certain Commercial HWC Fluids

Various types of commercial HWC fluids are available. A few are given here: (The list should not be taken as conclusive.)

1. **C X 4J** High-water based hydraulic fluid by Cincinnati Milacron Inc., USA. It is a synthetic chemical solution.
2. **EPPCO Hydraqua 541**. Concentrate by Expert Oil Co. USA. It is a micro-emulsion.
3. (a) **Plurasafe P 1000**—Hydraulic fluid from BASF, Michigan.
   (b) **Plurasafe P 1200**—Hydraulic fluid concentrate from BASF, Michigan.
   (c) **Plurasafe P 1200**—Hydraulic fluid concentrate from BASF, Michigan.
   (d) **Plurasafe P 1200**—Hydraulic fluid concentrate from BASF, Michigan.
   These are synthetic solutions.
4. (a) **Hydrolubric 120B concentrate**—Synthetic water additive solution type.
   (b) **Hydrolubric 141 concentrate**—Synthetic water additive microsolution.
   (c) **Hydrolubric 142 concentrate**—Synthetic water additive microsolution.
   These fluids are from E.F. Honghton and Co.
5. **Trim concentrate**—A chemical emulsion from Master Chemical Corporation.
6. (a) **Mobil Hydrasol B concentrate**—A micro dispersion from Mobil Oil Corporation.
   (b) **Mobil Hydrasol 78 concentrate**—A chemical solution from Mobil Oil Corporation.
7. **Petro-HWB Hydraulic fluid**—Micro-emulsion from Petro Lube Inc.
8. (a) **Fluid Power 95-5 dual concentrate**—A synthetic solution
   (b) **Fluid Power 95-5 dual plus concentrate**—A synthetic solution

(c) **Fluid Power 95-5 Tripurpose concentrate**—Micro dispersion synthetic solution.
(d) **Fluid Power 95-5 Qued concentrate**—Micro dispersion synthetic solution.
These are from Pillsbuty Chemical & Oil.
9. **Sunsoil high water based fluid concentrate**—A micro emulsion from Sun Refining and Marketing Co.
10. **Microlubric 19 concentrate**—A soluble extreme pressure mineral oil from Quaker Chemical Corporation.

2.10 COMMON FIRE RESISTANT FLUIDS

As discussed, various types of fire resistant oils are being used in hydraulic systems even though mineral oils cater to a majority of the applications. When a hydraulic system is operated near fire sources like an extrusion machine, boiler and steam turbine controls, welding, forging and foundry robots, heat treatment furnace and die casting machines, it is not advisable to use mineral oils due to the possibility of fire. In such applications fire resistant oils are the best solution. The three most commonly used fire resistant oils are:

(i) Water-in-oil emulsions (invert)
(ii) Water-glycol solutions, and
(iii) Phosphate esters.

Water-in-oil emulsion (invert emulsion) This is generally a mixture of water and oil in the ratio of 40 to 60. Water is dispersed into a continuous phase of oil in the form of tiny droplets of size varying from 3/4 to 2 micron approximately. In the case of fire due to ignition, the water forms steam which acts as a blanket cutting off the oxygen supply to the oil and thus prevents propagation of the flame. The ratio of water to oil may change in various proportions giving rise to various grades of water-in-oils. There is another variety of oils called oil-in-water emulsions where the percentage of water is much higher.

Disadvantages—A major problem with such oils is the depletion of water due to evaporation. Reduction of water from the emulsion reduces
(a) viscosity of fluid
(b) fire resistance
Hence it is better that the water content does not fall below 35% by volume.

Demulsification may be a regular problem with water-in-oil emulsions as the water when it gets condensed fails to re-emulsify and has a tendency to rest at the bottom of the tank as a separate layer. It is generally difficult to prevent this. If the reservoir is designed to create more turbulence, it may help avoid such stagnation of the oil and accordingly there is less chance of demulsification. One may note that removal of baffles in the oil-reservoir may improve chances of turbulence and thus reduce the chances of demulsification.

Periodic monitoring of water content should be a part of the routine maintenance activities of a hydraulic system.

The percentage of water content is checked by a refractometer or by the acid split test. The strength of the emulsion can be corrected by adding distilled or deionized water introducing it slowly at the pump's suction side so that internal churning can help it to emulsify. Water can also be added in the form of fine spray when the pump is working.

The problem of contamination is more severe in these emulsions than in mineral oil because an invert emulsion acts as a cleansing agent and the contaminants remain suspended instead of settling down. Moreover, filters finer than 10 micron rating may not be appropriate with this oil due to possible breakdown of emulsion into water and oil. Hence when the fluid is heavily contaminated with particles more than 10 micron sizes, it is better to discard it instead of filtering. Similarly when other lubricating oils get mixed with the invert emulsion, it becomes unstable breaking into water and oil. Hence such contaminated oils should be skimmed from the oil.

Water-glycols Water-glycol hydraulic oils are single phase solutions with a mixture of water and polyglycols in the ratio of 40 to 60. Their fire resistance property is similar to that of water-oil emulsion. As with invert emulsions, here also the water content needs to be checked regularly and should not be allowed to be less than 35% at any point of time. Viscosity of oil is inversely proportional to the water and hence loss in water percentage thickens the fluid and causes operational problems. Actual percentage of water can be found out by checking the viscosity. Besides water content, alkalinity of pH of the oil must be maintained. Due to oxidation, the alkalinity level goes down which may impair the corrosion inhibiting property of water oil. The normal alkalinity level is around 9.2 and should never be allowed to fall below 4. Initial pH level can be restored by the addition of the correct quantity of alkaline additive.

Phosphate esters These are synthetic fluids and used for high operating temperature (150°C) as compared to water-based fluids. The fire resistance property is due to their chemical composition. These oils develop acidity during usage due to hydrolysis or oxidation. Highly acidic fluids have a tendency to foam and entrap air. High acidity causes corrosion and accelerates the rate of hydrolysis leading to deposit formation. Acidity level of 0.3 mg KOH per gram is considered the maximum limit for reconditioning the fluid. Neutralization number is the measure of acidity. The fluid can be restored to its accepted level of acidity (around 0.05 mg KOH/g) by filtering it inside through the fuller's earth. The recommended quantity of earth/fluid ratio is 2–4 percent and the mesh size for the solid is 50/80. The oil can be maintained to have an acceptable range of acidity by incorporating by-pass filtration system using fuller's earth as a medium. Such a continuous treatment of the fluid not only controls the acidity but considerably reduces the consumption of fuller's earth as well. In such a system, when the acidity number of the fluid reaches 0.3, fuller's earth must be changed.

With a water-based fluid, it is advisable to put silica gel into the breather as moisture trap as water is the principal component for oxidation.

Costwise phosphate esters are 6–8 times costlier than mineral oils and hence any relaxation in the maintenance will be costly if it is to be replaced prematurely.

2.10.1 Hydraulic Oils used in India

In India, the state owned Indian Oil Corporation Limited has marketed various types of lubricants and oils for industrial use. Oils used for hydraulic systems are termed as circulating system oils. These are triple inhibited oils which contain suitable chemicals to enhance the working capability of the machines where these oils are to be used. As claimed by IOC, these circulating oils are suitable both for hydraulic systems and for enclosed gear boxes where extreme pressure oils are not needed. But it is cautioned that these oils are not to be used in steam or hydraulic turbines and similar critical situations.

This series consists of

1. Servosystem 311
2. Servosystem 314
3. Servosystem 317
4. Servosystem 321
5. Servosystem 328
6. Servosystem 526
7. Servosystem 533
8. Servosystem 553
9. Servosystem 563

Apart from these, Servosystem 517 oils are also used as hydraulic oils.

2.11 MAINTENANCE OF HYDRAULIC OILS

Ensuring good filtration and prevention of external as well as inhouse impurities are key steps in keeping the fluid clean. The cleanliness requirement of hydraulic oils has created new awareness about the maintenance requirement demanded by modern hydraulic equipment coupled with the high cost of petroleum products. Care of hydraulic oils starts before they are used into the system. The following points may be worth noting:

- The oil barrels during storage should be kept horizontal to prevent water or other contaminants collecting near the bungs.
- The gasket inside the bung should be kept wet so that it can be more effective as an airtight seal.
- The oil inside the barrel can get contaminated by ingress of airborne moisture or rust formation inside the barrel wall in the air space above the fluid. It is also known that the level of contamination of hydraulic oil as received by user is generally not better than SAE class 6 or NAS 10 which is not acceptable to many hydraulic equipment manufacturers. As per SAE class 6 level, the number of 10–25 micron nominal particle size per 100 ml of oil is around 40,000 which electro-hydraulic servo valves, proportional technology systems, precision hydraulic mechanisms and high pressure pumps cannot tolerate. Hence the oil from barrel needs to be pre-filtered before it is put to use in the system.

2.11.1 Pre-filtration

An auxiliary filtering unit capable of filtering 10 micron nominal particle size is used to pump the fluid from the barrel to the system tank. Though some electro-hydraulic servo valves can tolerate SAE class 6 contamination level, still the fluid should be prefiltered to SAE class 4 level. Otherwise the filter will clog within a few hours of operation and can damage the servo valve if the system is not built with a safety device. This brings us to the fact that oils should be pre-filtered at least to the level of cleanliness the filters are expected to maintain. This practice is mandatory where the system has electro-hydraulic servo valves.

2.11.2 Water Contamination

Water contamination of oil takes place due to:
  (a) Leakage of water from the oil cooler.
  (b) Moisture condensed from the atmosphere.
  (c) Violent churning and dispersion in the pumps and valves breaks the water into tiny droplets and help the contaminated water to form emulsion, which joins with the other deposits and forms sludge and a sticky substance impairing the performance of pumps, valves and cylinders.

Research study indicates that mineral oil fortified with zinc *dialkyldithiophosphate* (Zddp) as an antiwear additive (used in hydraulic oils) when combines with water, decomposes into metal particulates due to hydrolytic action. These metal particulates block the filter and cause filtration problems. Deposits from Zddp can even cause stickiness in the valve movement and clog fine throttles impairing the precision of hydraulic system.

How to Check Water Contamination: Periodic and routine check of water contamination and prevention of water ingress into the system should become a part of the maintenance schedule. When the level of water contamination is more than 0.1 percent, the oil should be centrifuged to separate the water.

Water contamination can be easily checked by taking a sample of oil. The sample should be taken when the system is running under equilibrium condition. Also, the sample tap should be protected to prevent any stagnant fluid getting into the sampling bottle. If the sample oil has a hazy or cloudy look it is an indication of free water present in the oil. A qualitative confirmative test is to put a small quantity of sample oil on a shallow pan and heat it. If audible crackling or popping occurs it is a clear evidence of water contamination.

2.11.3 Oxidation

A dominant factor which determines the service life of the oil is degradation through oxidation. Oxidation occurs when oxygen combines with the original hydrocarbon molecules and gives a chain reaction. Traces (even a few parts per million) of certain metals, particularly copper, iron, zinc, lead and water act as catalysts and markedly increase the rate of oxidation.

Temperature is another factor in accelerating the rate of oxidation. As a working rule oxidation rate is doubled for each 10°C rise in the temperature. An 'oxidation-inhibited' high viscosity index (HVI) oil may give a useful life of about 100,000 hours at 40°C even under the condition of aeration and catalysis. When the oil temperature is raised to 60°C, life drops to as low as 10,000 hours.

Oxidation produces soluble as well as insoluble products and these may form sticky substances and may deposit as gum or sludge in oil passages, various pump parts, valve spools and ports, etc. blocking them and restricting oil flow and making the machine run sluggishly. As a result of oxidation, the oil becomes progressively darker and gets acidic properties which ultimately deteriorates the life of the oil.

2.11.4 Total Acidity Number

Total acidity number (TAN) or the *neutralization number* is the measure of acidity of oil. It is defined as the number of milligrams of potassium hydroxide required to neutralize one gram of oil (mg KOH/g). Oil enriched with anti-oxidant additives has a long induction period. Such an oil takes a longer period to reach the TAN value of 1, but after that the increase in TAN value is very sharp and it quickly reaches the value of 2.

Though the permissible value of TAN is 2, a sharp increase in value is a better indicator for changing the oil.

Once the sharp increase is noticed, it is a warning to the plant engineer to prepare the machine for shut down for changing the oil. The value of TAN may be determined as per procedure given in IS-1448 or ASTM Standard D 974.

Increase in viscosity of oil as well as changes in oil-color may also indicate the time for changing the oil.

2.11.5 Steps to Combat Oxidation

Anti-oxidant additives are used in hydraulic oils to combat oxidation. Apart from this one can use other means to retard oxidation rate by:

- Installation of heat exchanger when oil temperature exceeds 60°C.
- Incorporating magnetic plugs and filters to flush out ferrous contaminants.
- Keep out copper and brass tubings from the system if the pressure rating is above 10 bar.

When magnetic plugs are used, provision should be made to facilitate frequent cleaning of the plugs, otherwise oil surge in the tank may dislodge the particles from the magnet and circulate them back into the system. Incidentally, use of magnetic plug reduces load on filters.

2.11.6 Change of Oil

When oil changes are effected, the system has to be thoroughly flushed to remove the old oxidized oil; partially oxidized oil is an effective catalyst for the oxidation of fresh oils. Flushing is done with same oil or with flushing oil at 40°–50°C and should be continued until the filter is seen without any trace of contaminants.

Typical flushing may range from 2 to 8 hrs. For large systems, say 2000 liters or more, flushing may continue for days. It may be desirable to flush even a new system before putting into use for the sake of reliability.

Keeping the oil clean not only extends the life of oil but provides trouble free operation of the hydraulic components and systems as a whole.

2.11.7 Comparison of Various Oils

We have seen from the earlier sections that different varieties of oils are being used in hydraulic systems all over the world. Their sheer variety may sometime create confusion for a designer to decide on the type of oil to be used for a particular system. In Table 2.5 a comparison of a few common oils is provided with respect to their physical properties, cost, etc. which may provide a general guideline to choose a hydraulic oil.

2.12 BIO-DEGRADABLE OILS

Eco-friendly oils developed so far are bio-degradable and therefore in the near future it is expected that such bio-degradable and bio-friendly oils will replace mineral based hydraulic oils throughout the bio-sphere to save the planet. The menace caused by petroleum based oils are well established nowadays such as the additives used in petroleum oils are more toxic than the oil itself.

Bio-degradable lubricants have found extensive use as engine oils, hydraulic oils in construction equipment and food processing machineries in recent times. The development of bio-degradable lubricant was based on the following principal environmental protection theories:

1. The prime ingredient should be bio-degradable in nature.
2. The associated materials (e.g. additives) would be environmentally non-toxic.
3. On natural degradation of the oil over a period of time, the oil should not produce any product which would be toxic in nature.
4. The oil should have no ill effect on the water and soil.
5. The natural gift of fish, insects and plant life should not get disturbed due to the toxicity of the chemicals present in such oil.

Three common types of such biodegradable oils are discussed in Table 2.6 to show their relative merit and demerits.

Hydraulic Oils and Fluid Properties 51

Table 2.5 Comparison of Oil Characteristics

Properties	Mineral Oil	Water Glycol	Water Oil Emulsion	Chlorinated Hydrocarbon	Phosphate Ester	Phosphate Ester Blend	Silicone
Sp. Gravity	0.864	1.060	0.916	1.43	1.275	1.15	0.93-1.03
Viscosity at 100°F (37.75°C)	Low-High 100-200 SSY	Low-Medium 200	Low-High 145	Low-High 145	Low-High 200	Low-High 250	Low-High
V.I.No.	70–100	150	—	30–40	40–50	Low	High
Flash Point (°C)	221°C	None	None	216°C	260°C	246°C	100-150°C
Max. Heating Temperature (°C)	107°C	65°C	65°C	150°C	150°C	150°C	315–370°C
Toxicity	None	None	None	Slight	Slight	Slight	
Pump life	Standard	Standard at low pressure	Reasonable	Comparable	Comparable	Comparable	Fair to Good
Lubricity Suitability To Roller Brg.	Very Good	Fair	Fair	Good	V.Good	V.Good	-do-
Common Seal Material	-do- Synthetic Rubber	Poor Synthetic Rubber	Fair Synthetic Rubber	V.Good Butyl and Silicon	V.Good Butyl Silicon	V.Good Butyl	Viton up to 230°C
Rust Prevention	Very Good	Fair	Fair to Good	Fair to Good	Fair to Good	Fair to Good	Fair
Effect on Standard Point	None	May soften	None	Incompatible	Incompatible	Incompatible	—
Particular Limitation.	Flammable	Poor lubricant	Shear breakdown	Seal compatibility problem	Seal compatibility problem	Seal compatibility problem	Very High Cost
Comparative cost	100	400	150-200	700	500	600	above 700
Comp. WT.	100	120	110	160	125	130	—

Table 2.6

Type	Advantages	Disadvantages	General points
1. Rapeseed oil linseed oil etc.	Soluble in mineral oil. • Better adhesiveness. • Very good lubricating property.	Reduced Thermal/Oxidation stability. • Limited pour point and boiling point • Incompatible with bearings which contains atomic lead.	• Not soluble in water. • Temperature range is from $-10°C$ to $70/80°C$. • Low cost • Degenerates through combustion. • Density 0.92
2. Synthetic lubester e.g. polyolester, diester, monoester etc.	• Soluble with mineral oil. • Better thermal stability than rapeseed oil. • Longevity higher than rapeseed oil. • Good pour point • Very good lubricating property.	• Costlier than rapeseed oil by at least 2 to 3 times.	• Not soluble in water. • Density 0.92 • May be applied from $30°C$ to $100°C$. • Degenerates due to combustion.
3. Polyalkalyne glycol. (PAG) e.g. ethylene oxide, Polyether etc.	• Very good longevity and higher life cycle.	• Poor rust prevention • Friction problem arises between some friction pairs of steel, aluminium. • Produces varnish.	• Cost between rapeseed oil and synthetic ester (twice that of rapeseed oil) • Density 0.99 • Can work between temperature range of $30°C$ to $100°C$ (up to $130°C$ some times) • Soluble in water

Review Questions

1. What are the broad tasks of a hydraulic oil in a hydraulic machine?
2. State the various properties of oil.
3. What criteria should be considered for selection of oils for a given hydraulic system?

4. What is bulk modulus? Is it a desirable property? Explain.
5. Why is pre-filtration advocated for pumping oil from a fresh drum to a hydraulic system?
6. What precautions are to be taken to save oil from contamination?
7. How does one check water contamination of hydraulic oils?

References

2.1 Fitch, E C, *Fluid Power and Control Systems,* McGraw-Hill Book Co. New York.
2.2 Palit, S R, *Elementary Physical Chemistry,* Book Syndicate Pvt. Ltd., Calcutta, 1938.
2.3 Radhakrishnan, M. "Fire Resistant Oils", *The Hindu*, Chennai, Dec. 19,1984.
2.4 Krist, Ing. Thomas, *Hydraulik,* Vogel Verlag, Wuerzburg, Germany, (pp. 69-70).
2.5 Parr, Andrew, *Hydraulics and Pneumatics*, Butterworths and Heinemann, Oxford, UK.
2.6 Nekrasov, B, *Hydraulics for Aeronautical Engineers*, Peace Publications, Moscow, (p. 20).
2.7 Biswas, B N, *Lubricants and Biodegradability*, Bharat Lube Industries Pvt. Ltd.

3
Filters and Filtration

To enhance the operational efficiency of hydraulic components and service life, filters play an important role. The oils used in hydraulic systems contain millions of foreign matters—both solid and liquid—which are quite undesirable in the system and therefore these need to be arrested. If one analyzes the causes of failure of most hydraulic systems, a majority of them will be due to solid particle contaminants in oil.

Filters are a medium in oil hydraulic systems used to arrest and purify the oils and thus get rid of those unwanted materials which are mostly in solid form like dust, dirt, worn out metallic parts and other particles. But these may also be in the liquid form like water, acids, paints, etc. Filters are thus an important part of a hydraulic system.

The quality of oil and the life and efficiency of a filter is another important parameter which one should consider while designing a system. The cost of repair and immature replacement of costly and expensive sophisticated filter elements in a hydraulic system can be avoided or minimized to a larger extent by constant and effective preventive maintenance of the oil and system.

Hence it is necessary to decide the type of filter and filtration method to be adopted in a hydraulic system at the design stage itself. It has been found from experience that the cost incurred through the use of optimum size of filters is compensated through less machine downtime and less maintenance. It has also been found that a filter with more surface area is more economical as the filter life is extended under identical flow conditions as may be clear from Fig. 3.1 if one compares two filters A and B (see the comparison of filters A and B). Various filters and filter terminology are used in the area of filtration science. Some of these are mentioned later on in this chapter.

A hydraulic system is adversely affected by contamination whether it is introduced into the system from the outside or generated from within the system. As contaminants are allowed to recirculate through a system, they wear critical component surfaces. Wear debris adds to the system contamination which, in

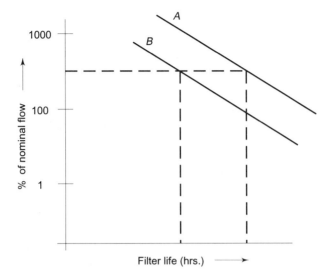

Fig. 3.1 *Enhanced life of filter with higher filter area*
Area of filter B > area of filter A

turn, generates additional contamination. This is sometimes called the "chain reaction" of abrasive wear.

Portable hydraulic servicing and filling equipment very often used by preventive maintenance operators, can transfer contamination to the system. The hose used to connect the hydraulic system to the portable equipment presents a potential contamination problem. Even if connection covers are used, flexing hose can add several thousands of particles per 100 ml of fluid. There are many such examples of how a system may get contaminated.

Maintenance procedures and personnel are other sources of contamination. A typical example is dropping a hose on the floor before installation and using the same again. Improper storing and handling of equipment such as failure to cap lines, pipes and component parts after cleaning and before assembly are some other examples.

3.1 NATURE AND EFFECT OF CONTAMINATION

An absolutely clean fluid is essential for reliable operation of hydraulically controlled systems. With more sophistication and increased pressure, it has become necessary to tighten clearances in valves and other moving parts. Abrasive particles are very often found to cause extreme damage to such devices. The common contaminants may be in the form of—

1. Particulate matter (metallic, fibrous, elastomer)
2. Water
3. Air
4. Chemical reaction products such as sludge
5. Other (non-compatible) hydraulic fluids.

3.1.1 Dirt Causes Downtime

The experience of hydraulic system designers and maintenance supervisors has shown that over 50% of the downtime of hydraulically operated machinery is caused by dirt in the hydraulic fluid.

Excessive fluid velocities add dirt to a system by erosion of pipe and component surfaces. Air entering the reservoir contains dirt and water vapour which causes the formation of rust on reservoir walls. An additional source of dirt is the dirt which is continuously being added to a system through maintenance practices. Every time a line is cracked to service a component, the system becomes more contaminated. It is also not uncommon for maintenance men to dump dirty fluid into a machine's reservoir by accident or due to lack of education.

3.2 SOURCES OF CONTAMINATION

Contamination may be classified as:
1. Built-in metal chips, welding scales, lapping compound, sand from casting or debris produced during the manufacturing process.
2. System generated—wear products generated inside the mating components may be categorized here.
3. Maintenance generated—contaminants generated by opening and closing a system and exposed often to dirty maintenance environments are examples of maintenance oriented contaminants. The new fluid added to the tank during filling-up operation is also another source of contamination.

3.2.1 Contaminants During Assembly Stage

1. Pipe scale—from pipes not cleaned before assembly.
2. Corrosion or rust—from ferrous metal components stored in unfavorable conditions.
3. Sand—residue foundry sand on castings, e.g. valve blocks.
4. Lints or fabric threads—from clothes used for plugging or cleaning components (only lint free clothes should be used for such purposes).
5. Swarf—produced by incorrect assembly tools or technique.
6. Adhesive particles—from surplus adhesive or joining compounds used on gaskets or static seals (e.g. thread sealants).

It is most important that care should be taken to ensure correct and clean assembly before a system is put into operation.

3.2.2 Contamination During Service

Once in service, likely sources of further contaminants are produced, e.g. metallic particles. Besides being harmful because of their abrasive and blocking character, finely divided metal can act as a catalyst promoting early breakdown of the fluid. Metallic particles are very likely to be developed in a new system due to initial pump wear. After a suitable 'running in' period sets in, the wear rate of pumps and other mating parts gets reduced due to the lubricating property of the fluid. Paint flake caused by the failure of old paint on the inside of tank or paints which are not compatible with the fluid are also quite harmful.

As the machine continues to operate, moving parts naturally begin to wear and generate dirt. Therefore every internal moving part in the system can be considered a source of future contamination, especially during the machine's "breaking in" or "running in" period. Component housings continuously experience hydraulic and mechanical shocks due to constant flexing from normal stresses and pressures. These actions cause metal and casting sand to break loose and enter the fluid stream.

Some other contaminants are:
- Acidic by-products—This is mostly caused by the process of oil oxidation. Normally, this is soluble but leads to corrosion problems.
- Sludge—This is also caused by oxidation and fluid-breakdown.
- Elastomeric particles—This is generated from seal wear.
- Airborne solids—They are inducted through the tank, joints, piston surfaces, etc. when leakage of oil takes place.

3.2.3 Damage Due to Solid Contaminants

It has been explained earlier that a lot of solid particles find their way into a hydraulic system. These particles are generally produced due to high mechanical stresses to which the system components are subjected. As during the operation of the machine, these particles are circulated unhindered throughout the system. They produce more solid particles and a chain reaction follows aggravating the situation and accelerating the failure of the system. Figure 3.2 depicts the scoring action of a solid contaminant on a finely machined valve spool or valve body. Solid particles in a system may cause:

1. Blockage of valve openings.
2. Jamming of pistons and spools.

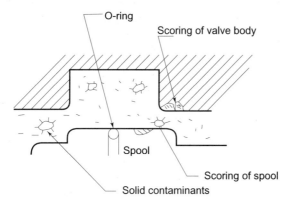

Fig. 3.2 *Damage due to solid contaminants*

3. Uncontrolled and increased leakage.
4. Changes in control characteristics, and
5. Ultimately component or system failure.

The amount of wear and tear of hydraulic components due to solid contaminants depends on the:

1. Material of the solid particles
2. Size of the solid particles
3. Nature of working clearance between mating parts of the components
4. Ratio of particle size to gap between the mating parts of the components
5. Shape of solid particles
6. The oil flow velocity
7. The working pressure of the fluid.

It is noted that hard and sharp particles may cause deep scratches, erosion, etc. and hence are more dangerous than soft and spherical particles. Soft and gelatinous particles may cause undesirable blockages that may lead to component failure. They may also interfere with lubrication by blocking lubricating passages which are generally very narrow and finely finished.

The nature and severity of damage that a solid particle can cause depends on the type of materials and their characteristics as may be evident from Table 3.1.

Table 3.1 Effects of Some Solid Contaminants

Contaminant material	Damaging effect
1. Carborundum	Severe damage
2. Steel, Iron, Brass, Bronze, Aluminium	Severe damage may take place
3. Laminated fabric fibers, seal residue, rubber particles, paint particles, oxidation products (gummy, etc.)	Slight damage only

3.3 EFFECT OF DIRT ON INDIVIDUAL HYDRAULIC COMPONENTS

It is accepted that 75% of the problems of a hydraulic system are dirt-related. Figure 3.3 shows graphically the particle concentration in the hydraulic fluid and the flushing time needed to clean the fluid. Due to the presence of particle contaminants, the fluid power component may have the following problems as given below:

(i) Cylinders
1. Excessive wear of cylinder rod, packing and seals.

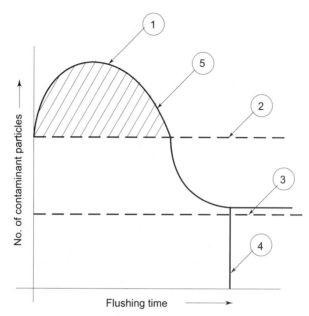

Fig. 3.3 *Particle concentration in hydraulic fluid*
1. Quantity of contaminants to be arrested.
2. Quantity of contaminants in new fluid.
3. Desired concentration of dirt particles.
4. Optimum flushing time.
5. Curve of system contaminantism.

2. Cylinder barrel scoring may damage the costly finished internal cylinder surfaces.

(ii) Hydraulic valves
(a) *Direction Control Value (D.C.V.)*
   1. Small and intricate orifices get plugged.
   2. Spool and housing lands wear—may cause excessive leakage.
   3. Spool may get struck resulting in solenoid failure, excessive shock load damaging the hose, pipe, fittings, etc.
(b) *Flow Control Value (F.C.V.):* 1. Flow orifices get eroded changing the flow characteristics and ability to regulate thus rendering the valve less responsive.
(c) *Check valves*
   1. Balls and seats-wear may take place resulting in leakage.
   2. Fluid may bypass check valves creating system malfunction.
(d) *Pressure valves*
   1. Erratic outlet pressure may cause valve malfunction.
   2. Sequence valves cannot close.
(e) *Pressure reducing valves*
   1. Erratic outlet pressure may cause valve malfunction.
   2. Valve may stick open or closed.

(f) **Relief valve**
    1. Chatter
    2. Excessive seat wear
    3. Change response time and amount of pressure overshoot
    4. Accumulated dirt makes pressure erratic.
  (g) **Servo valves**
    1. Sharp edges erode affecting metering and control.
    2. Nozzles become plugged generating undesirable signals.
    3. Nozzles may erode.
    4. Varnish deposits build up causing sticking and friction, sometimes called "sticktion".

(iii) Pumps
1. Excessive damage due to wear may take place in the pump and its parts because of contaminants including erosion and abrasion of valve plate.
2. The vanes may get stuck inside the slots creating erractic action and the vanes may wear down the cam ring necessitating untimely replacement.
3. The vane slots may wear out with rapidity.
4. Particle contaminants decrease life of bearings.
5. Wear and tear of the gears in a gear pump increases resulting in pump inefficiency.
6. Wear and tear of pistons and piston sleeves may increase.
7. In variable pump compensator controls problems like sticking, slow response, erratic delivery may take place.

(iv) Fluid
1. The wear debris may act as a catalyst to accelerate the process of oxidation and break down molecular structure causing gummy residue and varnish which are quite injurious to the system.
   This may ultimately lead to the following:
   (a) Attract additives and may change the composition of oil.
   (b) The fluid life gets shortened necessitating premature replacement of the fluid.
   (c) Lower down maximum operating temperature.
2. Hydraulic system malfunctions due to dirt may result in
   (a) Damage to equipment
   (b) Safety hazards
   (c) Scrapping of parts due to poor finish
   (d) Insufficient pressure
   (e) Increased downtime
   (f) Unnecessary maintenance and production cost
   (g) Higher oil cost with increased disposal problem

3.4 SYSTEM FAILURE

Presence of contaminants in the fluid may cause two types of system failures
  (a) Catastrophic failure

(b) Degradation failure

(a) Catastrophic failure—takes place when a component suddenly fails to function such as jamming of a check valve, or when a particle of a critical size gets into a specific location, for example, plugging of a fine orifice in a highly sensitive servo-valve.

(b) Degradation failure—a slow time consuming event because it is a wear process producing a slow deterioration of component performance over a period of time. A very good example is the slow increase in internal leakage in a pump as it wears.

3.4.1 Toleration Limit of Particle Size

There are two schools of thought as to the sizes of particles that cause degradation failure

One concept says that degradation wear is related to the dirt level of the system. Component wear is accelerated as dirt levels increase above the tolerable level of a component. A little increase in component life is obtained by reducing dirt levels below the level of component tolerance. This is because of the probability of a great number of particles with certain characteristics being present in the thin oil films of the system.

It is the experience of many filtration engineers that while catastrophic failure occurs due to particles above 5 µm dia. in the oil, degradation failure occurs due to particles under 5 µm size. However it is true that degradation failure is related to the dirt level of the hydraulic system. With increase in dirt level above the tolerable limit of a component, there is considerable acceleration of component wear. This means if the dirt level in a system can be contained to a lower level of component tolerance, there will be considerable increase in component life. It is more so when the hydraulic system is designed for high pressure as such a system is more sensitive to dirt levels than those designed for lower pressures.

The concentration of dirt particles is another factor which affects component wear. It is the expert opinion that the distribution of the particle sizes versus concentration (no. of particles) in the fluid follows a specific pattern—number of particles increasing exponentially with reduced particle size. A system design engineer is more concerned about those particles whose size is about the same as that of the oil thickness. It has been found that the thinner the oil film in a high pressure hydraulic system, the higher the probability of containing more smaller sized particles due to their specific distribution characteristics. These smaller size dust/dirt particles (specially those below 5 µm) are primarily responsible for degradation failure because the average typical clearance in the most modern hydraulic components is of the same size or smaller than 5 µm, providing enough scope for much smaller particles to enter the gap between the moving/sliding components and abrade/erode the surfaces which are critically finished and coated for easy sliding. The resultant worn out particles from such abrasions may contain both ferrous and non-ferrous metals critical enough to start a chain reaction of abrasion and formation of other undesirable products due to the presence of hard metallic and non-metallic worn out particles, e.g. (Fe, Zn, Cr, gummy and plastic/rubber materials etc.) which may act as a catalytic agent and oxidize the hydraulic oil under such conditions. The oxidation products contain both solids and acids. This is caused by water which invariably gets into the system as moisture which also corrodes the base metal.

3.5 CONTAMINATION LEVELS AND STANDARDIZATION

From the above discussion it is clear that it is the primary duty of all concerned personnel to maintain the contamination level in a hydraulic system to a tolerable limit. Various bodies, e.g. ISO, SAE, ASTM, CETOP, AIA, etc. have standardized level and class of contamination.

This standardization helps a hydraulic engineer to gain an idea about the quality of oil, i.e. the number and size of contaminants present in the oil. Accordingly, the engineer can use a filter to arrest the same, if needed.

The SAE standard 749D, Sept. 1963 is shown below in Table 3.2. Due to the small number of gradations the SAE classification is however rarely used.

Table 3.2 Contamination Level for Hydraulic Fluid (particle per 100 ml) as per SAE 749D

Range of particle size in micron	Contamination class						
	0	1	2	3	4	5	6
5–10	2700	4600	9700	24000	32000	87000	128000
10–25	670	1340	2680	5360	10700	21400	42000
25–50	93	210	380	780	1510	3130	6500
50–100	16	28	56	110	225	430	1000
100	01	03	05	11	21	41	92

Apart from the above SAE classification, there are other classification standards as shown below:

*ISO DIS 4406
*CETOP RP 70H
*NAS 1638
*MIL STD 1246A
*AIA—(Aerospace Industries Association of America)

The MIL STD 1246 A is used only for special cases.
Table 3.3 compares the different classification systems defining the quantity of particles of a certain size in 100 ml of oil.

Classification is done by counting and sizing the solid contaminants either under a microscope or an electronic particle counter if the dirt concentration is below 20 mg per liter.

Table 3.3 Various Contamination Standards and their Comparison

ISO DIS 4406 or Cetop RP 70 H	Particles per ml > 10 μm	ACFTD solids content mg/L	MIL STD 1246 A (1967)	NAS 1638 (1964)	SAE 749 D (1963)
26/23	140000	1000			
25/23	85000		1000		
23/20	14000	100	700		
21/18	4500			12	
20/18	2400		500		
20/17	2300			11	
20/16	1400	10			
19/16	1200			10	
18/15	580			9	6
17/14	280		300	8	5
16/13	140	1		7	4
15/12	70			6	3
14/12	40		200		
14/11	35			5	2
13/10	14	0.1		4	1
12/9	9			3	0
18/8	5			2	
10/8	3		100		
10/7	2.3			1	
10/6	1.4	0.01			
9/6	1.2			0	
8/5	0.6			00	
7/5	0.3		50		
6/3	0.14	0.001			
5/2	0.04		25		

Table 3.4 Contamination Data and Short Form Coding

	No. of particles per 100 ml			
	over 5 μm		over 15 μm	
Code	more than and up to		more than and up to	
20/17	500 k	1 M	64 k	130 k
20/16	500 k	1 M	32 k	64 k
20/15	500 k	1 M	16 k	32 k
20/14	500 k	1 M	8 k	16 k
19/16	250 k	500 k	32 k	64 k
19/15	250 k	500 k	16 k	32 k
19/14	250 k	500 k	8 k	16 k

	No. of particles per 100 ml			
	over 5 μm		over 15 μm	
Code	more than	and up to	more than	and up to
19/13	250 k	500 k	4 k	8 k
18/15	130 k	250 k	16 k	32 k
18/14	130 k	250 k	8 k	16 k
18/13	130 k	250 k	4 k	8 k
18/12	130 k	250 k	2 k	4 k
17/14	64 k	130 k	8 k	16 k
17/13	64 k	130 k	4 k	8 k
17/12	64 k	130 k	2 k	4 k
17/11	64 k	130 k	1 k	2 k
16/13	32 k	64 k	4 k	8 k
16/12	32 k	64 k	2 k	4 k
16/11	32 k	64 k	1 k	2 k
16/10	32 k	64 k	500	1 k
15/12	16 k	32 k	2 k	4 k
15/11	16 k	32 k	1 k	2 k
15/10	16 k	32 k	500	1 k
15/9	16 k	32 k	250	500
14/11	8 k	16 k	1 k	2 k
14/10	8 k	16 k	500	1 k
14/9	8 k	16 k	250	500
14/8	8 k	16 k	130	250
13/10	4 k	8 k	500	1 k
13/9	4 k	8 k	250	500
13/8	4 k	8 k	130	250
12/9	2 k	4 k	250	500
12/8	2 k	4 k	130	250
11/8	1 k	2 k	130	250

For a turbid fluid when the dirt concentration is above 20 mg/liter, the contamination level can be determined by gravimetric analysis, i.e. by weight.

3.5.1 ISO Code

Figure 3.4 shows plotting of particle size vs. number of particles. In ISO classification (ISO DIS 4406) the level of contamination of the fluid is defined by a two digit code—one being the number of solid particles above 5 μm in size and other being the number of particles over 15 μm in size in a 100 ml sample of fluid. The number of all particles larger than 5 μm in the oil sample is counted and given a code number and then a code number is given for all particles larger than 15 μm which is shown in Fig. 3.4 and Table 3.4.

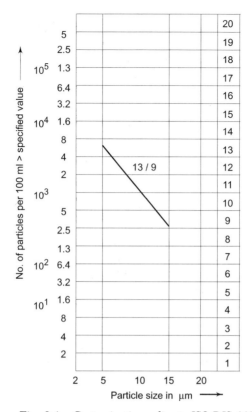

Fig. 3.4 *Contamination coding to ISO DIS 4406*

3.5.2 NAS 1638 Code

In this method of classification, particle sizes are divided into 5 ranges from 5 μm to over 100 μm size. The range and number of particles in each class is shown in Table 3.5.

Table 3.5 Contamination Level According to NAS 1638 (maximum number of dirt particles in 100 ml oil)

Class	Size range in mm				
	5–15	*15–25*	*25–50*	*50–100*	*> 100*
00	125	22	4	1	0
0	250	44	8	2	0
1	500	89	16	3	1
2	1000	178	32	6	1
3	2000	356	63	11	2
4	4000	712	126	22	4
5	8000	1425	253	45	8
6	16000	2850	506	90	16
7	32000	5700	1012	180	32
8	64000	11400	2025	360	64

(*Contd.*)

Class	Size range in mm				
	5–15	15–25	25–50	50–100	> 100
9	128000	22800	4050	720	128
10	256000	45600	8100	1440	256
11	512000	91200	16200	2880	512
12	1024000	182400	32400	5760	1024

3.6 FILTER RATING

Correct rating of a hydraulic filter is essential in order to
 (a) Prevent undesirable contaminants—solids and liquid particles into the system.
 (b) Enhance time intervals of carrying out preventive maintenance and increase ease of preventive maintenance.
 (c) Maintain timely switching functions of control valves as designed.
 (d) Increase life of filter, fluid and system components.
 (e) Maintain functional efficiency of the system and prevent operational disturbances due to filtration failure.
 (f) Minimize effect of pressure, flow and temperature variation of filter functioning.
 (g) Ensure high capacity of contaminant holding.
 (h) Ensure desired lubricity of oil and minimize chemical degradation of the fluid.
 (i) Reduce downtime due to maintenance shut down.
 (j) Ensure high reliability of filter and filtration
 (k) Ensure continuous availability of the system.

3.6.1 Filtration Performance

Dirt holding capacity and filtration capacity of filters are important characteristics that determine the application of filters. ISO has standardized a method called multipass test (ISO 4572) which helps to make a comparison of filter elements made by different firms but with identical filtration rating.

3.6.2 Multipass Test Rig

The diagram in Fig. 3.4 (a) shows the arrangement of a multipass test rig with the following components:
 (a) Hydraulic reservoir for test fluid
 (b) Pump
 (c) Heat exchanger
 (d) Flow meter
 (e) Cleaning filter
 (f) Differential pressure gauge
 (g) Electronic particle counter
 (h) Tank for hydraulic fluid to be injected
 (i) Test filter

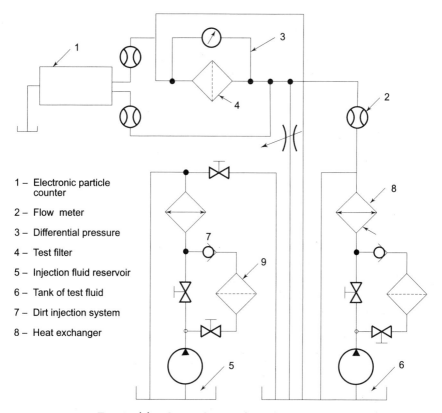

1 – Electronic particle counter
2 – Flow meter
3 – Differential pressure
4 – Test filter
5 – Injection fluid reservoir
6 – Tank of test fluid
7 – Dirt injection system
8 – Heat exchanger

Fig. 3.4(a) *Circuit diagram for multi pass test rig*

The test rig consists of two parts:
(a) The testing unit with test fluid, pump, coolant, test element and electronic particle counter.
(b) The dirt injection system.

In the dirt injection system, contaminated fluid with test dust is kept. The test is conducted after cleaning both the systems with ultra fine filters when the reference value of contamination particle count is achieved. The test filter is subjected to the fluid in which specific contaminated fluid is injected and both upstream and downstream fluid samples are collected and analyzed in the electronic particle counter. The pressure drop across the test filter is measured. The retention rate for filtration rating is defined by the degree of separation β_x where β_x denotes the β-rating and x denotes the particle size larger than x microns. β_x is defined as the number of dirt particles larger than a given particle size x counted upstream of the test filter divided by the number of dirt particles counted downstream of the filter element under the same pressure drop condition. β_x is a dimensionless number which is the degree of separation.

$$\beta_x = \frac{n_1}{n_0}$$

where β_x = degree of separation
  n_i = number of particle upstream or in $\geq x$
  n_o = number of particle downstream or out $\geq x$

Let us take a simple numerical example.
Assume a system requires 10 μm level.

Let n_i = 15000 particles where $x \geq 10$ μm/100 ml
  n_o = 200 particles where $x \geq 10$ μm/100 ml

$\therefore \quad \beta_{10} = \dfrac{n_i}{n_o} = \dfrac{15000}{200} = 75$

$\therefore \quad \beta_{10} = 75$ where 10 is particle size in μm.

Efficiency of a filter can be expressed by β ratio.

$$E_x = \text{Efficiency} = \dfrac{\beta_x - 1}{\beta_x} \times 100.$$

$\therefore$ If $\beta_x = 75$, it means that 98.5% filter efficiency is achieved which should be the minimum particle capture size.

Similarly $\beta_x = 2$ means 50% filter efficiency
  $\beta_x = 20$ means 95% filter efficiency.
  $\beta_x = 100$ means 99% filtration efficiency, i.e. absolute retention rate.

3.7 FILTER TERMINOLOGY

Before going into details of the types of filters and the cost of filtration, it may be better to introduce certain filtration terminology at this stage.

(a) *Filter and strainer* Filters and strainers are devices whose primary function is the retention of insoluble contaminants. Generally a strainer is a coarse filter. Various porous media or wire mesh screens are used as a filtration medium. The medium is mostly a screen or other forms of filtering devices that stop contaminants but allow the oil to move forward.

(b) *Mesh number* A simple screen or a wire strainer is rated for filtering fineness by a mesh number or standard sieve number. The higher the sieve number the finer the screen. However this method of filter sizing is now outdated and so more and more fine filtration is the need of the day, new and more accurate methods of filtration rating has come into existence.

(c) *Micron rating* Filters made of materials other than wire screens are rated in micron size. A micron is one-millionth of a meter. To have an idea about the size of a micron we consider a grain of salt which is about 70 microns in size. A bare human eye can see a 40 to 44 micron object. The equivalent size in microns for a few sieve numbers is given below for certain values of sieve numbers.

(d) *Equivalent sieve number and size in microns.* Equivalent size in microns for a particular sieve number is given below for a few values.

Table 3.6 Equivalent Micron Size for a Particular Sieve Number

Sieve number	Opening in micron
10	1650
20	830
50	297
70	210
100	149
140	105
200	74
270	53
325	44
	10
	5
	3
	0.5

(e) Particle count The dirt in a 100 ml sample is classified, counted, and put into a cleanliness class. A particle count determines the type (sand, steel, etc.) quantity (numbers), and quality (size) of dirt in a hydraulic system. In filter application, a particle count would give a good indication of the type of element and the degree of filtration required.

(f) Silt index A number which indicates the number of particles below 5 microns in a fluid sample.

3.8 FILTER DESIGN

To determine the correct type and size of filter for a specific hydraulic system, an engineer has to consider a number of factors for optimum filtration efficiency as well as system optimization.

The following criteria need attention:

(i) *Recommended values of filtration rating* The filtration rating should be appropriate to the system components to optimize dirt sensitivity. The following table (Table 3.7) is worth noting. The most sensitive component determines the designed rating.

Table 3.7 Recommended Filtration Rating for Fluid Power Components

System component	ISO cleanliness class	Absolute filtration rating recommended (μm)
1. Cylinder:	19/15	20
Servo cylinder	17/13	5
2. Pumps:		
Gear	19/15	20
Vane	18/14	10
Piston	18/14	10
3. Valves:		
D.C. valve	19/15	20

(Contd.)

System component	ISO cleanliness class	Absolute filtration rating recommended (μm)
Flow control valve	19/15	20
Pr. relief valve	19/15	20
Pressure valve	18/14	10
Proportional valve	18/14	10
Servo valve	17/13	5

(ii) *Flow rate through filter* The pore size, filtration area and possible location of filters may determine the total quantity of fluid passing through a filter. Sometimes this may be higher than maximum pump capacity, e.g. return of oil from several circuits through return lines may have a higher volume than the pump's flow rate.

(iii) *Pressure drop across the filter at normal viscosity and clean element* This is another important factor. Design values of pressure drop recommended (for pressure line filter) are given below:

(a) Without by-pass valve = 1.0 bar
(b) With by-pass valve = 0.5 bar
(c) Return line filter = 0.3 – 0.5 bar

However these values are not conclusive.

(iv) *Filter housing* The filter housing should have adequate fatigue strength and shock absorbing capacity.

(v) *Filter material compatibility* Compatibility of filter material is a very important parameter to decide the design of filter characteristics and their use.

(vi) *Operating temperature* The most important factor here is the viscosity index (VI) number. The fluid viscosity and VI number decide the filter size.

(vii) *Application* The location and application area—whether it is to be used in a laboratory or in a rolling mill determines the probable contamination based on which a filter has to be designed or selected.

(viii) *Filter mode* Varieties of filters with or without indicators, by-pass valve etc. are available in the market. While going for a specific design of a filter for a specific application, type of clogging indicators, e.g. visual, electric, electronic with or without by-pass valve are certain points to be considered as the overall efficiency of filtration is dependent on such factors.

The variation in cleanliness classes that can be achieved by using different recommended absolute filtration rating is shown in Fig. 3.5. The rapid rise of concentration of contamination and filtration classes that can be achieved with fine filtration can be observed from the diagram.

3.8.1 Estimation of Filter Size

The size of a filter and design of the fluid power elements have direct bearing on the fluid characteristics. Hence the following criteria need optimum attention while deciding the filter elements and filter size.

(i) *Fluid Viscosity*: The pressure drop across the filter element has to be converted to the value as operating viscosity by means of viscosity

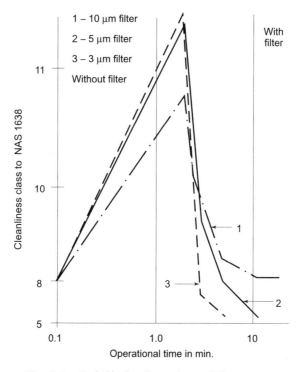

Fig. 3.5 *Probable cleanliness class and filter rating*

conversion factor in case the operating viscosity deviates from the reference value. Figure 3.6 can be used to find out the viscosity conversion factor which is needed to calculate the total pressure drop.

(ii) *Oil Density*: The density of oil is also an important factor to consider. The pressure drop across the filter housing depends on the fluid density, which can be calculated from the following equation.

$$\Delta P_h = \Delta P_r \cdot \frac{\rho_o}{\rho_R}$$

where ΔP_h = Pressure drop across the housing
ΔP_r = Pressure drop across the housing reference
ρ_R = Density of oil (refer data from oil manufacturer's catalogue)
ρ_o = Density of oil, operating value.

The size of the filter must also take into account the environmental condition around the machine and level of machine maintenance provided to the system. The permitted pressure drop across the filter may be calculated from the following equation.

$$\Delta P_t = (\Delta P_h + f_1 \cdot \Delta p_c) \cdot f_2 \text{ where}$$

ΔP_t = Total pressure drop
ΔP_h = Pressure drop across the filter housing
f_1 = Viscosity conversion factor (can be taken from Fig. 3.6)

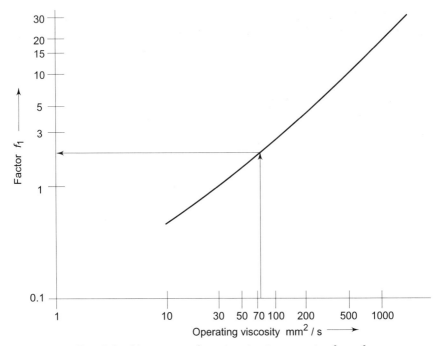

Fig. 3.6 *Monogram to determine viscosity conversion factor f_1*

f_2 = Environmental factor (may be taken from Table 3.8)
ΔP_e = Pressure drop across the element.

3.9 ENVIRONMENTAL FACTOR

As the atmospheric and environmental condition surrounding the hydraulic system has direct effect on the contamination level, the pressure drop across the filter is greatly influenced and the environmental factor must be taken into consideration while designing the filter. The value of the factor may be found out from Table 3.8. Here the amount of preventive maintenance also plays a major role.

Table 3.8 Values for f_2

Level of system	Contamination of machine surroundings		
	Low	Average	High
1. Good sealing of reservoir, regular preventive maintenance of machine system and filter, low dirt ingress, etc.	1.0	1.0	1.3
2. Irregular checking of filter	1.0	1.5	1.7
3. High dirt ingress, total lack of maintenance	1.3	2.0	2.3

Low—Machine in air conditioned and confined room condition; Average—Machine in hot and humid open air condition; and High—Machine in foundries, steel mills, mining and mobile machinery.

Though the designed value of pressure drop for a filter size could be calculated as shown here, most designers prefer to use various monograms available in the industry to determine the pressure drop for a specific filter size against a specific flow rate.

The graphical representation shown in Fig. 3.7 can be used to find the pressure drop across the filter housing. Figure 3.8 shows the pressure drop across a clean filter with 4 different sizes of filters when used with an oil having kinematic viscosity of 30 cSt.

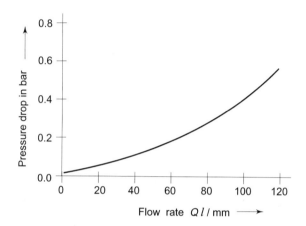

Fig. 3.7 *Pressure drop across filter housing*

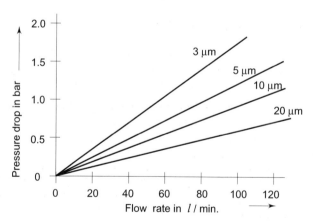

Fig. 3.8 *Pressure drop across filter element*

3.10 TYPES OF FILTER

Two main types of filters basically are used in a hydraulic system. They are:
 1. Surface filter

2. Depth filter

Surface filter Surface filters are nothing but simple screens which are used to clean oil passing through their pores.

The screen thickness is very thin and dirty unwanted particles are collected at the top surface of the screens when the oil is passed. A common example of a surface filter is a strainer.

Depth filter They are thick walled filter elements through which oil is made to pass and thus the undesirable foreign particles are retained. The capacity of depth filter is much higher than surface filters as much finer materials have a chance of being arrested by these filters. Whereas in the case of surface filters nothing finer than the rated size of the pore could be arrested.

Dirt holding capacity of surface filters compared to depth filters is shown in Fig. 3.9.

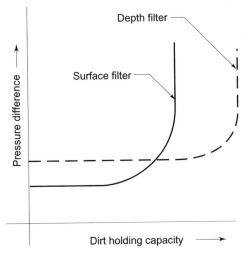

Fig. 3.9 *Comparison of dirt holding capacity of surface and depth filter*

Other classifications Filters are also classified as
 (a) Full-flow filter
 (b) By-pass filter

Full-flow filter In a hydraulic system it is necessary that all the flow must be through the filter element. This means that oil must enter the filter element at its inlet side and must be sent out through the outlet after crossing the filter element fully. In case of full flow filtration, the filter is sized to accommodate the entire oil flow at that part of the circuit.

By-pass filter Sometime the entire volume of oil need not be filtered and only a portion of the oil is passed through the filter element and main portion is directly passed without filtration through a restricted passage as shown in the circuit diagram in Fig. 3.10(a).

Filters and Filtration **75**

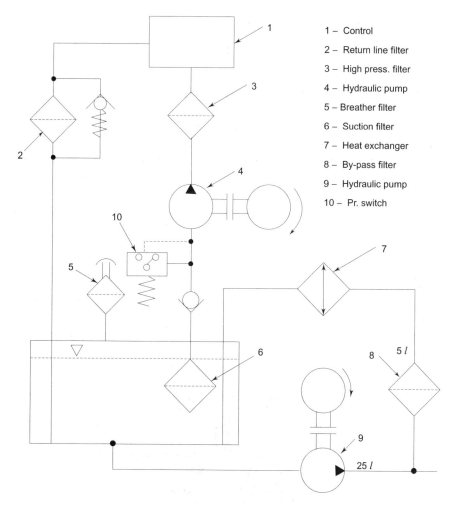

Fig. 3.10(a) *Positioning of filter and use of by-pass filter*

Due to pressure difference between the inlet and the restricted passage, a portion of the oil is drawn towards the filter cartridge and flows back to the restricted passage through the throat. 99% of the dirty material is found to be arrested by the filter cartridge of such filters.

The advantages of using a by-pass filter are:
1. Filtering is independent of system in operation or under shut down.
2. Filter elements can be replaced without shutting down the main system.
3. Less down time.
4. Less maintenance.
5. Inexpensive filter elements.
6. High dirt retention capacity due to low pulsation free constant volume flow.

The sizing of a by-pass filter is based on:
1. Filter area
2. Expected volumetric flow through the filter.

In non-life threatening or non-damaging applications, some designers do not prefer a by-pass valve. They want 100% filtration. If the filter ultimately clogs and the clog indicator is ignored, the system just stops operating. In such an eventuality the machine operators or maintenance mechanics have to change the clogged filters.

A pressure line non-by pass filter contains a built-in valve that blocks flow through the filter as the element clogs and differential pressure increases. However the dirty element is not exposed to high differential pressure as no flow passes through the filter. A filter with no-by pass valve built in is shown in Fig. 3.10 (b).

Filter area: The filtration rating must be determined first before the required filter area can be calculated. The filtration rating depends on the hydraulic components and probability of contamination generation. The sizing of the by-pass filter is calculated keeping in view the volumetric flow through the filter and filter area.

The following equation may be used to determine the volumetric flow and area of volumetric flow.

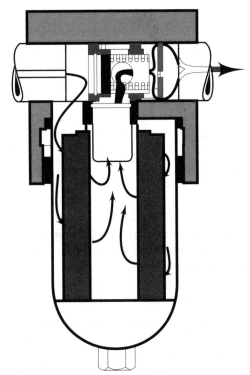

Fig. 3.10(b) *Filter with no-by pass valve built in*

$$Q_v = Q_e \cdot f_1 \cdot f_2$$

where Q_v = Volumetric flow for filter design
Q_e = Effective volumetric flow
f_1 = Viscosity conversion factor
f_2 = Environmental factor

The formula to calculate filter area (A) is given below:

$$A = \frac{Q_n \cdot f_1}{q}$$

where A = Required filter area
Q_n = Volumetric flow through by pass filter
q = Specific area loading
f_1 = Viscosity conversion factor

The specific area loading can be calculated from Table 3.9.

Table 3.9 Specific Area Loading

Filtration rating $\beta_x \geq 100$	Specific area loading liter/min/cm²
3 μm	0.0025
5 μm	0.0035
10 μm	0.005
20 μm	0.005

3.11 FILTER CONSTRUCTION

Strainers on the delivery line in the tank are without a housing or casing. All other types normally comprise a filter element fitted inside a pressure-tight casing. This may be designed for in-line mounting or with a tee-head and a detachable housing enclosing the element. The former can provide the more compact unit but tee-head construction is generally preferred since this allows the element to be removed for inspection, cleaning or replacement without break in the line connection.

Filter heads may also incorporate:
(i) A differential pressure indicator to measure the pressure drop.
(ii) A by-pass valve to by-pass the filter element when the pressure rises to a point to damage the element.

3.11.1 Types of Filter Media

Filtration can be defined as the removal of solid particles from a liquid or gaseous stream by means of a porous medium. There are numerous types of filter media available in the market. They can be either fixed pore or non-fixed pore type. Fixed pore size media are preferable to the non-fixed pore type for several reasons as given in Table 3.10.

Table 3.10 Fixed Pore vs. Non-fixed Pore

Fixed pore	Non-fixed pore
• Woven mesh • Sintered powder • Membrane (Mixed ester and polymeric) • Resin impregnated fiber matrix	• Fiber glass, unbounded • Cotton wound • Sand beds • Diatomaceous earth, • Fullers earth • Poorly bonded paper • Waste (Shredded Newsprint, Rags, Wood chips)

Table 3.11 Filter Element, Application and Characteristics

Type of element	Average filtration rating	Specific points to note
Felt	25–50 microns	Subject to element migration unless resin-impregnated.
Paper	Down to 10 microns	Medium dirt holding capacity, low permeability, low element strength suitable for suction/return time filter/fine filter. Subject to element migration. Disposable, low cost, low pressure drop.
Fabric	Down to 20 microns	Higher permeability than papers. Higher strength with rigid back up mesh.
Wire gauge	Down to 35 microns	Suitable for suction strainer
Wire wound	Down to 25 microns	Better mechanical strength.
Wire cloth	Down to 10 microns	Expensive, high strength and free from migration.
Glass fiber	Used as random layers with binding agent. Very fine filter for precision components.	Fine filtering, good dirt holding capacity, good chemical resistance, absorption of particles over wide pressure drop range, compatible with all fluids.
Edge type (Paper disc)	10 to 1 micron	High resistance to flow, clogs readily. Degrees of filtration variable with compression.
Phenolic resin impregnated paper	Fine filter, disposable low cost.	Same as above. Fluid compatibility should be a deciding factor.
Edge type (metal)	Down to 25 microns.	Very strong self-supporting suitable for high temperature, at high pressure.
Sintered woven wire cloth (Metal granules	10-20 microns	High strength suitable for high temperature, complete freedom from element migration, high

(*Contd.*)

Type of element	Average filtration rating	Specific points to note
sintered together. The diameter of granules determines the filtration rating)		cost, low dirt capacity. Sensitive to pressure shock, high pr. drop Good mechanical strength, suitable for high pressure & temperature, low dirt holding capacity. However element migration not completely eliminated under severe condition.
Sintered porous metal (may be used as protection filter).	Down to 2 microns	Low manufacturing cost, very high strength, high pressure drop, suits low flow rates.
Sintered porous metal with woven wire reinforcement.	—	Low manufacturing cost, very high strength, high pressure drop.
Sintered PTFE	5–25 microns	High cost, subject to element migration, strength improved with reinforcement.
Sintered Polythene	30 microns	Low resistance to flow and temperature above 60°C
Sintered metal felts	Down to 5 microns	High cost, but freedom from element migration. Difficult to clean. Elements usually replaced.
Membrane Filter	Down to sub-micron	Low mechanical strength, poor dirt capacity, used in oil-reclamation.
Magnetic filter	Ferrous particles only	Little or no resistance to flow.
Filter clothes	Down to 10 microns	May be used in air-breather.
Metal non woven type	Stainless steel wire random layering fine and very fine.	Suits high operating temperature, high pressure drop, used for all fluids. Very expensive, limited cleaning, good fatigue properties. Good compatibility.
Activated Earth		Limited application in hydraulics. Use as a depth type filter.

3.11.2 Constructional Feature of Filter Media

Various types of cellulose, glass, plastics, papers and metals are used for construction of filters. Generally filters used in hydraulic systems are classified as surface filters and depth filters. As the dirt holding capacity and filtration capacity of surface filters and depth filters vary according to their construction, the material used and constructional features also vary to some extent. The materials used and types of construction preferred for both these two types are given in brief in Table 3.12.

Table 3.12 Surface Filter

Material	Construction	Advantages	Disadvantages	Application
1. Square mesh stainless steel, galvanised iron and phosphor bronze	Wiremesh	Easy to clean elements. Low pressure drop.	Low filtration rating (< 10 μm) different small free filter area.	Lubricating oil filters, coarse filters, suction filter, etc. May be used for filtering water, resistant fluids, etc.
2. Wire gauges, stainless steel gauges	Braided mesh	Easy to clean. Low pressure drop and very high pressure difference	Multipass not possible	Coarse filters Protection filters.
3. Stainless steel	Split tube with triangular section wire wound around at different pitch angles	Easy to clean can be used for corrosive media, water, fire resistant fluids, etc.	High filtration rating (> 50 μm) small free filter area.	Suitable for back flushing or as coarse filter.

3.11.3 Filter for Fire Resistant Fluids

To ensure perfect compatibility of the filter material with fire resistant fluids the following materials need be used for filters.

Filter Element: Glass fiber non-woven, metal non-woven, stainless steel wire mesh

Filter Housing: Steel, C.I with phosphated or electro-less nickel plated surface.

Filter for fire resistant fluids should have a larger area than those for mineral oil based filters because of higher component wear, possibility to have more micro-organisms.

Recommended Filtration Rating and Class   For general purpose system–10 to 25 μm absolute; 18/14 cleanliness class
For proportional valve   –10 μm absolute ;   17/13 cleanliness class
For servo system   –5 μm or less absolute; 18/14 cleanliness class

3.11.4 Dirt Capacity and Service Life

Dirt capacity is the amount of dirt that can be allowed to the filter test system until the terminal ΔP is reached.

On the other hand, service life is defined as the length of time that a filter can last in an actual system before the terminal ΔP is reached. The following variables generally determine the dirt capacity and service life of a filter.
1. Ingestion rate
2. Generation rate
3. Flow rate
4. Terminal ΔP
5. Film integrity
6. Particle size distribution.

3.11.5 Gaseous Contamination

Very often air and moisture entraining into the hydraulic oil may create an undesirable effect in the smooth functioning of the system. The specific point worthy to note here is that air may exist in the oil from free solution to free state condition continuously changing state with change of pressure in the system. Catastrophic failure of system component may take place if free air contained in a hydraulic fluid operating at 200 bar enters a low pressure region e.g. when the air from valve enters the return line the sudden expansion of air may damage the related parts. Air may also cause erroneous response of the control elements resulting in system malfunction. Apart from this the most common problems caused by presence of air in oil-system ranges from sluggish operation, system instability, poor servo-response, pump cavitation, oil oxidation, undesirable shock to cylinder movement, etc.

3.12 LOCATION OF FILTER

It is difficult to decide the location of the filter in a hydraulic system. Various locations are used to arrange a filter either in the return line or in the pressure line but each has its own advantages or disadvantages.

However, in general one may use a filter either in the intake side of the pump or in the outlet side of the pump. As discussed earlier, sometimes filters are fitted in the return line of the system. One may also note here that in some sensitive circuit design, filters are used to protect sensitive valves for which in-built filters are used as in the case of servo-valves.

In general the following locations are preferred:
(a) *Return line filter*–It may reduce the amount of dirt ingested through the cylinder and seals from reaching the tank.
(b) *Intake filter*–They are fitted before the pump so that they can prevent random entry of large and other contaminants like large chips into the pump and thus prevent damage to it.
(c) *Pressure filter*–A pressure filter is used sometimes at the pump outlet to prevent entry of contaminants generated in the pump, into other components like valves, etc. and thus help in avoiding the spread of such undesirable elements into the whole system. This will thus protect valves, hydromotors, cylinders, etc.

(d) A last chance (or final control) filter to keep out of a component the large debris that can cause the component to fail through in-built arrangement or by additional protective design.

Each location is important and hence many use all four in the system. However, a decision by the designer to use filters in all possible locations should be taken after careful consideration of the pressure and flow parameters.

Wherever we locate the filter in the system, we must understand that there are certain pockets where the maximum amount of contaminants are concentrated. One such area in the overall hydraulic system is the reservoir. Hence a maintenance person has to take necessary steps so that regular cleaning of hydraulic reservoir is not neglected.

3.13 PHYSICAL MEASUREMENT OF CONTAMINANT SIZE AND FLUID ANALYSIS

3.13.1 Fluid Sampling

Fluid analysis is a check on the fluid to insure that the fluid conforms to specifications. It checks if the composition of the fluid has changed, and determines its overall dirt contamination.

To take a fluid sample, select a fitting or pipe union on the pressure side of the system as a sampling point. Cleanse the sampling point with a solvent such as kerosene. Allow the system to reach its operating temperature. While the system is at low pressure, crack the fitting or pipe union and allow the fluid to flush the sampling point. Fill a clean sample bottle with the fluid. For analysis of tank oil, static oil may also be collected from tank. Method of collection is shown in Fig. 3.11.

3.13.2 Points to Note While Collecting the Sample

1. The sampling device should be cleaned and flushed out with clean solvent.
2. The sample collected should be stored in a clean bottle after removing the last trace of cleaning solvent.
3. While taking the sample from a dynamic portion of oil the m/c should run for a few minutes before the same is collected.
4. Some amount of oil (not less than 2 to 3 liters) should pass through the sampling device before the actual sample is drawn.
5. The sample bottle must be sealed with a protective foil which should only be lifted while putting the sample.
6. Sample should be collected by only trained personnel.

The oil sample is analyzed in order to determine:
   (a) The solid particle contamination
   (b) The oxidation level of oil
   (c) The usefulness of the filters used
   (d) The flushing time while commissioning a system
   (e) The state of oil and system components

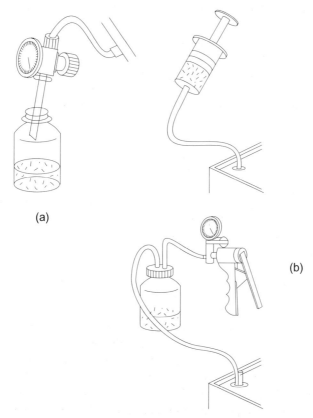

Fig. 3.11 *Collection of fluid sample*
(a) moving fluid
(b) static fluid

Though the best method of checking the quality of oil sample is by an electronic particle counter, due to cost and other reasons it may not be economical to keep it in all workshops. A microscope analysis can give a general estimate of particle size. The state of oil condition can however be found out most economically by using the blotting paper test and along with a microscope the estimation is near perfect.

3.13.3 Checks Performed to Insure that Fluid Conforms to Specifications

Viscosity   The internal resistance of a liquid to flow.

A change in viscosity may mean a decrease in machine performance because of increased leakage, higher pressure drops, lack of lubrication, and/or overheating. It may also mean than an incorrect fluid was used to fill the machine's reservoir.

A change in viscosity ± 10% from original specifications or from a previous sample is considered excessive.

In the application of a filter, the viscosity must be known to determine the initial clean pressure differential across filter element and housing.

Specific gravity   The ratio of weights between water and an equal volume of another liquid.

A change in specific gravity may mean an increase in the lift of a suction line and may also indicate that an incorrect fluid was used to fill the machine's reservoir.

A change in specific gravity ± 1% from original specifications or from a previous sample is considered excessive.

In the application of a suction filter, specific gravity must be known to determine the lift or head in the suction line.

3.13.4   Checks Performed to Determine if the Composition of the Fluid Has Changed

Neutralization number   A number which indicates the amount of potassium hydroxide needed to neutralize all the acids in a fluid.

A check on the neutralization number is performed to determine if additives have deteriorated or oxidation is excessive.

Water content   Water in hydraulic fluid reduces lubrication and affects additives and promotes oxidation. In filter application, water content is a consideration in element and housing compatibility.

Aniline point   The lowest temperature at which a liquid is completely miscible with an equal volume of freshly distilled aniline.

3.13.5   Checks Performed to Determine the Fluid's Overall Dirt Contamination

Particle count   The dirt in a 100 ml sample is classified, counted, and put into a cleanliness class.

A particle count determines the type (sand, steel, etc.) quantity (numbers), and quality (size) of dirt in a hydraulic system.

In filter application, a particle count would give a good indication of which type of element and the degree of filtration required.

Silt index   A unit-less number which indicates the number of particles below 5 microns in a fluid sample.

The silt index gives a verification of varnish and shows the silting tendencies of the fluid. In the filter application, a high silt index may indicate the need for fine filtration or the need to use coarse filter elements.

Gravometric determination of dirt contamination by weight may be done to show the dirt level by weighing the dirt in each sample. It is quite helpful to decide the nature of filtration needed.

3.14 MAGNETIC FILTER

Magnetic filters in a hydraulic system may range from a simple plug to long magnetic pole suspended inside the reservoir. They are normally fitted in tanks to conventional filters incorporating a permanent magnetic element.

A magnetic filter will attract and collect only ferrous metal particles, such as wear particles. A proportion of non-magnetic particles may also be retained, although performance in this respect is unpredictable.

Magnetic filters can only be regarded as secondary filters for removing ferrous particles or initial wear products. They cannot be used as alternative to other types of filters.

3.15 OPTIMUM FILTRATION

Filter ratings: To have an optimum filtration accuracy, the filter rating should be properly decided in a hydraulic system. By the term cut-off rating, we mean the nominal diameter of the smallest particle size retained by the filter, this value is normally expressed in microns. In the case of a filter medium with non-uniform pore size, the cut-off can only be a nominal rating. If the medium has an exact and consistent pore size, then the cut-off can be specified as an absolute rating on the basis that no particle larger than the cut-off size can be passed.

In practice, it has been experienced that the absolute rating may not at all be met as larger particles of non-spherical shape may well be passed, depending on the actual shape of the particle and the mechanical construction of the filter medium. Another important point to note here is that certain filter media, e.g. papers, felts and clothes have generally variable pore-size and thus no absolute rating at all. The effective cut-off is determined by the random arrangement involved and the depth of the filter.

3.15.1 Degree of Protection Required

This is a most interesting point. Logically it would appear that lower the cut-off rating of the filter element, the better, particularly as a solid contaminant more than 2 microns in size can be regarded as damaging and blocking. But this is open to several practical objections as shown here.

(i) high cost of such filters
(ii) need of frequent replacement of elements
(iii) possible restriction of flow-rate
(iv) more pressure drop, etc.

Table 3.13 Filtration Requirements (systems)

Type of system	Suitable cut-off microns
Low pressure hydraulics (industrial)	150–100
70 bar industrial hydraulics	50
100 bar industrial hydraulic	25

(Contd.)

Type of system	Suitable cut-off microns
150 bar industrial hydraulic	15
Push-pull movement	15–10
Machine tool with precision feed	10
150–200 bar heavy duty hydraulics	10
With electro hydraulic servo valves	5–2½
With precision servo controls	2½

Table 3.14 Filtration Requirements (components)

Components	Suitable cut-off (microns)
Sliding components (general)	Less than 100% of the working clearance
Fine orifices etc. (general)	Less than 100% bore opening
Elastomeric seals	25–30
Relief valves	15–10
Flow control valves	15–10
Low-gain servo-valves	10
High-gain servo-valves	5 or better

3.16 AUTOMATIC PARTICLE COUNTER AND ITS PERFORMANCE CHARACTERISTICS

Principle of Operation

The automatic particle counter is basically an electro-optical scanner with an electronic counter. It utilizes the principle of visible light scattering/total absorption.

A schematic block diagram of the measurement system employed is shown in Fig. 3.12. Whenever a particle traverses the illuminated zone of the optical sensor, a part of the light beam is interrupted by the particle and either scattered or absorbed. The quantity of energy thus diverted is related to the diameter of the particle and its effective cross-sectional area. The output is in the form of electrical pulses which are amplified, counted and displayed in a digital format. The digital output is further recorded with a printer.

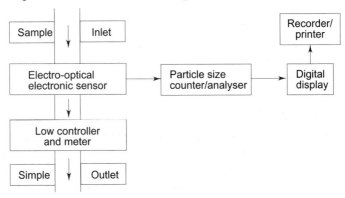

Fig. 3.12 *Schematic block diagram of automatic particle counter*

3.17 CONCLUSION

Oil contamination studies do help us to understand more about hydraulic systems. In particular it helps to know the performance characteristics of the filters in the system.

In addition, it gives us an indication of the problem as well as clues for our further studies.

Some typical clearances between various matching components are shown in Table 3.15. The values projected are of general nature and should not be taken as conclusive and specific.

Table 3.15 Typical Critical Clearance Fluid System Components

Item	Typical clearance	
	Microns	Inches
Gear Pump (Pressure Loaded)		
Gear to side plate	1/2–5	0,000,02 - 0.0002
Gear tip to case	1/2–5	0,000,02 - 0.0002
Vane Pump		
Tip of vane	1/2–1*	0.000,02 - 0.000,04
Sides of vane	5–13	0.000,2 - 0.000,5
Piston Pumps		
Piston to bore**	5–40	0.000,2 - 0.001,5
Valve plate to cylinder	1/2–5	0.000,02 - 0.000,2
Servo Valve		
Orifice	130–450	0.005 - 0.018
Flapper wall	18–63	0.007 - 0.002,5
Spool sleeve (R)**	1–4	0.000,04 - 0.000,15
Control Valve		
Orifice	130–10,000	0,005 - 0.40
Spool sleeve (R)**	1–23	0.00,04 - 0.000,90
Disc type	1/2–1*	0.000,02 - 0.000,04
Poppet type	13–40	0.000,5 - 0.001,5
Actuators	50–250	0.002 - 0.010
Hydrostatic bearings	1–25	0.000,04 - 0.001
Antifriction bearings	*1/2–	0.000,02–
Slide bearings	*1/2–	0.000,02–

* Estimate for thin lubricant film.
** Radial clearance.

Review Questions

1. State the nature and effect of contaminants in a hydraulic system.
2. State the type and sources of contaminants in a hydraulic system.
3. What is filtration rating? How do you define absolute rating? Does it differ from nominal filter rating?
4. What is a by-pass filter? State its advantages and disadvantages.

5. How does one test a hydraulic filter? Describe the function of a test rig for testing a filter. Show its circuit diagram.
6. What type of filters are used for fire resistant fluids?
7. Write a short note on toleration limit of particle size.
8. What are the various standardization systems used for contaminant levels and contaminants class? Describe briefly.

References

3.1 McNiele, L S, *Simplified Hydraulics*, McGraw-Hill Co, New York.
3.2 'Microscopic sizing and counting particles', Aerospace fluids on membrane filters, ASTM D2390-6ST, August 1, 1969.
3.3 'Detection and analysis of particulate contamination,' Milipore Corporation, ADM-30.
3.4 'Basic Fluid Power Research Programme,' Oklahoma State University, USA.
3.5 'Operational manual of liquid borne particle Monitor,' Royco Instrument, California, USA, 1970.
3.6 Reik, Martin, *Filtration in Hydraulic System.*
3.7 Publication from Pall Industrial Hydraulics Limited, Europa House, Havant Street, Portsmouth, UK, 1983.
3.8 Publication material from m/s Regeltechnik Friedrichs hafen GmbH, Germany.
3.9 O + P Discussion: Current status of Filtration in Hydraulics, Oelhydraulic und Pneumatik, Germany.
3.10 Filtration for Profit Making Design: A report published in Hydraulics & Pneumatics, March, 1978. Vol-31, No. 3. USA, PHP-10.

Hydraulic Pumps—Construction, Sizing and Selection

A basic hydraulic system consists of the following major components
1. Reservoir—an oil supply tank
2. Pump and prime mover—the source of power
3. Pressure relief valve—for system safety
4. Control valves—for direction and other controls
5. Cylinder/hydro motor—the working element.

Diagrammatically we can depict the system as shown below:

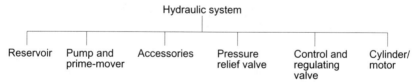

Though each of the items is equally important for the system, the pump plays a specific and unique position in the system. It can be compared to the heart of a human being. The main purpose of the pump is to create the flow of oil through the system and thus assist transfer of power and motion.

The pump is coupled to and driven by the prime mover of the system, which is most often an electric motor. The inlet side of the pump is connected to the reservoir; the cutlet or pressure side is connected to the control valve and thus to the rest of the system.

4.1 PUMP CLASSIFICATION

Basically pumps can be classified as positive and non-positive pumps as shown below.

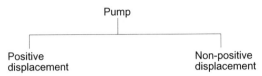

They are further classified as shown in chart on page 91.

4.1.1 Positive Displacement Pump (PD)

Positive displacement pumps are those whose pumping volume changes from maximum to minimum during each pumping cycle. That is, the pumping element expands from a small to a large volume and is then contracted to a small volume again.

Positive displacement pumps are used where pressure is the primary consideration. In these pumps the high and low pressure areas are separated so that the fluid cannot leak back and return to the low pressure source. The pumping action is caused by varying the physical size of the sealed pumping chamber in which the fluid is moved. As fluid moves through the pumping chamber, volume increases and is finally reduced causing it to be expelled—alternately increasing and then decreasing the volume. Since the volume per cycle is fixed by the positive displacement characteristics of the pumping chamber, the volume of fluid pumped for a given pump size is dependent only on the number of cycles made by the pump per unit time. Gear, vane, piston, screw pumps are some examples of such pumps.

In such pumps the flow enters and leaves the unit at the same velocity, therefore, practically no change in kinetic energy takes place. These pumps provide the pressure with which a column of oil acts against the load and are hence classified as hydrostatic power generators.

Advantages of positive displacement pumps
1. PD pumps are widely used in hydraulic system.
2. They can generate high pressure.
3. They are relatively small and enjoy very high power to weight ratio.
4. They have relatively high volumetric efficiency.
5. There is relatively small change of efficiency throughout the pressure range.
6. They have greater flexibility of performance under varying speed and pressure requirements.

4.1.2 Non-Positive Displacement Pumps (NPD)

Pumps where the fluid can be displaced and transferred using the inertia of the fluid in motion are called non-positive displacement pumps. Some examples of such pumps are centrifugal pumps, propeller pumps, etc.

They are usually used for low pressure (up to 40 bar) and high volume systems. Because of their fewer number of moving parts, NPD pumps cost less and operate with less maintenance. The chamber in NPD pumps is connected so that as pressure increases, fluid within the pumping chamber circulates. These pumps

Hydraulic Pumps—Construction, Sizing and Selection 91

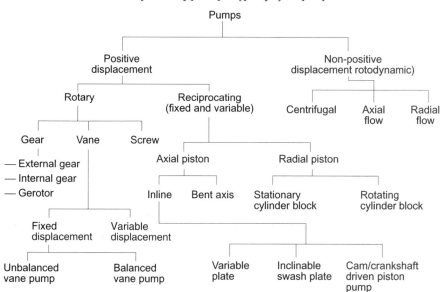

Classification of principal types of hydro-pumps

make use of Newton's first law of motion to move the fluid against system resistance. The action of the mechanical drive in the pumping chamber speeds up the fluid so that its velocity accounts for its ability to move against the resistance of the system. In centrifugal pumps, rotational inertia is imparted to the fluid, whereas in propeller pumps, the inertia imparted is translational. For a given pump size, the volume of fluid pumped is dependent on the speed of the rotating member and the resistance the pump must overcome.

Advantages of non-positive displacement pumps
1. There is a low initial cost.
2. There is minimum maintenance cost.
3. They can be operated quietly.
4. They are capable of handling almost any type of fluid for example, sludge and slurries.
5. Simplicity of operation.
6. High reliability.

Because the inlet and outlet passages are connected hydraulically, centrifugal pumps are not self-priming and must be positioned below the fluid-level. Most have large internal clearances—though some use flexible impeller blades that reduce internal clearances to increase pumping efficiency.

The impeller imparts kinetic energy to the fluid. Hence centrifugal pumps are hydro-dynamic or hydro-kinetic power generators.

4.1.3 Construction of Non-Positive Displacement Pump

Centrifugal Pumps A simple line diagram of a centrifugal pump is shown in Fig. 4.1. Construction: The primary parts of a centrifugal pump are:

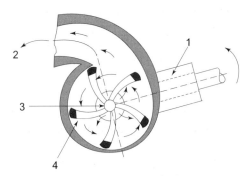

Fig. 4.1 *Sketch of a centrifugal pump*
1-Inlet, 2-outlet, 3-eye, 4-impeller blades

1. Inlet port
2. Involute shaped pumping chamber
3. Drive shaft
4. Impeller
5. Outlet port.

Operation Rotation of the impeller imparts a centrifugal force to the fluid causing it to be directed outward from the center of the impeller. The blades are curved opposite to the direction of rotation to increase efficiency. Defuser blades are sometimes attached stationery to the pump housing and they redirect the fluid in such a way as to reduce the velocity and internal clearances and increase the ability of the pump to develop more pressure. As stated earlier, centrifugal pumps are used primarily for low pressure hydraulic applications only. However, staging in series may be used to increase system pressure significantly. This creates high pressure keeping the flow rate the same through each of the pumps placed in series.

4.1.4 Construction and Principle of Positive Displacement Pump (PD)

As defined earlier a PD pump:
  (a) has to generate a continuous flow of liquid and also
  (b) has to supply the necessary force (pressure) to the flow of liquid.

Positive displacement pumps may be either (a) reciprocating or (b) rotary type. These pumps can be also divided as:
  (a) fixed delivery, i.e. pumps with fixed displacement (constant delivery) in which the flow volume is constant and when the speed is constant. These pumps have no control of rate of flow.
  (b) variable displacement pump: In variable displacement pumps the output volume, i.e. the flow rate of the pump is variable, i.e. in these pumps the flow rate is controlled to vary in a certain range at constant speed as per the demand of the system.

Reciprocating pump Reciprocating pumps are generally cylinders with a piston operating as a pump. They are available in various sizes, shapes and driving mechanisms. Due to high pressure they are used for specialized applications. Such pumps are capable of developing very high pressure if the reciprocating principle is used in a rotary form. A pump using this principle is known as a *piston pump*.

Rotary pump Rotary pumps are the most common in oil hydraulic systems when low to medium pressure is the prime consideration. Although there are many varieties of rotary pumps, out of those, the three prime types are: (1) gear, (2) vane and (3) screw. Although they overlap to some degree in both performance and application, they are completely different in structure and usually serve quite different functions.

Some advantages of rotary pumps over reciprocating pumps are
1. Power to wt. ratio is satisfactory
2. Simpler construction
3. Direct coupling to motor(el)
4. Continuous pulsation-free supply of oil

4.2 GEAR PUMPS

The basic gear pump consists of two meshed gears, a case or housing to encompass the gears, and two cover plates that enclose the ends of the gears. Each gear is mounted on a shaft which is supported on bearings in the end covers. One of these shafts, the drive shaft, is coupled to the prime mover, i.e. the drive motor. Two ports—inlet and delivery ports are provided. Their exterior position on the pump is relatively unimportant, but they are most often located on opposite sides of the gear case. It is absolutely essential, however, that inside the pump these ports open directly on opposite sides of the mesh point of the gears. For best performance, the ports should be as large, straight, and as unobstructed as possible. A diagram of an external gear pump is shown in Fig. 4.2.

Spur gears and helical gears both are used. Helical gears run diagonally across the width of the gear at an angle (helical angle) determined by design. The larger the pitch angle, the smoother is the pump operation. However when the gears are in mesh, and the helical angle is such that one end of a gear root is on the suction side while the other is still on the delivery side, then an undesirable leakage path along the root results. Thus there is a limit to the helical angle. Helical gears are quicker but are more expensive.

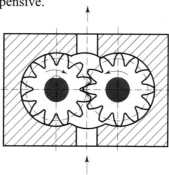

Fig. 4.2 *Gear pump*

The flat end faces of the gears are closed off by wear plates. Leakage between wear plates and the gear end faces must be reduced to a minimum. On the other hand, clearance is necessary to permit the gear to rotate without undue friction. Most designs, therefore, use floating wear plates. Discharge pressure is applied to the outside of the plates. Thus frictional losses become approximately proportional to pressure, with the seals getting tighter with higher pressures and counteracting the tendency of increased volumetric losses.

4.2.1 Pumping Action

The pumping action in a gear pump takes place as follows. A tooth space of one gear is filled by a tooth of the other gear. As the meshed gears start rotating one tooth space after another is evacuated, and in the resulting spaces a vacuum is created. Atmospheric pressure in the tank forces the oil into the tooth space from the port located at that point. This tooth space full of oil is carried around the periphery of the gear until the teeth again mesh and the oil is forced out of the space by the meshing tooth and flows out of the delivery port located at that side of the mesh point. The elimination of this space prevents oil from crossing over to the intake side, and thus the pump dispels one toothful of oil per tooth per revolution. Close fitting covers and other parts prevent leakage of oil through the mating parts. The line contact of the gear teeth over one another prevents flow through the mesh and the close-fitting of the housing prevents flow back around the periphery. From the suction side to the delivery side the pressure develops tooth by tooth as shown in Fig. 4.3 (a) which may be taken as linear in characteristics. The fact that the gear teeth can force fluid out of the tooth spaces against resistances placed down stream is the essence of the pumping concept and is the measure of success of the gear pump. Without a restriction downstream, it

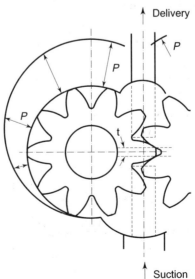

Fig. 4.3 (a) *Development of pressure in a gear pump*

can be seen that the pump would merely transfer fluid from one side to the other at zero pressure. Positive pressure is created only when a restriction is encountered.

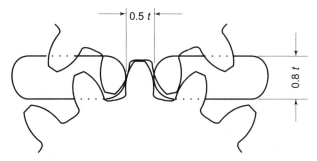

Fig. 4.3(b)   *Recess for relief for trapped oil in gear pump*

4.2.2 Trapped Oil

In gear pumps it is also noted that as one gear meshes into the other, a remainder of the fluid tends to get trapped between the tooth tip of one gear and the tooth-root of the other. The compression of the oil in such an eventuality would result in very high force noise. One method for relieving entrapped fluid is to provide shallow recess. The recess is only required on the discharge side, but if the pump can be run in either direction, suction and delivery being reversible, two recesses are to be provided as shown in Figs 4.3 (a) and (b). Figure 4.3 (a) also shows the gradual rise of pressure in a gear pump. Where the design is such that the idler gear has a bushing insert which revolves on a stationery shaft, the method as per the sketch may be used. Radial holes are drilled through tips and roots and roots of the idler gear and of the bushing as shown in Fig. 4.4 (a). Grooves are cut in the stationery shaft which communicate with the holes and provide a return passage for the entrapped fluid. Another modification for relieving entrapped oil consists in making the shaft hollow and discharging entrapped oil through a

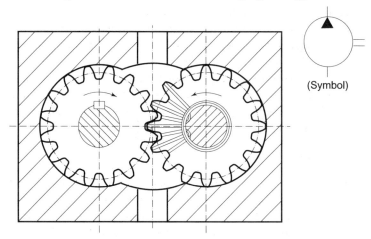

Fig. 4.4(a)   *Gear pump: Pressure balancing by drilling holes*

separate connection. The pump has then two deliveries—one from the pressure side and a minor one that delivers entrapped oil for a separate low flow rate application.

4.2.3 Pressure Balancing

The principle of fluid mechanics states that:
   (a) Pressure acts equally in all directions.
   (b) Pressure exerted upon one point or area of a fluid body immediately translates to all portions of the body and acts with equal intensity throughout the body.
   (c) Pressure always seeks to move the fluid from a high pressure to a low pressure point.
   (d) Pressure acting upon an area becomes a force.
   (e) Equal pressures acting in direct opposition to one another cancel or neutralize each other.

Pressure balancing in a gear pump is the proper utilization of all the above facts. As a pump operates, it carries oil from a point of less-than-atmospheric pressure (vacuum) to one of high pressure (discharge). This action creates a pressure gradient around the perimeter of the gear inside the pump. Due to this undesirable gradient, the confining surfaces tend to distort. When the pressures are relatively low, the strength of the pump materials is sufficient to resist; but as the pressure increases, the materials measurably distort. We can say, then, that this distortion is the result of the unbalanced pressure force. If this force is balanced (neutralized), distortion can be eliminated. Since this unbalanced force can be quite detrimental, bearings have to take up the unbalanced forces. The usual approach is to rely on roller bearings rather than to complicate the design by any of the force balancing methods available. One method balances these forces by connecting ducts as described in Section 4.2.2. Difficulty in such a design is not only the cutting of additional passages but also the need for preventing increased leakage that can occur between gear tips and housing wall. The leakage path here is considerably shorter than in conventional gear pumps and hence increased leakage is expected. The gain in torque efficiency is thus cancelled by loss in volumetric efficiency.

Because of pressure balancing as shown in Fig. 4.4 (a), a number of holes result all of which pass through the shaft center, but are not allowed to run into each other. Both driver and idler are treated this way. The result is good balancing, increased internal leakage and added cost.

4.2.4 Wear Plate

Wear plates are placed at the ends of the pump gears between the gears and the cover plates. The first function of the wear plates is to provide contact surfaces of superior bearing quality. Most wear plates are made of bronze, but aluminium is also commonly used. Many pumps have no wear plates and use the ground surfaces of the cast iron cover plates as bearing surfaces as cast iron is an excellent bearing material itself.

Through clearances around the wear plates and holes (such as gear-shaft holes) through the wear plates, oil finds its way between the wear plates and the cover plates. This oil exerts pressure against the wear plates and this pressure is the maximum in the system. In the area of the discharge port, the pressures on both sides of the wear plate are equal; the wear plate is in neutral atmosphere and remains flat. However, in the area of the inlet port, the high pressure on the 'underside' is substantially greater than the pressure on the gear side. This creates an unbalanced force which deflects the plate against the gear. Therefore in a relatively shorter time, the gear teeth may mill away the bearing surface, increasing the clearance across the ends of the gears and reducing the pump's performance.

If one can isolate the areas of varying pressures this will result in neutralizing the pressure on both sides of the place and deflection of the wear plate may be eliminated. This is done by sealing off gears with differing pressures under the wear plates so that fluid and pressure transfer from one area to another cannot occur. The sealed areas can then be vented to the area in the gear chamber directly above so that the pressures on both sides of the wear plate become substantially equal. It may be mentioned here that this pressure balancing cannot exactly follow the pressure gradient of the pump chamber. However any pressure separation is helpful, and reasonable approximations can be made. In a gear pump the inlet area is the most critical, elementary pressure balancing consists of merely sealing the areas beneath the wear plates at the intake port. More elaborate balancing separates the entire area under the gears into many small areas, each one vented to the gear chamber directly opposite. Balancing provides a reasonably neutral atmosphere for the entire wear plate. Thus it can remain flat and provide a uniform bearing surface for the gears, maintain design clearances, improve operating efficiency, and extend service life. In Fig. 4.4 (b) it is shown how pressure build up in the suction side can be eliminated to balance the pressure.

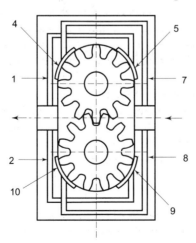

Fig. 4.4(b) *Pressure balancing*
5, 4, 9, 10—4 Chambers diagonally opposite
7, 8—Duct line connected to suction line
1, 2—Duct line connected for delivery line

Here high pressure built up during operation in the suction chambers 5 and 9, which communicates through clearances between the gears and housing on the delivery side, is reduced by the drain of some fluid back through ducts 1 and 2. The pressure built up in the other two chambers equalizes or balances the one sided pressure on the delivery side thus relieving the gears of unilateral pressure.

4.2.5 Manufacturing Requirements

Clearance range: 0.025 mm diametral and end clearance in small pumps of 25 mm diameter gears and 0.1 mm to 0.125 mm end clearance and 0.150 mm diametral clearance on pumps with 90 mm diameter gears.

Gears Commonly $14\frac{1}{2}°$ or $20°$ involute teeth. The minimum number of teeth on $14\frac{1}{2}°$ pressure angle should be about 16. One should keep in mind that pump gears are different from transmission gears. Transmission gears have sliding action but for pumps the gears are required to have rolling action in order to have higher longevity and perfect pumping action. In a gear of a given diameter, the teeth and their intermediate spaces become larger as the number of teeth is reduced. This on the other hand means that amount of oil carried in the space is increased resulting in higher supply per rotation of the gear. However when conventional $14\frac{1}{2}°$ involute gears are used and if the number of teeth is reduced to say, 10, the resultant undercutting of the teeth makes them dimensionally and structurally weak. They also leave unprotected pockets on each side of the mated gears tooth when the gears are meshed creating undesirable thrust. If the involute angle is increased, the shape of the gear teeth becomes more triangular. With such gears, the undercutting is significantly reduced and number of teeth can be reduced than 12 or 10 thus providing robust and strong teeth but the roots and unproductive pockets also get reduced. Most gear pumps are of this type having an angle of 23 to $28\frac{1}{2}°$ in most cases. However in general, gear pumps having 11 to 15 teeth through 8 or 9 gears are also used.

4.2.6 Manufacturing Range

They are generally available in the following ranges:
- For a continuous pressure of 200 bar
- Minimum pressure range of 10 to 100 bar
- Minimum speed of rotation from 400 to 500 rpm
- Maximum speed of 3000 to 6000 rpm
- Maximum flow rate of 300 l/min
- Minimum flow rate of 3 to 100 l/min

The volumetric efficiency is generally low to medium in most gear pumps. The oil is carried around the periphery from the suction to the delivery side. Some volume of oil (10%) is trapped in the clearance spaces of the meshing gears as described earlier. This trapped volume of oil creates a heavy load on the bearings. The effect can however be minimized by providing relief grooves. The characteristics of volumetric efficiency are illustrated in Fig. 4.5.

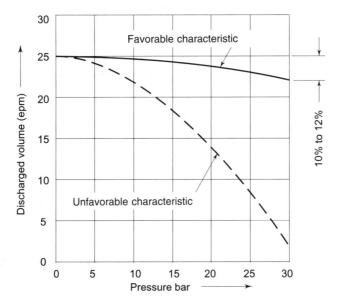

Fig. 4.5 *Characteristic of gear pumps*

4.2.7 Multigear Pumps

In such a pump there are 3 or 4 gears in one housing. The center gear is connected to the motor shaft. The two outside gears are driven, and thus turn in the direction opposite to each other. One advantage of this type of pump is that it has two independent outputs, each of which may do a particular job. The two outputs may also be connected together as one. One disadvantage is the short sealing range of the center gear thus limiting the system pressure. A simple design of a multigear pump is shown in Fig. 4.6.

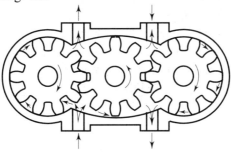

Fig. 4.6 *Three gear pump (gear on gear type)*

4.2.8 Herringbone Gear Pump

A herringbone gear pump is made up of two helixes going in opposite directions. A helix is a curve produced when a straight line moves up or down around the surface of a cylinder. Most helical or spur gear pumps are reversible. But a herring-bone gear pump is not. When the herringbone gear pump is turned in the wrong direction, oil is held in the middle 'V' and a back pressure will result.

Helical gear pumps are quiet even at high speed and deliver large volumes of hydraulic oil at high pressure. A herringbone gear pump operates in the same way on a helical or spur gear pump but the space at the center of the gears helps to make the pulsations in the discharge small. Also end thrust is reduced by the opposed helical gears forming the herringbone.

4.3 INTERNAL GEAR PUMP

Figure 4.7 illustrates an internal gear pump. The operation of internal gear pumps is similar to external gear pumps. But they are very efficient and produce less noise. The internal spur gear drives the outside ring gear which is set off-center (eccentric) to each other. Between the two gears on one side there is a crescent shaped spacer which is a stationery part of the housing around which oil is carried. The inlet and outlet ports are located in the end plates between where the gear teeth mesh and unmesh at the two ends of the crescent shaped spacer.

4.3.1 Operation

Oil is transferred from the inlet to the outlet port. The internal gear drives the external gear and effects a fluid tight seal at the place where the teeth start meshing. Rotation causes the teeth to unmesh near the inlet port, the cavity

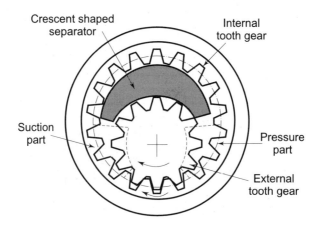

Fig. 4.7 *Internal gear pump*

volume increases and suction occurs. Oil is trapped between the internal and external gear teeth on both sides of the crescent-shaped spacer and is carried from the inlet to the outlet cavity of the pump. Meshing of the gear teeth reduces the volume in the high pressure cavity near the outlet port and exits from the outlet port. Some internal gear pumps produce high pressure (up to 400 bar) but do not create much noise.

4.3.2 Drive

In some internal gear pumps, the drive shaft is coupled with the external gear and the internal gear is driven; but in some other types the drive shaft may be connected with the internal gear and the external gear may be driven. When the external gear is the driven gear, by rotating the crescent shaped separator and the external gear assembly by 180°, the input shaft rotation may be reversed.

This is a fixed displacement pump and is used from low pressure system (sometime as low as 15 bar pressure only) to also very high pressure.

4.4 GEROTOR PUMP

A schematic diagram of a gerotor pump is shown in Fig. 4.8. It resembles the internal gear pump described earlier. The main parts of a gerotor pump are:
(1) Outer ring
(2) Outer gerotor
(3) Inner gerotor

The inner gerotor is the driver and the outer is idler. There is only one shaft involved which means a certain simplification in design. The number of teeth of the outer and inner gerotor vary. The outer gerotor has always one tooth more. The axis around which the inner element rotates is offset from the axis of the outer gerotor, which driven by the inner gerotor, rotates inside the ring. If the number of teeth for the inner and outer gerotors are 4 and 5 respectively, while the inner gerotor makes one revolution, the outer one makes only 4/5 or 0.8 revolution.

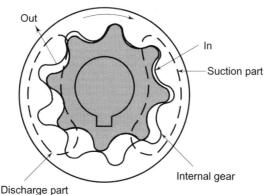

Fig. 4.8 *A gerotor pump*

Disadvantages   Gears must be made to high precision. Otherwise internal leakage becomes excessive, though the gerotor pump is more compact than external gear pumps.

Ratings: Continuous pressure = 125 bar
Max. speed = 2000 to 3600 rpm
Max. delivery = 200 l/min.

4.4.1 Lobe Pump

Here two lobes are used to rotate one against the other, creating two chambers between the lobes and the wall of the pump chamber. The elements are rotated, one being directly driven by the source of power, the other through timing gears. As the elements turn, fluid is caught between each lobe and the wall of the pump chamber and carried around from suction to discharge. This type of pump is used for pumping gas, air, liquid with low pressures and higher flow rate.

4.5 SCREW PUMPS

Screw pumps are one of the most reliable pumps and perform very efficiently at a considerably low level of noise, a factor which makes them most suitable for certain types of applications. These pumps transfer fluid by displacing the oil axially through a closely fitted chamber formed between the screw recess and housing wall. They are fixed displacement pumps and generally flow rate can be varied by varying the drive motor speed only. The only moving parts are the screws. A two element rotary type of screw pump and its cut section view are shown in Figs 4.9 (a) and (b). Mostly they are available either as one, two or three screw designs. In some special designs, one can use four or five spindle also. Number of threads and size determine the flow quantity of the pump.

In the screw pump with two screws, timing gears are required to maintain a running clearance between the two meshing screws. The pumping action in the pump comes from the "sealed chambers" moving from the inlet to the outlet. This is like a nut moving along a threaded rod when the rod is rotated. The liquid does not rotate but moves linearly. As in case of gear pumps, here also the meshing of

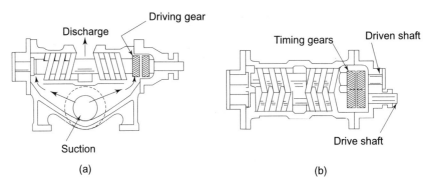

Fig. 4.9 *(a) Two element rotary screw pumps (b) Plan cut away view*

the thread flanks forces fluid to flow in the space between the rotor and housing wall. The volume of fluid moves forward uniformly with the rotation of the screw along the axis and is discharged to the delivery port.

Sometimes the screw may encounter an undue thrust load toward the inlet port of the pump due to undesired pressure peaks which can be balanced by the use of right and left handed threads so that the pump inlets are on each end with the pump outlet being at the middle of the screw. However side loading which is very common in gear and vane pumps are generally non-existent in the case of screw pumps. Therefore one main bearing to support the driving rotor is generally adequate for screw pumps to withstand the sideload, if any. In Fig. 4.10, a three spindle screw pump is shown. In this design the drive motor is connected to the middle screw and on its both sides two idler screws are provided. The idler screws are driven by the pressure of the liquid and therefore no timing gears are required here. The optimization of a three-spindle screw pump performance can be bettered by improving the manufacturing tolerance of its various parts, i.e. the profile of spindle thread and the housing unit. The manufacturing of a three spindle screw pump is costly and therefore it is used for any special application which demands low flow rate, no pressure pulsation and maximum noise elimination. They are used for the hydraulic systems of lift, elevator, hydraulic press, ship winch, crane, wood working m/c, machine tools, etc. Mostly they operate up to 250 bar pressure (and sometimes as high as 400 bar) at 1000 cm^3 per min. Pumping of higher viscosity oil may derate the pump considerably.

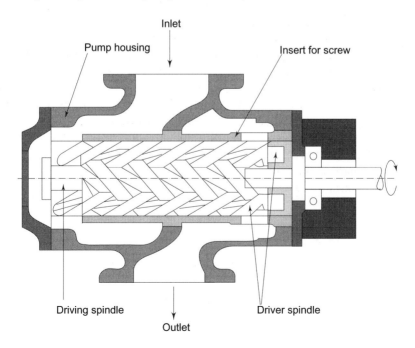

Fig. 4.10 *3 Spindle screw pump*

Advantages of screw pump
1. These pumps are most reliable.
2. They can operate at very high speeds—say up to 3500 rpm and sometimes at even higher speeds.
3. The oil supply is pulsation free, continuous.
4. There is no oil churning, pump turbulence, etc.
5. These pumps are very quiet because of rolling action of the spindle rotor. This eliminates pulsation and vibration, hydraulic whine, noise, etc.

Disadvantages
1. Because of its specific profile, the manufacturing of a screw pump is quite difficult especially if one wants to maintain a close tolerance.
2. Pressure rating of the screw pump is viscosity dependent. They can easily pump oil up to 20 cSt viscosity. Pumping of higher viscosity oil may derate the pump characteristics.
3. Increase of viscosity may decrease the pump efficiency considerably.
4. Overall volumetric and mechanical efficiency in screw pumps is relatively low.

4.6 VANE PUMPS

Classification Vane pumps are classified as:
(a) Fixed displacement, unbalanced pump
(b) Fixed displacement, balanced pump
(c) Variable displacement, unbalanced pump
(d) Variable displacement pressure compensated vane pump.

Pumping action in a vane pump is caused when the vanes are allowed to track along a ring—mostly called a cam ring.

Constructions
Figure 4.11 illustrates a normal vane pump unbalanced type. This type of vane pump consists basically of the following essential parts:
(a) Driven rotor with slots for vanes
(b) Sliding vanes in the slots
(c) Stationary circular ring
(d) Port-plates with kidney shaped inlet and delivery port.
(e) Housing.

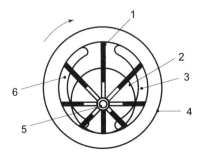

Fig. 4.11 *Fixed displacement unbalanced type vane pump*
[1-Vane, 2-Rotor, 3-Delivery port, 4-Pump body ring, 5-Shaft, 6-Inlet port]

Sometimes the rotor, vanes, ring and wear plates may form a cartridge unit and can be easily replaced.

Principle of operation The rotor axis is positioned eccentric to the circular ring inside which it rotates. Since the vanes are free to slide in their slots, they move outward due to centrifugal force and as the vanes make contact with the inner ring wall, a positive seal takes place between the vane tip and the cam ring. Thus a number of chambers are formed between the vanes and cam ring. The chamber changes their volume continuously because the vanes follow the inner contour of the ring.

A port plate which fits over the ring, rotor and vanes is used to separate the incoming oil from the outgoing oil. The discharge and suction side of the pump are sealed from each other at any time by at least one vane. To obtain such sealing, it is necessary that the track between the two ports is slightly wider than the space between two vanes. Hence with six vanes the angle between the end of suction port and beginning of delivery port must be at least 60° and the angle between end of delivery and the beginning of suction port should be same. This leaves a maximum of 120° of port length with a six vane pump.

Side load on shaft A pump during its operation experiences two different pressures—the working pressure at the pump outlet—and the pressure at the pump inlet which is less than the atmospheric pressure. This means that one half of the pumping mechanism is less than atmospheric pressure while the other half is subjected to the full system pressure. This creates an undesirable side loading of the rotor shaft causing damage to the intricate parts of the pumps. Such unbalanced forces reduce the life cycle of the pump considerably. Hence vane pumps with eccentrically placed circular rings are seldom used and a balanced vane pump is designed with a cam ring with an elliptical contour. An elliptical cam ring is shown in Fig. 4.12.

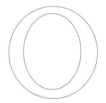

Fig. 4.12 *Elliptical cam ring*

4.6.1 Balanced Vane Pump

In a balanced design vane pump a circular rotor with vane slots is concentrically positioned with the axis of an elliptical cam ring as shown in Fig. 4.13 (a). This creates two inlet and two outlet chambers such that the two pressure quadrants oppose each other and the two inlet chambers are directly opposite to one another. Therefore the forces acting on the shaft are fully balanced and side loading of the shaft as narrated earlier is eliminated completely. A schematic diagram of this balanced vane pump is shown in Fig. 4.13 (b).

In actual design both the inlet ports are connected together. Similarly both the outlet ports are also connected together so that in the pump housing there is only one pump inlet and one delivery port.

These type of pumps are constant volume or fixed displacement pumps and can work up to 175 bar pressure. They are relatively quiet and of simple construction. The vanes move to and fro in the slots depending on the cam ring contour they come in contact with while traversing it radially.

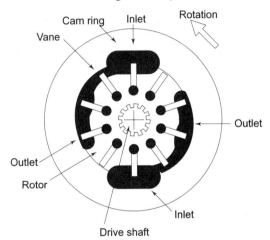

Fig. 4.13(a) *Fixed delivery balanced vane pump*

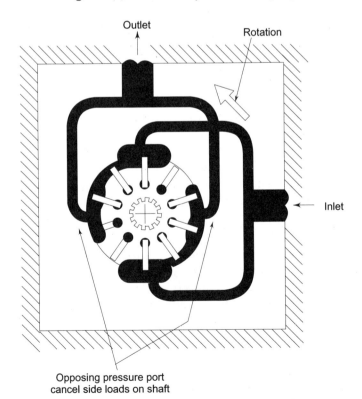

Fig. 4.13(b) *Balanced vane pump with two opposing inlet and outlet*

4.6.2 Vane Loading

Without a positive sealing, proper function of the vane pump is quite difficult. Generally the vanes are displaced from the slots by centrifugal forces as soon as the rotor starts rotating and achieve a positive sealing. That is why designers prescribe that the minimum operating speed of vane pumps should not be below 600 rpm. The leakage of oil is dependent on the geometric configuration of vanes and cams. Tighter seal between the vane tip and cam ring wall is desired to minimize the leakage. Industrial vane pumps are designed such that a portion of the system pressure is directed to the underside of vanes, the higher the system pressure, the more force is developed to push the vane out against the cam ring. This hydraulic loading of the vane develops a tight seal, but with too great a force the vane and cam ring would wear excessively and the vanes would be a source of drag and the pump will malfunction due to mechanical failure of pump parts. The scoring action will be more pronounced if the vane is of a straight rectangular cross-section type as shown in Fig. 4.14 (a).

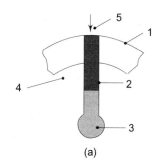

Fig. 4.14 *(a) Vane loading in a vane pump [1-Cam ring, 2-Vane, 3-Oil pressure, 4-Rotor 5-Heavy load here]*
Courtesy: M/s Parker-Hannifin Corporation, Ohio, USA, Bulletin No. 0107-B1. 1970

4.6.3 Solution Against Excessive Vane Loading

As a compromise between best sealing and least drag and wear, vanes may be designed with only partial loading.

A vane with a bevelled edge as shown in Fig. 4.14 (b) may be also used.

The use of vanes with a chamfer or bevelled edge eliminates high vane loading. The complete underside vane area is exposed to the system pressure as well as a large portion of the area at the top of the vane. This results in a balance. The pressure which acts on unbalanced area is the force which loads the vane.

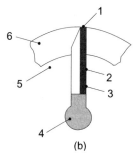

Fig. 4.14 *(b) Vane loading eliminated with chamfered vanes [1-Force and sealing takes place here, 2-Vane with chamfer, 3-Unloaded area 4-System pressure, 5-Rotor, 6-Cam ring]*
Courtesy: M/s Parker-Hannifin Corporation, Ohio, USA.

Making vane tips round or bevelled depends on materials and experiences. Making vanes thinner reduces the area against which the pressure is applied. This decreases thrust, friction and also rigidity of the vanes.

4.6.4 Types of Vanes

For high pressure vane pumps, one generally uses the following types of vanes as shown in Figs 4.15 (a), (b), (c), (d) and (e).
  (a) Dual vanes
  (b) Pin vanes
  (c) Angled vanes
  (d) Intra vanes
  (e) Spring loaded vanes

A dual vane, as the name suggests, consists of two vanes in each slot of the rotor as illustrated in Fig. 4.15 (a). With this arrangement, the vane is considerably balanced with a possibility of positive sealing. An intra vane consists of a smaller vane within a larger vane with a bevelled edge. The delivery pressure (i.e. system pressure) from the pump is directed above the smaller vane resulting in less vane loading. Similar to this design, in a pin vane the pressure is

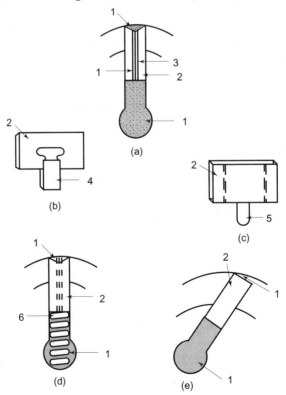

Fig. 4.15 *Types of vanes: (a) Dual vane (b) Intra vane (c) Pin vane (d) Spring loaded vane (e) Angular vane*
[1-System pressure, 2-Vane, 3-Chamfer, 4-Intra vane, 5-Pin, 6-Spring]
Courtesy: M/s Parker-Hannifin Corporation, Ohio, USA.

directed underside the pin which forces the vane out against the cam ring. In a spring loaded vane, a spring is used to force the vane against the cam ring along with system pressure (Fig. 4.15 (d)). Whereas in Fig. 4.15 (e), the vanes are positioned at an angle in the rotor. This may reduce the loading on the vane without any mechanical means.

4.6.5 Cartridge Assembly

Very often in vane pumps, the pumping mechanism is designed and formed as an integral assembly which normally is called a "cartridge" assembly. This consists of vanes, rotor and cam ring sandwiched between the port plates. Such an assembly is easy for maintenance and servicing when the system malfunctions. Another important advantage of using such an assembly lies in its easy replaceability when it is old and needs replacement due to wear and tear. A new cartridge assembly can easily be fitted in place of the old one while the old one can be serviced if needed. Another advantage is that while fitting the new cartridge, the pump volume can be increased and decreased by using a cartridge having identical external dimension but changing the internal dimension to suit the current and anticipated pumping requirement, if needed. Such a cartridge is shown in Fig. 4.16.

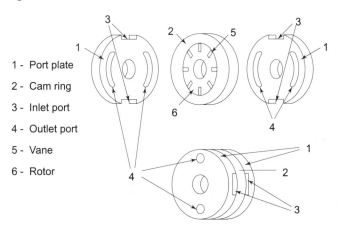

1 - Port plate
2 - Cam ring
3 - Inlet port
4 - Outlet port
5 - Vane
6 - Rotor

Fig. 4.16 *Vane pump cartridge assembly*

4.6.6 Variable Displacement Vane Pump

Very often in a hydraulic system the flow rate of the pump needs to be variable. This can be easily achieved by varying the rpm of the electric motor. But it has been seen that this is economically not feasible and hence is not practicable. The other way is to vary the pump displacement and this can be easily effected. The displacement of a vane inside the pump and therefore its delivery is proportional to the eccentricity between the rotor axis and the cam ring. Changing the geometric position of the ring relative to the rotor center will change the delivery volume as per system need.

Construction The schematic design and constructional feature of a variable displacement vane pump with the method of adjustment is shown in Figs 4.17(a), (b) and (c). The main parts of such a pump are:
(1) Mardened cam ring
(2) Rotor
(3) Vanes
(4) Screw for position adjustment
(5) Thrust (Needle) bearing
(6) Stop

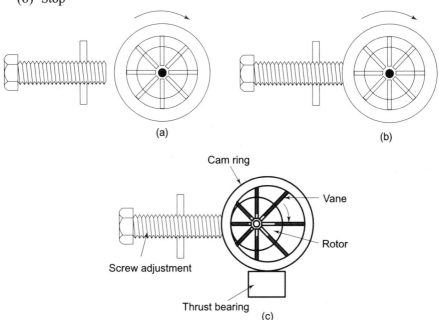

Fig. 4.17 *Variable volume vane pump with a, b, c method of adjustment*

4.6.7 Principle of Operation

The rotor containing the vanes is positioned eccentric or off-center with regard to the cam ring (circular in construction—not cam shaped) by means of the adjusting screw. Hence when the rotor is rotated, an increasing and decreasing volume can be created inside the cylinder bore.

If the screw is adjusted slightly so that the eccentricity of the rotor to the cam ring is not enough, the flow will be less whereas with higher eccentricity the delivery volume will be increased. With the screw adjustment backed completely out, the cam ring naturally centers with rotor and no pumping will be there as the eccentricity will be zero.

In Fig. 4.18 three extreme different positions of the eccentricity (e) are shown schematically, i.e. when eccentricity (e) is positive, i.e. for $e = +e$, Q is maximum. But when eccentricity (e) is 0, then Q becomes zero as both rotor and cam ring

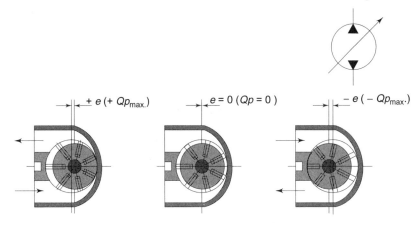

Fig. 4.18 *Vane type pump with adjustable flow rate Q*

become concentric. When eccentricity e is negative, i.e. $e = -e$, the direction of flow gets reversed. The hydraulic symbol is also shown here.

4.6.7 Pressure Compensated Vane Pump

In certain hydraulic system design, it is desired that when the predetermined system pressure is reached, the pump should stop pumping further oil to the system. This is possible if one uses a pressure compensated pump. Generally variable displacement vane pumps may be used for such purposes. The pressure compensated variable vane pump consists of an additional spring which is adjusted to offset the cam ring as shown in Fig. 4.19 (a). As the pressure acting on the inner contour of the ring is more than the pressure exerted by the spring, the cam ring becomes concentric to the rotor and pumping action stops; except for leakage oil, no flow is permitted to the system. The flow–pressure relationship of a variable displacement pressure compensated vane pump is shown in Fig. 4.19 (b). The point at which pressure is set is also shown graphically. In Fig. 4.19 (a), the vanes are angular which may be noted in comparison to Fig. 4.19 (c) which illustrates another design of a pressure compensated vane pump where the vanes are normal straight vanes only.

In such a pump the pumping mechanism has to act very fast and any build-up of fluid inside the housing is undesirable. The leakage is generally hot. If this were diverted to the inlet side, the fluid would get progressively hotter. Externally draining the housing alleviates the problem. The external drain of a pump housing is called "case drain". The symbol of such a pump is shown in Fig. 4.19 (d).

As has been expressed earlier, vane pumps are moderate to high in overall efficiency and power rating. A good vane pump may have a volumetric efficiency as high as 90%. The characteristics of a vane pump are shown in Fig. 4.19 (e). Fig. 4.19 (f) shows a two stage vane pump. The oil is drawn in during suction in the first stage pump and the output of this pump enters the inlet port of the second

112 *Oil Hydraulic Systems: Principles and Maintenance*

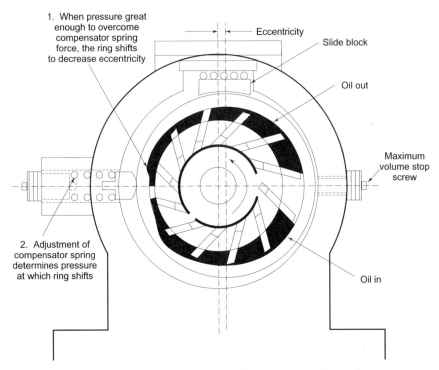

Fig. 4.19(a) *Pressure compensated variable vane pump with angular vanes [Courtesy: M/s Vickens Sperry Rand, USA]*

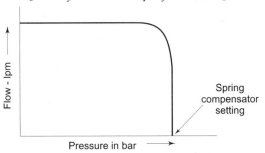

Fig. 4.19(b) *Flow-pressure relationship of a pressure-compensated vane pump*

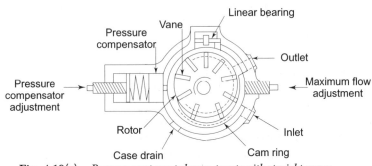

Fig. 4.19(c) *Pressure compensated vane pump with straight vanes*

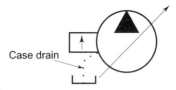

Fig. 4.19(d) *Variable volume pressure compensated pump symbol*

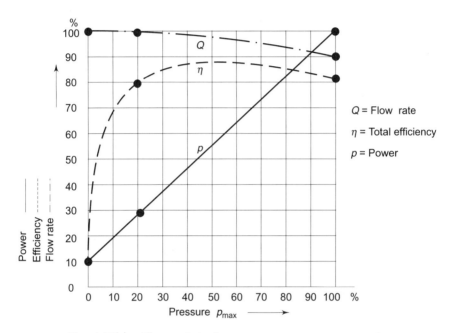

Fig. 4.19(e) *Characteristic of a vane type pump at constant speed*

stage pump which forces out the oil through its outlet port as shown through the arrow. A dividing valve with a spool area in the ratio 1:2 is maintained between the two pumps to control the flow and pressure.

4.7 PISTON PUMPS

In a piston pump, pumping is effected by a reciprocating piston in a finely machined and polished cylindrical bore in a cylinder block. The piston while reciprocating allows oil to be drawn inside the bore as it is retracted and the oil is expelled from the bore during the forward stroke of the piston. All the piston pumps available in the industry can, basically, be classified as:

  (i) Fixed or constant displacement pump
  (ii) Adjustable or variable displacement pump.

As per geometrical and physical arrangement of pistons they may be classified in three groups:

114 *Oil Hydraulic Systems: Principles and Maintenance*

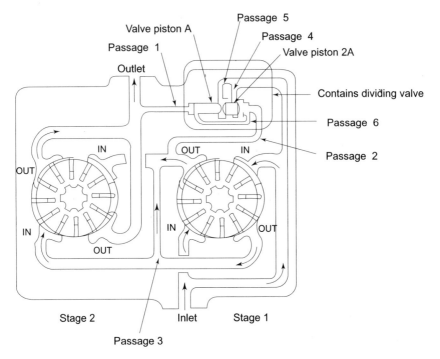

Fig. 4.19(f)   *Two stage vane pump*

1. In-line crank shaft driven pump
2. Axial piston pump
3. Radial piston pump

Apart from the above, some axial piston pumps are also available in bent axis form and are therefore termed as bent axis pumps.

A simple piston pump as shown schematically in Fig. 4.20 (a) consists of a finely machined and finished cylinder barrel and a plunger (piston) which is the moveable member inside the housing. The shaft of the plunger is connected to a prime mover which is mostly an electric motor—which is used along with a cam to produce the constant reciprocating motion of the piston plunger. In a simple design of such a pump the inlet and outlet ports are controlled by ball valves which allow oil in one way only. As the plunger makes an outward motion, oil enters into the increasing chamber inside the barrel. When the chamber is filled, the plunger is pushed in, the ball in the inlet gets closed and the ball at outlet port unseats and the oil is forced out. Such continuous cycling of the piston results in supply of oil in pulses only as may be evident from Fig. 4.20 (b). Pulsation creates undesirable effects and in order to eliminate and minimize the effect of oil pulsation and also to increase the flow rate capacity in piston pumps a number of cylinders and pistons are used in parallel and operated by a crank shaft.

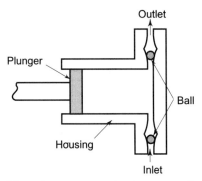

Fig. 4.20(a)   *Schematic principle of operation of a piston pump*

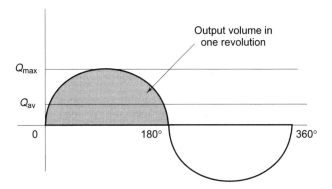

Fig. 4.20(b)   *Oil flow in pulses in a single piston pump*

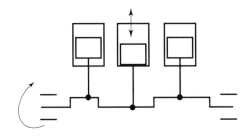

Fig. 4.20(c)   *Crankshaft driven piston pump*

4.7.1   Crankshaft Driven Piston Pump

Figure 4.20 (c) illustrates a crankshaft driven piston pump. Crankshaft drives are well known from automotive and steam engines. They require a parallel arrangement of cylinders and hence may require more space. This specific factor may limit their use in modern hydraulic systems which need to be more compact. However, there are still many applications where one or two-cylinder piston pumps are used, e.g. in certain mobile applications like small fork-lifters, simple loaders, etc. A typical 24-cylinder—12 cylinders in two opposite rows each—crankshaft driving two pistons, has been found to work at a very high pressure of 400 to 450 bar with a flow rate of 40 liter per min. with prime mover running at 1200 rpm.

4.7.2 Axial Piston Pump

All axial piston pumps may be classified as:
1. In-line axial piston pumps
2. Bent axis piston pumps.

Axial (in-line) pumps are also categorized as:
(i) Swash plate axial piston pumps, and
(ii) Wobble plate axial piston pumps.

All these pumps may be either fixed capacity pumps or variable capacity pumps.

4.7.3 In-line Piston Pump

A simple exploded view of an in-line axial piston pump is shown in Fig. 4.21 (a). Here the pistons are arranged axially parallel to each other around the circumferential periphery of the cylinder block. The pistons are driven to and fro inside a number of bores in the cylinder barrel. In such pumps either the cylinder barrel or a plate (swash plate) is rotated which makes the pistons have to and fro motion. Controlled by ball valves, the oil is sucked in or pumped out, the flow rate of the pump at a given rotational speed remaining constant. However the flow rate can vary if the speed of rotation of the prime mover is altered or the angle between the axis of the plate and the barrel is varied.

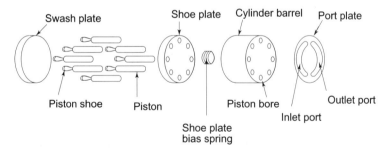

Fig. 4.21(a) *Exploded view of axial piston pump*

In axial piston pumps, as the pistons are arranged inside the piston block parallel to each other (i.e. in line), they are termed as in line axial piston pumps. The piston, cylinder barrel, swash plate, piston shoe and their mutual geometrical positions are shown in Fig. 4.21 (b).

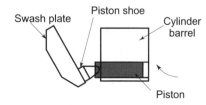

Fig. 4.21(b) *Schematic arrangement of swash plate, piston and piston-shoe*

In-line piston pumps are available either as a
(i) fixed cam or swash plate or as
(ii) rotating swash plate.

Three different designs of axial piston pumps are illustrated in Figs. 4.22 (a), (b) and (c). In Fig. 4.22 (a) we can see the cylinder body containing the axially placed pistons is made to rotate against a cam plate. The cam plate or swash plate is kept fixed and positioned at an angle with the axis of the cylinder block. The shoe plate to which the shoes are fitted is part of the rotating group which includes the drive shaft, cylinder block and pistons. As the cylinder barrel is rotated, the piston shoe follows the surface of the swash plate. Since the swash plate is at an angle, the piston has to reciprocate within the cylinder bore and thus oil is sucked during one half of the circle of rotation and during the other half of rotation, the oil is forced to the outlet port.

The cam plate is also known as *tilting plate* and the swash plate as the *reaction plate*. The shoes are flexible links connected to the shoe plate.

In contrast to the above, a schematic diagram is illustrated in Fig. 4.22 (b), the location of the fixed swash plate and the cylinder block is reversed, the piston

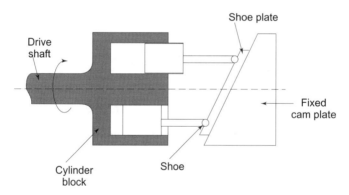

Fig. 4.22(a) *Swash plate stationery, cylinder block rotates*

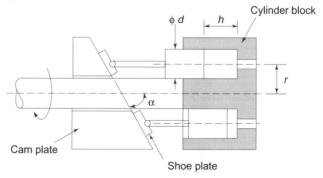

Fig. 4.22(b) *Stationery swash plate, cylinder block rotates.*
Position of cylinder block and plate interchanged

arrangement remaining the same as before. The pumping action takes place as the rotating cylinder block causes the pistons to reciprocate as they follow the angle of the stationery swash plate. While the pistons retract inside the cylinder bores and the cylinder block passes over the kidney shaped inlet port in the valve plate, the fluid is forced into the cylinder bore through the valve opening. Further rotation causes the pistons to force out the oil.

The theoretical delivery from the pump depends on the number of cylinders (n), the speed of pump (N), the diameter of bore (d), and the stroke length of the piston (h). The stroke length (h) is again dependent on the angle of the swash plate with the axis of the cylinder block.

For a fixed capacity axial piston pump the angle is mostly 30° for majority of swash plates.

The theoretical volume (q) of such a valve may be calculated from the following formula:

$$q = n \cdot N \cdot h \cdot A, \quad \text{where } A = \text{area of piston bore.}$$

In Fig. 4.22(c), the swash plate rotates with the drive shaft while the cylinder block is kept fixed. The swash plate in such pumps is called a *wobble plate*. The

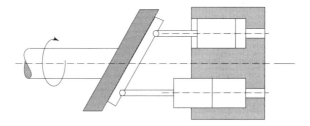

Fig. 4.22(c)   *Wobble plate axial piston pump*

shoe plate is thus prevented from rotating and the swash plate rotating on the surface of the shoe plate, produces a motion on it that moves the pistons in and out while reciprocating inside the bore. Swash plate in Figs 4.23 (a) and (b) are sometimes called cam plates or tilting blocks and in (c) they are called wobble plates while shoe plates are referred to as reaction plate. Figure 4.23 (a) illustrates a wobble plate pump while in Fig. 4.23 (b) the schematic arrangement of wobble plate, pistons and other mechanical parts of such a pump are shown.

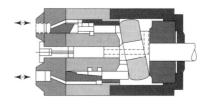

Fig. 4.23(a)   *A wobble plate axial piston pump*

Hydraulic Pumps—Construction, Sizing and Selection **119**

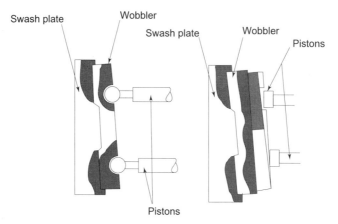

Fig. 4.23(b) *Mechanical arrangement of swash plate, wobbler and pistons*

Function of wobble plate pump As stated earlier, in a wobble plate pump, the cylinder block is fixed and the swash plate rotates. It is necessary to provide inlet and outlet valves for each piston. Though it means additional parts, still it is advantageous because of the following reasons:

(a) The outlet valves which are check valves, protect the pump mechanisms from heavy back pressure surges.
(b) The other advantage is that the flow can be split. For example, in a 10-piston pump, pistons 1, 3, 5, 7, 9 can be connected to one load and pistons 2, 4, 6, 8 and 10 to a second load. This will reduce the chances of damage to some extent as both systems would completely remain separated but will supply the oil to the single outlet port only.

But because of outlet check valves, pressure oil can not be fed back to the cylinder.

4.7.4 Variable Displacement Axial Piston Pump (Swash plate design)

It is clear from the above discussion that the stroke length of a piston is determined by the swash plate angle. With a large swash plate angle, the pistons have a larger stroke whereas smaller swash plate angle only permits shorter stroke length. When the swash plate angle becomes zero, no displacement takes place at all. This is illustrated schematically in Figs 4.24 (a), (b) and (c).

Hence piston displacement and volume flow rate in swash plate pump designs can be varied by changing the angle of the swash plate which can be effected either mechanically or by the action of the yoke actuating piston operating from the fluid supplied through the pressure compensating valve. However the maximum angle is generally limited to $17\frac{1}{2}°$ from various design considerations. By varying the angle, pump output and flow can be changed from zero to maximum. These pumps are also available in pressure compensated design as the flow generation and pressure development can be biased against a preset pressure as per the design need of the system.

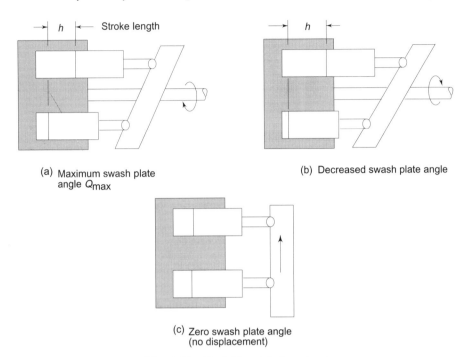

(a) Maximum swash plate angle Q_{max}

(b) Decreased swash plate angle

(c) Zero swash plate angle (no displacement)

Fig. 4.24 *Variable axial piston pump*

4.7.5 Pressure Compensated Variable Displacement Axial Piston Pump

A schematic diagram of such a pump is shown in Fig. 4.25 (a). In this design the swash plate is connected mechanically to a piston which senses the system pressure. This piston is called a compensator piston and is biased against a spring.

Initially, the return spring positions the yoke to full delivery. As the system pressure increases, the compensator valve spring of the piston moves to allow fluid to act against the yoke actuating piston. The system pressure is dependent on the setting of the compensator spool spring and adjustment. When the pressure is high enough to overcome the valve spring, the spool is displaced and oil enters the yoke piston. The piston is forced by the oil under pressure to decrease or stop the

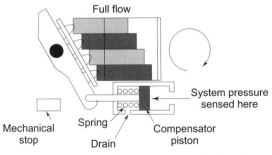

Fig. 4.25(a) *Pressure compensated piston pump*

pump displacement and no flow takes place as shown in Fig. 4.25 (b). If the pressure falls off, the spool moves back, oil is discharged from the piston to the inside of the pump core, and the spring returns to the yoke to a greater angle. The symbol of such a pump is shown in Fig. 4.25 (c). A cut section of this pump is illustrated in Fig. 4.25 (d).

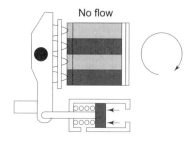

Fig. 4.25(b) *Pressure compensated axial piston pump-pump stops flow [Courtesy: M/s Parker Hannifin, USA]*

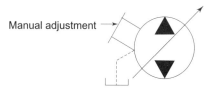

Fig. 4.25(c) *Variable volume overcenter pump symbol with manual adjustment*

Fig. 4.25(d) *Pressure compensated piston pump*

4.7.6 Bent Axis Axial Piston Pump

In case the drive shaft, pistons and cylinder block shown in Fig. 4.24 (c) all rotate around a common axis, there will be no reciprocating motion of the pistons and therefore there will be no pumping action. But as shown in Fig. 4.26 (a), if the axis of the cylinder block is made to form an angle (α) with the axis of the drive shaft, there will be reciprocating movement of the pistons. Let the distance between the center line of block and center line of the piston be r. As the entire unit rotates, the piston is forced into a reciprocating motion relative to the cylinder block. The amount of this motion is indicated by h.

As tilt angle α increases, stroke h also increases.

$\therefore$ $\qquad h = 2\,r\tan\alpha$, where $r = D/2$.

Theoretical displacement is the product of stroke, piston area, and number of pistons.

$\therefore$ $$Q = n \cdot h \cdot \frac{\pi}{4}d^2 = 1.57\,d^2\,n\,r\,t \text{ and } = \frac{\pi}{4}d^2\,n\,r\tan\alpha$$

where, d = piston diameter
$\qquad h$ = stroke
$\qquad \alpha$ = tilt angle
$\qquad n$ = number of pistons
$\qquad r$ = distance between cylinder block center and piston center = $D/2$.

From the diagram in Fig. 4.26(a) we find that in a bent axis pump, the rotating group consists basically of a cylinder block, pistons, universal link which keys block to drive shaft, shaft bearings and drive shaft. The cylinder block is supported by the cylinder bearing sub-assembly which is then free to rotate on the bearing.

As the drive shaft rotates it causes rotation of the cylinder block and reciprocation of the pistons inside the cylinder bore starts. The rotating cylinder block bores are open to inlet and outlet slots of the valve plate, which is fixed. The angle between the drive shaft and the cylinder block can be fixed or may be made variable.

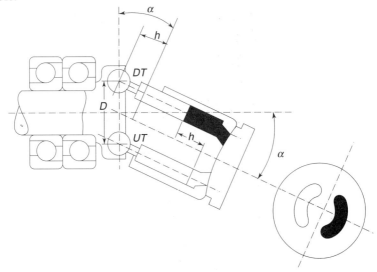

Fig. 4.26(a) *Bent axis axial piston pump*

Hydraulic Pumps—Construction, Sizing and Selection   123

The mechanism to control the angle between drive shaft and cylinder block consists of a yoke, valve block and actuating control. Positioning the yoke can be done by using hand wheel or by other methods.

With bent axis pumps, an oil film is needed between cylinder block (which rotates) and valve plate (which is fixed) to reduce friction. A pressure gradient is formed on the discharge slots of the valve plate, as shown in Fig. 4.26 (b).

The consequence of the pressure gradient is a separating force between the cylinder block and valve plate.

However, if the bore is restricted at the outlet as shown, then the separating forces are neutralized. This is a rather universal method with both bent axis and in-line pumps. A possible disadvantage is a slight pressure drop in the restriction.

Another form of balancing is shown in Fig. 4.26 (c). Here the restriction is avoided, and the bore goes straight through the cylinder block. Two balancing pools are provided. At the discharge port, the balancing pool is connected to zero pressure and at the suction port, the pool is at system pressure.

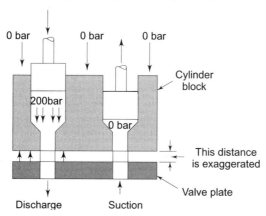

Fig. 4.26(b)   *Pressure gradient on valve plate*

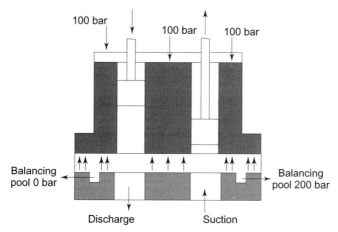

Fig. 4.26(c)   *Balancing of pressure gradient*

The maximum tilt angle is generally kept up to 30°.
Fixed displacement bent axis pumps of most manufacturers have angles = 15°, 20°, 25°, 30°.

4.7.7 Radial Piston Pump

They are available in two basic types:
(a) Radial piston pump with stationary cylinder block
(b) Radial piston pump with rotating cylinder block

In a stationary cylinder block type radial piston, pump, the cylinder block is kept stationary. Reciprocating motion is imparted to the pistons by a rotating cam. The outer bed of each cylinder is connected to the inlet and discharge check valves, which are opened by the suction of the receding pistons and pressure of the advancing pistons builds up respectively. A schematic diagram of a radial piston pump is shown in Fig. 4.27(a).

The rotating cylinder block and the pump housing with the track ring are positioned eccentrically which allows the radially placed pistons to make to and fro movement.

In a radial piston pump with rotating cylinder block the rotating cylinder block is connected to the drive shaft. The outer ring may be either fixed or rotating with the cylinder block. Along with the rotating cylinder, the pistons confined by thrust ring, reciprocate into the slots provided in the cylinder block.

The cylinder block and the outer ring are eccentrically placed. The pistons are thrown out by centrifugal force and pushed back because of the eccentricity between rotor-ring and cylinder block.

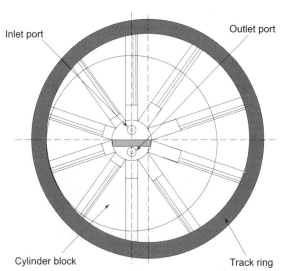

Fig. 4.27(a) *Schematic view of radial piston pump*

The principle of operation of a constant delivery radial piston pump is shown in Fig. 4.27(b). These pumps are quite robust and work at a high rpm with very high flow rate. The pressure sometimes is as high as 450 to 500 bar.

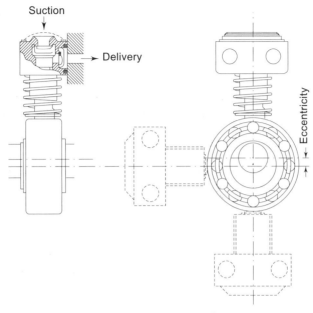

Fig. 4.27(b) *Principle of operation of a constant delivery radial piston pump*

The shape and internal construction of 7-piston radial piston pump are shown in Figs 4.28 (a), (b) and (c). For a radial piston pump one should note the following:

Fig. 4.28 7-*Piston pump*
(Courtesy: National Hydraulics Co. Ltd., Canada)

1. Separate and specific arrangement for lubrication of bearing is preferred.
2. Suction and pressure ports are valve controlled.
3. Pump pre-filling is advisable to avoid cavitation.
4. These pumps are found to suit majority of the hydraulic oils used in the market.
5. Operating parameters are found to be quite high, e.g. pressure rating—from 200 to 450 bar or more, flow rate—30 cm^3 to 80 cm^3/rotation, noise level—about 75 to 80 dB (A).
6. In most designs, pump rotation can be from right or left.

The piston type of pumps—both axial and radial—are developed and used for increasing pressure ranges. In order to enhance their longevity and improve their reliability especially at high pressures, special care is being taken by pump manufacturers to use appropriate combination of materials for various pump parts in order to have:

1. Higher safety even against dry running condition.
2. Improved protection of various pump mating surfaces from mechanical distortion and damage.
3. Higher resistivity against shock load, impact, vibration and oscillatory forces.
4. Improved ability to reduce friction between its sliding surfaces even in case of its parts having too pronounced surface asperities due to poor manufacturing.
5. Higher resistance against high temperature application.
6. Enhanced applicability under wide temperature variance (– 70° to 300°C).
7. Easy machinability including weldability during repairs, if needed.
8. Improved compactness.
9. Higher manufacturing precision.
10. Less and easy maintenance.

For a radial pump it is imperative that the side load capability of the pump is enhanced. In order to improve the pump-longevity a radial piston pump is supported on heavy duty roller bearings so that it can stand the extreme working pressure.

4.7.8 Energy Saving

Under the present day energy crisis condition, it is imperative for every one of us to take measures to conserve energy. Although the idea to keep the use of energy in hydraulics as low as possible is not new at all, but in the past not much efforts were made to optimize the energy saving parameters as overall efficiency was of more concern to the system builders at that point. This has now changed to a great extent. One of the main reasons is the exorbitant cost of energy. An important debate today is the future availability of energy. This means that not only the cost but also the consumption of energy has become an important deciding parameter in the modern industrial and social context. One way to save energy is to improve the efficiency of pumps and reduce their chances of leakage. The need of the hour is to develop energy-efficient pumps as well as hydraulic systems.

4.8 SELECTING AND SIZING OF HYDRAULIC PUMPS

During the last few decades, the fluid-power components and the oil-hydraulic systems as a whole have undergone a tremendous transformation in design, construction and application. It is needless to say that modern hydraulic systems are capable of a very high degree of efficiency in energy transmission and conversion. A properly engineered hydraulic system may attain a conversion efficiency even as high as 99%. Various types of basic fluid power elements have been developed in this period and the quality of each type has been progressively improved upon towards more sophistication as a result of which more and more hydraulic components with higher degree of quality and reliability are now readily available from the shelves of component dealers. This is true for all fluid power components and the hydraulic pump is no exception.

Amongst all members of the hydraulic family, probably the most important is the pump. Nowadays varieties of pumps with various configurations are developed and available from different hydraulic manufacturers. This particular phenomenon has posed a very interesting but peculiar problem to the hydraulic system designer. A fluid power designer today finds himself in a fix while selecting and sizing his pump. As far as functionability, reliability and maintainability of the designed system vis-a-vis the needed parameters of the desired system is concerned, a modern designer is confronted by the question as to which of the various types of pumps is most suited to the required application. Not only has he to choose between various basic types, e.g. a gear or a piston pump, but also to select between the different design characteristics of each type of pump to suit the size, function and control of the designed system. A hasty decision in this regard may lead to a wrong proposition and hence before jumping to a conclusion, a detailed study of the pump and the system characteristics and their mutual compatibility is a must.

The present day energy crisis which has engulfed the entire world with an unprecedented oil scarcity makes it obligatory for the fluid power system designer to look at the design parameters with an energy conscious eye to obtain an optimum return of the inputs to the system both energywise and costwise. Though each of the fluid power elements of a hydraulic circuit has its own individual contribution to the optimal system efficiency and reliability, the major responsibility lies in the correct choice of the type and size of the pump to match the overall system characteristics and reliability. It is important for the system designer to scrutinize very intimately some of the most relevant pump selection criteria before taking any decision.

Examples of certain common types of hydraulic pumps have been taken up here to make a comparable study of a few of their physical and functional characteristics. Pumps which are most commonly used include gear, screw, vane and piston types. One difficulty in presenting a consolidated comparison table for all the characteristic properties of pumps or any other components is that they vary too much typewise, size wise, construction, make and conditionwise, which means that pumps of similar type of design and size, but of different make may behave variedly under identical conditions as far as efficiency, leakage, maintenance, life-expectancy or other properties are concerned. Hence it is quite obvious that a designer has not only to compare the functional characteristics of the various types of pumps of similar size at a particular working speed or pressure, but he has also to make a detailed study of the similar pump of identical size but of different make under similar working parameters so that the final choice is found from every engineering aspect. Certainly it is quite a difficult and laborious job when one considers the varieties of interrelated factors. Some of the most common and important factors are taken up here to project a probable comparison between the pumps.

4.8.1 Pressure and Pressure Pulsation

Pressure is a fundamental pump selection factor. For the majority of common industrial applications, the system pressure is lower or medium except in certain places where very high pressure may be needed. In general, for up to a pressure of 100 bar any pump may be suitable, but in high pressure applications, piston type pumps are the natural choice. Internal gear pumps are also being used increasingly for moderately high pressure system. Screw and vane pumps are very efficient in low pressure applications whereas external gear pumps are used from very low to medium pressure range. Pressure pulsation is another important factor which needs careful consideration while selecting a pump. In Fig. 4.29, a comparison has been drawn between an external and internal gear pump to show the range of

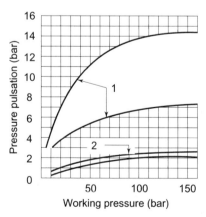

1. External gear pump
2. Internal gear pump

Fig. 4.29 *Comparison of pressure pulsation between external and internal gear pump*

pulsation of pressure over the working pressure at 1500 rpm. From the diagram it is seen that the rate and range of pressure pulsation of the external gear pump is much higher than the internal gear pump. In case of the internal gear pump the pulsation image is very narrow and almost constant.

4.8.2 Flow Rate and Power of Hydraulic Pump

Flow Rate Pumps have to generate hydraulic power and also supply a volume of oil (Q) at a certain pressure (P) to the system.

The theoretical volume of a pump can be calculated from the formula.

$$Q_{th} = i \cdot v \cdot n = cn$$

where Q_{th} = theoretical flowrate
  i = number of pumping chambers inside the pump per revolution
  v = volume of each pumping chamber, dm^3
  n = pump rpm 1/min
  $c = iv$ = pump constant (flow rate) for each revolution of the pump dm^3/revolution

The flow rate capacity of any pump is dependent mainly on two factors—the geometric size of the pumping chamber and the rotational speed of the pump. The maximum flow rate will depend on the maximum possible speed of the pump, which is limited by the structural stability of the pump body and other constructional features. To size a pump, the theoretical pump displacement rate, Q_{th} per revolution, may be calculated as shown in Table 4.1.

4.8.3 Pump Displacement Formulae

Some useful formulae for calculation of pump displacement, flow rate, etc. for various types of pumps are given in the box below:

Table 4.1 Useful Formulae for Pump Displacement

Useful formulae	Nomenclature	
	D—Vane rotor diameter cm	Screw, cm (where a, the distance between
External gear pump—two gears	D_p—gear pitch circle diameter where	the two screw centerlines, and d_o are
$q_t = 2\pi D_p bm$	$D_p = mZ$, cm	interrelated as
Screw pump—two spindles	D_c—center distance	$\cos \theta/2 = a/d_o$)
$q_t = [\pi(d^2_o - d^2_i)/4 - (d^2_o/4)$	between piston in	e—eccentricity of
$(\theta\pi/180 - \sin\theta)] h$	cylinder block,	vane/rotor piston
Variable vane pump—single cell	axial-piston/bent-axis	cylinder block, cm
$q_t = 2eb (\pi D - sZ)$	pump, cm	h—pitch of screw, cm
Balanced vane pump—two cells	P—pump output power,	l—vane stroke, cm
$q_t = 2\pi l b(D + 1)$	kW	m—gear module, cm
Radial piston pump	P_i—input power to pump,	n—rpm
$q_t = (\pi/2)d^2 eZ$	kW	p—pressure, bar
Bent-axis piston pump		

(Contd.)

$q_t = (\pi/4)d^2 ZDc \sin \alpha$
Swashplate, axial piston pump
$q_t = (\pi/4)d^2 ZD_c \tan \alpha$
Hydraulic power
$P = (pQ)/612$
Theoretical hydraulic power
$P_t + (pQ_t)/612 = (PQ)/612\, \eta_v$

Input power
$P_t = P_t/\eta_m = (pQ)/612\, \eta_v\, \eta_m$
$= (pQ)/612\, \eta_o$
Volumetric efficiency
$\eta_v = Q/Q_t\,(100)$
Overall efficiency
$\eta_o = P/P_t$
Theoretical pump displacement
$Q_t = (q_t)n(10^{-3})$

P_t—theoretical power of pump, kW
Q—actual flow from pump, l/min
Q_t—theoretical pump flow, l/min
Z—number of gear teeth/vanes/pistons
a—axial distance between screw spindles, cm
b—width of vane/gear teeth, cm
d—piston diameter, cm
d_f—core diameter of screw, cm
d_o—outer diameter of

q_t—theoretical pump displacement, cm^3/rev
S—vane thickness, cm
q = angle of d_o overlap as measured at a screw centerline, degrees (Fig. 7.2)
a—offset angle, bent-axis pump/tilt angle, swashplate pump, degrees
η_m—mechanical pump efficiency, %
η_o—overall pump efficiency, %
η_v—volumetric pump efficiency, %

To calculate angle of screw overlap (θ), the following formula can be used (shown in Fig. 4.30).

$$\cos \frac{\theta}{2} = \frac{a}{d_o} = \frac{R+r}{2R}$$

4.8.4 Power and Pump Efficiencies

By definition, power is work done per unit time.

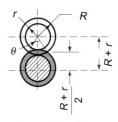

Fig. 4.30 Screw overlap angle (θ)

$$\therefore \quad P = \frac{\text{Force} \times \text{distance travelled by oil}}{\text{time}}$$

where P = power

$$= \frac{F \times S}{t}$$

$$= \frac{F \times S \times A}{t \times A} = \frac{F}{A} \times \frac{S \times A}{t}$$

$$= p \times \frac{V}{t} \quad \text{where } F = \text{Force, kgf}$$

$$= p \cdot Q$$

$$= \text{Pressure} \times \text{flow rate}$$

where,
A = Area of cross section of pipe, cm^2
S = distance travelled, cm
t = time taken to travel, min
$\dfrac{F}{A}$ = Pressure = p, bar
$S \times A$ = Volume of oil = V, cm^3

$$Q = \text{Flow rate} = \frac{V}{t} = \text{l/min}$$

∴ Power = Pressure × flow rate

A practical formula will be more clear from the following example.

Example 4.1 *If a hydraulic system working at a pressure of 30 kgf/cm² has flow rate of 8 l/min, calculate the power.*

Here $p = 30 \text{ kgf/cm}^2$, $Q = 8 \text{ l/min}$

∴
$$\text{Power } P = p \cdot Q = 30 \frac{\text{kgf}}{\text{cm}^2} \times 8 \frac{\text{dm}^3}{\text{min}}$$

$$= \frac{30 \text{ kgf}}{\text{cm}^2} \times \frac{8 \times 1000 \text{ cm}^3}{\text{min}}$$

$$= 30.8 \frac{102 \cdot \text{kgf} \cdot 1000 \text{ cm}}{102 \cdot \text{min}}$$

$$= 30.8 \cdot \frac{102 \cdot \text{kgf} \cdot 10 \text{ m}}{102 \cdot 60 \cdot \text{sec}} \text{ kW}$$

$$= 30.8 \cdot \frac{1}{612} \text{ kW} = \frac{30 \times 8}{612} \text{ kW}$$

∴ Theoretical power output of a

$$\text{pump } P = \frac{pQ}{612} \text{ when } P = \text{Pressure in } \frac{\text{kgf}}{\text{cm}^2}$$

Q = Flow rate in l/min.

∴
$$\text{Power, } P = \frac{p \cdot Q}{612}$$

4.8.5 Pressure, Flow Efficiencies

These are fundamental pump selection factors. The formulae are shown in Table 4.1. In general, for pressure up to 100 bar, any pump may be suitable. For high pressure applications, piston pumps are a natural choice. Internal gear pumps also are used increasingly for moderate pressure systems. Screw and vane pumps are very efficient in low pressure applications.

The flow capacity of any pump depends on the geometric size of its pumping chamber and its rotational speed.

A pump's maximum power rating depends upon pressure, flow, speed and the mechanical strength of the pump body. Some modern hydrostatic transmission pumps can supply power to nearly 1500 kW, although most industrial application pumps produce in the 350 kW range.

Actual pump efficiency depends on pressure, flow, speed, design and manufacturing clearances. Volumetric efficiency depends mostly on leakage losses caused by change in viscosity or excessive mechanical clearance. Pump overall efficiency is determined by mechanical and frictional losses and volumetric loss.

In general piston pumps have higher overall efficiency than gear and vane pumps. The efficiency of a pump is greatly influenced by:
(a) size, geometrical clearance, and accuracy of the fit of the internal mating parts of the pump.
(b) oil viscosity and properties of oil e.g. lubricity and the effect of operating temperature on the oil viscosity.
(c) operating pressure, and
(d) pump rotational speed.

While selecting a pump, it is important to remember the permissible clearances of all pump components to achieve higher volumetric efficiency. Table 4.2 shows a few of the critical clearances.

Table 4.2 Critical Pump Clearances

Type of pump	Type of clearance	Clearance, micrometres (μm)
Gear	Radial, between teeth tips and housing	20-25
	Axial between gear face and end covers	30-50
	Gear shaft and bearing	10-20
Piston	Bore and piston	10-20
Vane	Vane and slot walls	10-40
	Non-parallelism between slot walls	15-30

4.9 OIL-COMPATIBILITY

Rules of Thumb Pump–oil compatibility depends upon clearances of mating parts, oil viscosity, lubricity and wear rate of pump parts, operating temperature, and pressure. Some rules for oil compatibility are given below.
1. *Gear pump*—General kinematic viscosity range is 30 to 70 cSt with a working temperature of 10° to 30° C.
2. *Screw pumps*—Suitable for nearly all types of fluids which present pumping problems due to minimum lubricity: water–oil emulsions, and synthetics such as phosphate esters. They accept oil from 20 to 70 cSt without much difficulty.
3. *Vane pumps*—Suitable for oil from 30 to 50 cSt above 70 bar at 10° to 35°C with tolerable efficiency.
4. *Piston pumps*—Accept oil viscosities of 60 to 70 cSt most efficiently. They are suitable for most of the fire resistant oils.

Actual pump efficiencies vary extensively. Table 4.3 indicates various parameters and capacities that can be expected from common types of hydraulic pumps.

Table 4.3 Pump Capacities

Type of pump	Pressure, bar		Flow rate, liter/min		Speed, rad/s		Efficiency, %		Maximum power, kW†
	Maximum	Common median	Maximum	Common median	Maximum	Common median	Volumetric	Overall	
External gear	300	10-100	400	3-100	350	75-200	65-85	60-70	100
Internal gear	350-400	300	450	200	300	200	98	85-90	125
Fixed vane	175	125	200	5-150	300	150-200	85-90	75-85	50
Variable vane	125	75		100-150	300	180	90	85	30
Screw	175	30-60	7500	3-1000	300	150-300	95	75-85	50
Variable axial piston	700	300-350	1000	*	350-400	200-300	to 95	85-90	150
Bent axis piston	700	250-300	800	*	400	200-300	>95	>90	125
Radial piston	1000	300-400	2000	*	350	150-200	>95	>90	350

*Moderately higher than other pumps
† General industrial requirement is 25–30 kW

4.9.1 Life Expectancy

Modern internal gear pumps run silently at moderately high pressures, have life expectancies to 20,000 working hours, and have volumetric efficiencies in the range of 97%. They are more costly than external gear pumps.

For low and moderately high pressures and efficiencies, external gear pumps are probably much cheaper than others, but with pressure and time, their noise levels may rise very abruptly.

Radial piston pumps have a very long life expectancy and are well suited for high pressure applications. The life expectancy of axial piston pumps may be as high as 40,000 working hours when operated at 200 to 250 bar. This may decrease to less than 15,000 working hours when run at 300 to 350 bar.

Table 4.4 Weight-to-Power Ratio

Type of pump	Ratio	
	Common median, kW	Maximum possible, kg/kW
External gear	0.5-0.75	1.5
Single-stage vane	0.2-1.25	5
Variable axial piston	0.3-0.6	6
Bent axis piston	0.3-2.5	10
Radial piston	0.6-10	20

4.10 SIZE

Modern hydraulic pumps are very compact and have better weight to power ratios than other energy conversion systems. This ratio can be important, specially in aviation and mobile applications. The ratio depends mostly on the type of pump and the materials used for the pump and its parts. Of the common pumps, vane pumps probably have the best weight-to-power ratio, around 0.2 kg/kW. Piston-pump ratios vary from about 0.3 to 0.6 kg/kW, while most gear pump ratios vary between 0.5 to 0.75 kg/kW. (Please refer to Table 4.4.)

The cost of any pump depends on many factors; probably the most important is the sophistication of the design which varies widely from pump to pump. However a survey shows that

(a) an average gear pump is the least expensive
(b) piston pumps are generally more costly than other pumps
(c) rotary pumps are more advantageous than piston pumps, considering cost and performance for low pressure applications
(d) internal gear pumps are more costly than vane pumps in certain cases, while
(e) screw pumps are generally more costly than both.

4.11 NOISE

Generated noise levels vary with the pump, pump component materials, pump mountings, methods applied to eliminate vibration, rigidity, manufacturing and fitting accuracies of the pump elements, size and flow capacity pressure, speed of rotation, pressure pulsations, and the other components connected in the circuit. Experience shows that external gear and piston pumps are the noisiest while screw pumps are very quiet with vane and internal gear pumps somewhere in between. Noise levels developed in typical pumps are shown in Fig. 4.31 (a).

Any pump which generates noise above 90 dB (A) is a loud pump; those around 60 dB (A) or less are considered quiet running. Pumps of different types with identical displacement, pressure and speed, working under identical conditions will generate different noise levels.

The rise in noise intensity generated in a positive displacement pump with an increase in pump speed is higher than with an increase in pressure or displacement as may be seen from Table 4.5. A pump should not be selected solely because of cost, pressure or size. Other considerations are important, such as overall pump-system compatibility, pump reliability and its life expectancy. Experience has shown that screw pumps are very efficient when used within a pressure range of 20 to 30 bar. They are the most silent and pulsation free, and their reliability factor is very high while used with proper viscosity.

Vane pumps also produce less pulsation and noise. They are better suited than external gear pumps for medium pressures in stationary applications. Their overall efficiency is less than piston pumps.

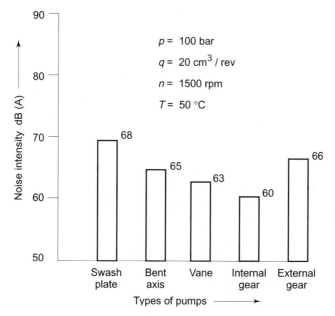

Fig. 4.31(a) *Noise developed in typical pumps*

The comparative noise behavior of two pumps with 32 l/min and 20 l/min capacity respectively working at 1500 rpm with oil of viscosity of 32 cSt can be seen in Fig. 4.31(b). The noise level of a pump kept in a noise isolating room is found to be less by almost 18 dB (A) compared to the noise level measured at site for a pump installed on a C.I. oil reservoir.

The pattern of rise of noise level depends on the pump construction, flow rate, speed, pressure, etc.

In the machine building industry, hydraulics alone offers the possibility of generating high power densities within a very small volumetric space which no other energy system is generally capable of. Positive displacement pumps and motors have specific advantages over others in this aspect. Necessarily, therefore, they tend to be the loudest generators of noise in hydraulic system.

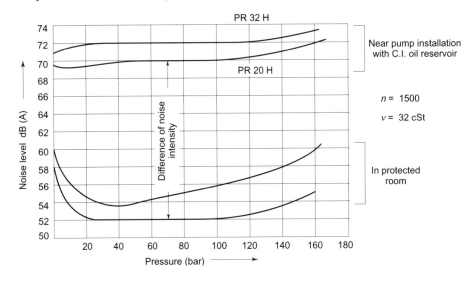

Fig. 4.31(b) *Noise intensity in protected room and near pump installation measured at 1m away*

The rise in noise level is considerably influenced by the rotational speed (n), operating pressure (P) and volume of oil per revolution of the pump. This may be clear from the graphical representation in Figs 4.32 (a) and (b). In Fig. 4.32 (a), the rise in noise level of axial piston pump is illustrated with increase in rpm, operating pressure and volume of oil per revolution. In comparison to an axial piston pump, a vane pump produces less noise when n, p and v are increased by same amount under similar working parameters as may be seen from Fig. 4.32 (b). Noise level also increases with the increase in power rating of the pump as may be seen from Fig. 4.33 where a variable axial piston pump is found to generate more noise level at higher power rating compared to low power rating. A fixed displacement pump generates less noise intensity than a variable displacement pump under similar working parameters and size. It may be noted

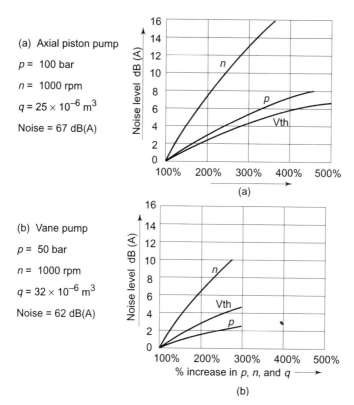

Fig. 4.32 *Rise of noise level with pressure, flow and rpm*

that the rise in noise intensity by a positive displacement pump with increase of pump speed, is higher than that with increase of pressure or displacement volume. Sometimes it may be advisable to use two or more lower capacity pumps instead of using a single larger capacity pump in which case the effective noise level is reduced. Pump ripple is one major cause of pump noise. Let us see here what causes pump ripple and how to measure it.

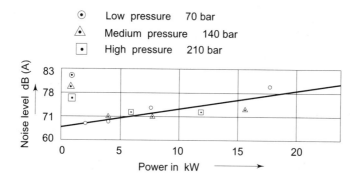

Fig. 4.33 *Dependence of power and noise intensity*

Table 4.5 Noise vs. Speed, Pressure and Displacement

Type of pump		Variable axial piston			External gear			Internal gear		
Speed and noise	Initial n-1000 rpm			p-200 bar			q_t-25 cm³/rev			
	1000 rpm noise level, dB (A)	71			65			63		
	Speed increase, %	50	100	200	50	100	200	50	100	20
	Noise level increase, dB(A)	4.0	7.5	13.0	6.0	11.5	15.0	2.5	5.5	10
Pressure and noise	Initial p-50 bar			n-1500 rpm			q_t-25 cm³/rev			
	50 bar noise level, dB (A)	69			67			63		
	Pressure increase, %	100		200	100		200	100		200
	Noise level increase, dB (A)	2.5		5.5	3.0		5.5	2.5		3.0
Displacement and noise	Initial q_t-25 cm³/rev			p-100 bar			n-1500 rpm			
	25 cm³/rev noise level, dB (A)	75			67			67		
	Displacement increase %	100		200	100		200	100		200
	Noise level increase %	4.0		10.0	6.5		10.0	6.0		8.0

4.12 PUMP RIPPLE

Positive displacement pumps (PD) by design can not deliver absolute constant flow. Similarly PD motors do not rotate at constant speed. Small variations of flow that take place during pumping are called *ripple*.

Generally ripple is not a very serious problem unless one wishes the pump noise to be minimum. However, it is better that maintenance and service personnel should have a clear idea about the mechanism of ripple.

4.12.1 Cause and Magnitude of Pump Ripple

All PD pumps are designed to displace a discrete volume of flow. For example vane pumps transport fluid volume in the compartments between two vanes. As each of these fluid compartments enters the pump discharge side, its volume starts reducing while the fluid is squeezed out. If the discharge volume is plotted against the angular position of vanes, the shrinkage of the flow, produces a sinusoid curve as these variations in output follow a sinusoidal pattern. This is the case with most vane and piston units. In the case of gear pumps too, a sinusoid may also be assumed. In Fig. 4.20 (b) a piston pump with a single piston is shown to have a delivery curve as two pulses during a complete rotation. Figure 4.34 shows delivery pattern of piston pumps with multi-pistons (4, 5 and 6 pistons respectively). With more pistons, the delivery schedule of the pump does not remain as simple as shown in Fig. 4.20 (b).

4.12.2 A Piston Pump with 6 Pistons and its Ripple

It is assumed that the zero base line separates suction from discharge and since only the discharge is of interest here, the sinusoid below zero is ignored. The total delivery can be obtained by adding at each instant the delivery of each piston as shown in Fig. 4.34 (c).

Calculation of amount of ripple Instead of plotting sinusoids, instantaneous delivery and ripple can also be calculated.

In the diagram of single piston pump (Fig. 4.20 b) the instantaneous displacement for any given moment is

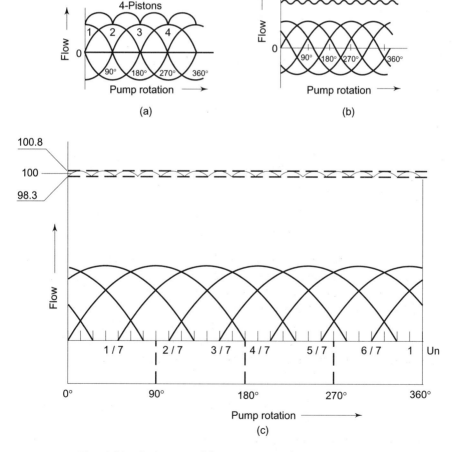

Fig. 4.34 *Piston pump delivery pattern with 4, 5 and 6 pistons*

$Q_i = Q_m \sin \theta$ where Q_m = maximum flow rate and θ = angle at which pistons are located.

The average value $Q_{av} = \dfrac{Q_m}{\pi}$

The theoretical delivery of the pumps with n pistons is

$$Q = Q_{av} \times n$$

∴ for a single piston, the instantaneous volume

$$Q_i = Q_m \sin \theta = \pi Q_{av} \sin \theta$$

$$= \dfrac{\pi Q}{n} \sin \theta.$$

The angle θ is measured from the point where the piston begins to discharge. Obviously there are other pistons ahead which are already discharging. For example for a nine piston pump there would be one at $\theta = 40°$, the second at $80°$ or $\left(40° + \dfrac{360°}{n}\right)$, the third at $\left(40° + 2 \cdot \dfrac{360°}{n}\right)$ and so on.

∴
$$Q_i = \dfrac{\pi Q}{n}\left[\sin \theta + \sin\left(\theta + \dfrac{360°}{n}\right) + \sin\left(\theta + \dfrac{360°}{n} \times 2\right)\right.$$
$$\left. \cdots + \sin\left(\theta + \dfrac{360°}{n} \cdot 8\right)\right]$$

or
$$Q_i = \dfrac{\pi Q}{n} \sum_{}^{k=n-1} \sin\left(\theta + \dfrac{360°}{n} \cdot k\right) \text{ where } k = n - 1.$$

If we divide $Q_i^{k=0}$, by the theoretical volume Q, then

$$\dfrac{Q_i}{Q} = \dfrac{\pi}{n} \sum_{}^{k=n-1} \sin\left(\theta + \dfrac{360°}{n} \cdot k\right)$$

The factor is the proportion by which instantaneous delivery differs from average delivery of a pump. Let us consider a nine piston pump with one piston at $20°$ within the discharge area. The other pistons will be at $60°, 100°, 140°, 180°, 220°, 260°, 300°, 340°$, but only piston $20°, 60°, 100°$ and $140°$ contribute to the delivery. At $180°$, the delivery is zero and the pistons at $220°, 260°, 300°, 340°$ are on the suction side, which means that all these angles of which the sinus is zero or negative, should be neglected as they do not add to the delivery volume at the instant.

∴ here $\dfrac{Q_i}{Q} = \dfrac{\pi}{4} [\sin 20° + \sin 60° + \sin 100° + \sin 140°] = 0.99$

This means that at this instant the delivery is 99 percent of the average delivery.

In order to obtain maximum deviations from average delivery, it is necessary to find the angles θ for which θ_1 becomes longest. For example, with a 6-piston pump, maximum deviation is 1.047 times average delivery, which means that at this point the deviation is 4.7 percent of the average delivery. These maximum deviation constitute the ripple flow that takes place during pumping.

4.13 CHECKLIST

Although a majority of hydraulic designers need not know how to design a hydraulic pump, they are required to know the selection process. This pump selection checklist can help:

- safe and maximum system working pressures
- allowable pump speeds
- rated pump performance
- system flow requirement
- reciprocal relationship of pressure, speed, and flow
- suitability of variable displacement control
- pressure surge tolerance
- leakage losses
- volumetric and overall efficiency
- contamination tolerance
- operating reliability and durability
- life expectancy at various loads, speeds
- oil characteristics and their relationship to rate of pump wear
- pump noise generation under different speeds, pressures, and flows
- system temperature
- maintainability
- servicing and spares availability
- filtration requirements
- drive type and mounting
- special coating on sliding surfaces
- suction conditions
- manufacturing characteristics: component clearances and fit
- compactness and power-to-weight ratio, and
- cost and economic factors for overall system compatibility.

Review Questions

1. State the role of a pump in a hydraulic system. Classify pumps. What is a positive displacement pump? Why is it called "positive" displacement?
2. What is a centrifugal pump? How does it differ from a positive displacement pump?
3. Differentiate between a swash plate and a wobble plate.
4. How does an external gear pump differ from an internal gear pump? What type of gears are generally used in gear pumps? State them.

5. What is pump ripple? Why does pump ripple occur in a pump? What is the advantage of using an odd number of pistons in a piston pump compared to an even number of pistons?
6. Which pumps are noisier—external gear, internal gear or piston pumps? Which pump generates the least noise?
7. Which are the most important factors one should consider while selecting a hydraulic pump for a specific application? State them.

References

4.1 Werner Holzbock G., *Hydraulic Power and Equipment*, Industrial Press Inc., 200 Madison Avenue, New York, USA.
4.2 Yaple F.D., *Hydraulic and Pneumatic Power and Controls*, McGraw-Hill Book Co., New York, USA.
4.3 Dr. Fitch E.C., *Fluid Power and Control Systems*, McGraw-Hill Book Co., New York, USA.
4.4 Addision, Herbert, *Treatise on Applied Hydraulics*, Chapman and Hall, London, UK.
4.5 Publication from m/s Vickers Serry of India Ltd., Bombay, India.
4.6 Herion Taschen Buch from m/s Herion Werke, kg, Stuttgart, Germany, 1969.
4.7 Publicity Material and Training Literature from m/s Bosch GmbH, Germany, 1970/71.
4.8 A. Duerr and O. Wachter, *Hydraulic Werkzeug Maschinen* Carl Hanser Verlag, Muenchen, Germany, 1968 p. 61.
4.9 M/s Parker Hannifin Corporation, *Industrial Hydraulic Handbook*, Ohio, USA, 1973. (p. 11-10, p. 11-5).
4.10 Publicity material of m/s Rexroth Maneklal Industries Ltd., Ahmedabad, India.
4.11 Andreas Breuer Sterken, *Geraeusch Minderung an hydraulischen Komponenten*, Oil Hydraulic u. Pneumatic Zeitschrift furer Fluid Technic, Germany, May, 1992.
4.12 *Laermbekaemfung in der Hydraulic* published in Constructions Jahr Buch, 1986/87 from Oil Hydraulic u. Pneumatic, Aachen, Germany.
4.13 Laermarn Konstruieren '77, VDI Bericht Nr. 273, Juli, 1977.
4.14 Majumdar, S.R., "Selecting & Sizing Hydraulic Pumps" Published in *Hydraulics & Pneumatics*, USA, April 1985.

5
Direction Control Valves

IN a hydraulic system, the pump generates the flow of oil which is to be fed to the cylinder or other actuators. The pressure energy is fed to the actuator through a number of control blocks called *valves*. Hence the valve is nothing but a device which is necessary to control the oil energy.

Various types of valves are used in hydraulic systems to control or regulate the flow medium. Basically valves are expected to control:

(a) Blocking or stopping of flow

(b) Direction of flow

(c) Pressure of flow media

(d) Flow quantity

(e) Other special functions.

In keeping with the above functions, different types of valves have been developed and used in hydraulic circuit design.

5.1 WHAT TO CONTROL?

This is an important query for a hydraulic expert. What are the various parameters that need to be controlled in the case of oil energy? The answer to this query can be given as follows:

(i) In a hydraulic machine the actuators make to and fro motion. Hence the direction of oil feeding the cylinder or motor needs to be reversed. This means that the direction of oil flow is an essential aspect of hydraulic operation. This is achieved by direction control valves.

(ii) Oil energy does the work due to pressure in the oil system. In a mechanical or hydraulic system there may be a need to increase or decrease the oil pressure depending on the specific requirement. Hence a group of valves called *pressure control valves* have been designed.

144 *Oil Hydraulic Systems: Principles and Maintenance*

(iii) The speed of the actuators needs to be altered as per operational requirement. For this purpose hydraulic systems use flow control valves.

Other than this, special valves are needed for specific operations and functional requirements.

Now let us discuss briefly the functional principles of control valves used in a hydraulic system.

5.2 DIRECTION CONTROL VALVE

It is important to know how to reverse the direction of a hydraulic actuator. What mechanism is needed to start, stop or reverse a cylinder movement? Change in the direction of the cylinder or a motor is effected by use of the direction control valve. The name of the valve implies the function it performs. Direction control valves are employed in a hydraulic system to determine the direction of the fluid in the hydraulic circuit. Sometimes they are used also as a selector switch.

However, other than the above, many special constructional features inside a valve body are also possible to make or break the oil supply to a system.

5.2.1 Construction

Two types of construction are generally used for common direction control valves, e.g.:

(i) Seat valve or poppet valve
(ii) Spool valve or sliding valve

However other than these, many special constructional features are also possible.

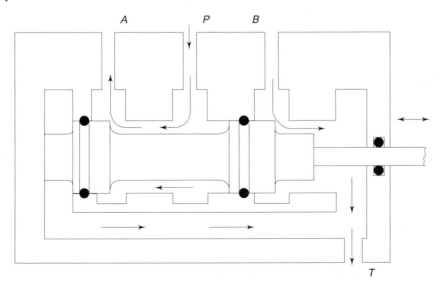

Fig. 5.1 *Basic design of a spool valve*

In direction control valves, the flow paths may connect a supply port to an outlet port (P to A or B) or may allow a pressurized port to unload to the tank (A or B to T). (See Fig. 5.1.) In certain cases all the flow paths through the valve may be blocked. This is effected by shifting the internal parts of the valve. The internal part may be

(a) A poppet lifting off or moving between sealed seats
(b) Sliding plates moving across ports
(c) Sliding spools with dynamic seals moving in a bore
(d) Spools moving in a bore with stationery seals
(e) Lapped spools moving in mating lapped sleeves

The general construction of valves is simple in nature having either a spool or poppet. But there are more complicated valves with multispools generally used for specific applications.

5.2.2 Poppet Valve

Here a poppet, a ball or a similar item like a plate is made to sit over a specially constructed, finely machined and polished seat. The advantage of this type of valve is that it has a high response and relative insensitivity to contamination. They are suited to high pressure duties and so may be preferred for specific applications. However, they are less suitable for large valve sizes but since the opening load becomes excessive they may be more suitable for indirect actuation. The lift required is dictated by the poppet angle.

The advantage of the poppet valve, with ball, plate or cone as the valving element, lies in its

(i) Absolute sealing ability in the closed position
(ii) No leakage or very little leakage takes place

But the disadvantages are

(i) While cracking initial movement of poppet is not helped at all due to the leakage oil as in spool valve since axial pressure compensation is virtually impossible.
(ii) The internal design is a bit complicated and sometimes difficult for fine finishing of the seats, etc.

5.2.3 Spool Valve

It may be either with a rotary spool or a sliding spool.

(i) Rotary spool   With a rotary spool the hydraulic fluid is directed through longitudinal grooves machined into a rotable piston. As a rotational movement is necessary to effect changeover, this type of spool is used predominantly in manually operated valves.

Rotary valves rely on close contact being maintained between a rotating port plate and a back-up member and are rather more difficult to produce adequate sealing for high pressures. They are more appropriate for lower pressures.

Sliding rotary valves or poppet valves may be preferred to spool valves for a particular application although they tend to be more expensive to produce and are prone to higher leakage rates unless manufactured to the highest standards of precision and accuracy.

(ii) Sliding Spool With the sliding spool type of valve, there is a small piston called a spool inside the valve casing, which slides inside the casing thereby opening or closing the ports made specially inside the valve body and according to the position of the axially displaced piston (spool), the ports get oil supply. Sliding spool valves are generally balanced permanently. During operation, only spring force, frictional force and flow forces are to be overcome.

The linear control motion is especially suitable for operation by a solenoid as well as hydraulic, pneumatic or other mechanical means. Due to functional requirements, the sliding piston is fitted with a definite play inside the valve housing. The pressure difference and viscosity of oil influences the play of the valve fitting. A diagram of a spool type valve is shown in Fig. 5.1.

5.2.4 Principle of Operation

In spool type valves, the valve spool is to slide in a matching bore thereby connecting or breaking various ports of the valve. In seat type valves, mostly balls or poppet are used as the valving element.

Spool control may be effected by a spring, by a mechanical system, pressure or by electrical means. The spool is the most favored type of valve for selectors particularly for industrial application. Spools themselves are generally classified by the porting conditions. Possible variations in the porting conditions are numerous and need judicious care for application which will be discussed later in this chapter.

The following parameters are normally to be taken into consideration while designing a valve:

   (i) pressure force
  (ii) valve size
 (iii) valve body and sleeve and their rigidity
 (iv) friction forces
  (v) material of internal parts
 (vi) valve fitting and permissible clearance
(vii) sealing arrangement
(viii) leakage and resultant pressure drop, etc.

5.3 OPERATING METHODS AND PARAMETERS

The operation of directional valves may be manual, mechanical or by pressure. Many times electrical signals are also used. The general description "mechanically controlled" is usually given to valves operated by cams or similar other mechanical means. The details of valve actuation possibilities are described below.

Valve actuation Various control techniques are nowadays used for actuation of the valve spool. But in the most normal case, the spool movement of direction control valves can be actuated by application of a direct force on the valve-spool or poppet. The actuating force could be applied:

(a) Manually
(b) Mechanically
(c) Hydraulically
(d) Pneumatically
(e) Electrically
(f) Electro-pneumatically
(g) Electro-hydraulically or
(h) Other means e.g. remote control

(a) Manual actuation A manually actuated valve as shown in Fig. 5.2 uses the muscle power or spring force to actuate the spool. Common manual actuators include levers, push buttons, pedals, etc.

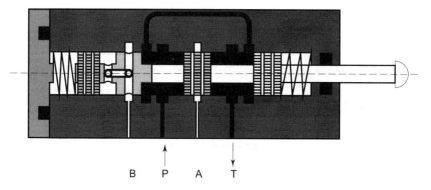

Fig. 5.2 *Manual actuation of valves*

In a lever operated valve, the angular movement of the lever is transmitted to a tappet and from there to the spool. The detents in the individual positions are achieved with the use of balls which are pressed into annular grooves in the tappet by springs. Pedal operation is used when hands of the operator need to be kept free. Sometimes both hands are needed for controlling the valves.

(b) Mechanical Various mechanical devices are used to control a direction control valve, e.g. roller, plunger, roller tappet, spring, etc. The roller tappet is pushed in by a cam or a similar device and presses on the spool. A very common mechanical actuator is a plunger. With a roller at its top, the plunger can be actuated by a cam which can be attached to the cylinder. The mechanical

actuation is used when the shifting of a directional control valve must occur at the time a cylinder reaches a specific position. A spring is very often used to reset the valve position for both manually or mechanically actuated valves described below.

Spring offset: A two position valve is generally actuated to shift it to its extreme position by using any of the actuating system as discussed. But the spool is made to return to its original position by using a spring when the earlier actuating force is withdrawn. Such a direction control valve is known as a spring offset valve.

(c) Pneumatic In order to have a higher force of actuation and eliminate mechanical control, direction control valves can be actuated using oil or air pressure. The pressure is applied at the two extreme spool lands if it is a 4-land spool or to separate pilot pistons if it is a 2-land spool as stated below. The pneumatic pressure acts on a piston with a large effective area, which in turn transfers the actuating force to the spool. The control is achieved by 2-way pneumatic valve. The connections are made with Z and Y ports of the valve. However with application of air pressure, care needs to be taken to see that no leakage of pressure air takes place to the main valve.

(d) Hydraulic The hydraulic pressure may act directly on the end face of the spool. The pilot ports are located on the valve ends. Hydraulic pilot valves are used to provide pilot pressure against the main valve spool ends. However, pilot pressure could be of lower value if needed, compared to the main system pressure. Pilot valves are actuated by mechanical or electrical means in most cases.

(e) Electrical actuation Electrically actuated valves use a solenoid to operate the valve spool as shown schematically in Fig. 5.3 (a). The solenoid can be either AC or DC.

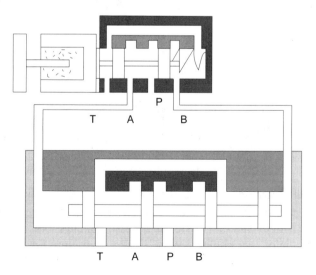

Fig. 5.3(a) *Indirect control*

The armature plunger of the electromagnet presses on the spool when the electromagnet is excited. In the case of AC solenoid, the ferromagnetic core is made of stacked iron laminations to reduce the heat effect. The advantage of a solenoid lies in its less switching time—specially for AC solenoids. However the disadvantage is that they are more sensitive to mechanical loads.

Electrically actuated valves made it possible to couple hydraulically powered elements with the advantages of open and closed loop control technology forming electrohydraulic driven systems. An electromagnet is the main control element which is coupled to the valve-spool to actuate the valve operating at very high pressure. Over the years this concept and technology has improved to a great extent with higher dynamic performance. Naturally keeping in tune to the advancement of electrical and electronic systems, apart from electromagnets, newer and more advanced form of control techniques are being used in hydraulic systems today, e.g. relays, torque motors, linear motors, proportional magnets, etc. More details on this will be dealt with in subsequent chapters.

Electromagnets are basically of two types:

(a) Dry type
(b) Wet type

A simple line sketch of an electromagnet is shown in Fig. 5.3 (b).

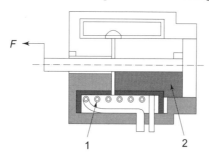

Fig. 5.3(b) *Electromagnet*
1 – Armature winding; 2 – Iron core

The actuation time for an AC solenoid is generally 25 to 30 milliseconds. They are available in 120 or 220 V, 50 Hz. The average life cycle is 7200 to 7500 switching operations/hour before failure whereas a DC solenoid works at low voltage, e.g. 12/24/36 V and the switching time is about 40 to 50 milliseconds. The average life is about 15000 operations per hour.

(f) Electro-hydraulic/electro-pneumatic actuation This is a combination of electric and hydraulic or pneumatic control method. The valves like 3/2, 4/2 DC valves are actuated by solenoids which in turn control the main valves. These valves are therefore called pilot valves and the controlled valve is known as the main valve.

5.4 CONTROL TECHNIQUE

The control technique of DC valves may be classified in two groups:
(a) Direct controlled units.
(b) Indirect controlled units.

Direct controlling means that the actuation is carried on directly on the valve-spool.

The indirect controlling mechanism unit contains two parts:
(i) One pressure oil controlled DC valve.
(ii) One direct controlled DC valve generally termed as a Pilot Valve mostly using solenoid as the actuating element.

The pilot valve is necessary to control and actuate large valves as otherwise the size of the solenoid in relation to the actual valve would be relatively large.

5.4.1 Pilot Operated Direction Control Valve

Directional control valves may be direct acting or pilot operated from a remote point. As stated earlier they may be operated manually, electro-hydraulically and pneumatically. As already mentioned, the valving action is commonly accomplished by the action of the valving element energized by the actuator.

As the hydraulic machines are worked on moderate to high pressure sometimes bigger size direct acting control valves may require higher amount of actuating force. In such cases in pilot controlled direction control valves, the main valve spool is operated with oil pressure which is directed to the spool by the pilot valve mostly by a 4-way valve solenoid actuated for 4-way or 5-way main valves.

Operating principle

The outlet ports A and B of the pilot valve are connected to the end faces of the main spool. The ports P and T are connected as required, with the pilot passages X and Y or with the ports P and T of the main valve.

(a) Internal pilot fluid supply   (P)
(b) External pilot fluid supply   (X)
(c) Internal drain fluid supply   (T)
(d) External drain fluid supply   (Y)

In the case of external pilot fluid supply, the required fluid for moving the main spool must be introduced from an external pressure source through Port Z, whereas in the case of internal supply this can be taken from the pressure port P of the main valve. The pilot fluid can either be drained externally through port Y or internally through port T of the main valve, as desired.

The type of pilot fluid flow is determined by appropriate installation or omission of a ball type plug or threaded plug. The ball type plug is accessible after removing the pilot valve, the threaded plug after removing a screw plug.

5.4.2 Pressure Loss

According to the rated size of a valve, the maximum oil flow (Q_{max}/min) is fixed. As the oil is flowing inside the valve and pilot valve housing, there is always a pressure loss which is dependent on the oil quantity and the viscosity of the oil.

5.4.3 Construction of 2, 3 and 4-way Direction Control Valves

The term 'way' is used to mean flow path through the valve, including reverse flow. One-way, two-way, three-way and four-way valves are common. One-way valves will allow flow only in one direction and does not permit return flow.

A two-way direction control valve consists of two ports or openings P and A which are connected or disconnected by the moving spool. In one extreme spool position, the flow from P moves to A and in the other extreme position no flow is allowed from P to A and the flow stops. A simple diagram with the positionwise symbol of the two positions (open and closed) is shown in Figs 5.4 (a) and (b).

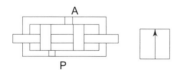

Fig. 5.4(a) *Two way DC valve and symbol of P → A*

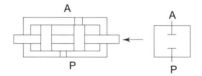

Fig. 5.4(b) *Two way DC valve with P ⊣ A*

A three-way direction control valve is shown in Fig. 5.5.

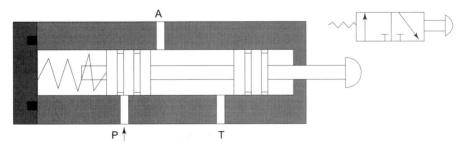

Fig. 5.5 *3/2 Direction control spool valve manually actuated spring return with symbol*

It has three ports or openings P, A and T, i.e. pump port P, actuator passage A and tank port T. In one extreme position of the spool, the pressure oil flows from P to A to move the actuator, the tank port T remaining closed. In the other extreme spool position the oil from P gets closed to A and the oil from the actuator is allowed to pass through A to T and then to the tank. Therefore the valve alternately connects or disconnects oil supply to the cylinder by the sliding spool. As the valve has 3 ports (openings) and the valve takes two distinct positions, the valve is called a 3-way 2-position direction control valve, i.e. 3/2 DC valve.

A 4-way direction control valve is shown in Fig. 5.6 (a). The valve consists of 4 ports, P, A, B and T. Here when the valve is spring centered, i.e. not manually actuated, P connects to A and B to T. But when the spool is manually actuated, the spool shifts compressing the spring, thereby opening P to B and A gets opened to T. As soon as the manual force is relased from the spool, the compression force of the spring brings the spool back to its original position. This valve is called a 4/2 direction control valve as the valve has 4 ways (Ports or openings) and it takes up two distinct positions. Figure 5.6 (b) shows the symbol of 3/2 DC valve as well as the 4/3 DC valve.

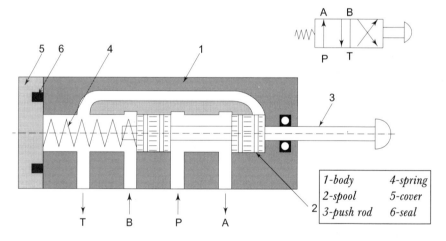

Fig. 5.6(a) *4/2 Direction control spool valve manually actuated spring return with symbol*

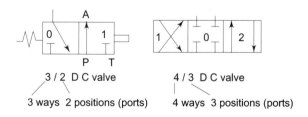

Fig. 5.6(b) *Valve position*

The spool of the direction control valve consists of a shaft with two, three or four lands as shown in Figs 5.7 (a) and (b). Most directional control valves use only 2-land or 4-land spool, the normal practice being to use a 2-land spool valve for low-rating valves and 4-land spool for higher rated valves:

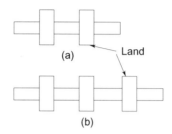

Fig. 5.7 *Sliding spool (a) Two land spool (b) Three land spool*

5.4.4 Position of the Valve

Valve position does not refer to the location or orientation of the valve but rather to the position of the internal moveable parts of the valve and the resultant flow conditions determined by that position. Normally, direction control valves have 2 or 3 positions. They are:

0 position or neutral position
1 operating position
2 operating position

Valve positions and nomenclature of direction control valves are shown schematically in Fig. 5.6. In the same diagram the nomenclature of a 4/3 direction control valve is also shown.

It is necessary to differentiate between neutral and operating positions. In DC valves with spring return, the neutral position is defined to be the position to which the valve returns after the actuating force has been removed. The neutral position is indicated by the number 0. The operating position on the other hand has the numbers 1, 2, etc. If a valve offers two useable flow combinations or positions it is called a 2-position valve. Some valves have three positions with a neutral position. Directional control valves without a neutral position retain their position after the actuating force has been removed.

The starting position is defined as the position taken up by the valve after being fitted into an installation, when the pump is switched on.

5.4.5 Valve Spool

We have discussed sliding spool used in a direction control valve earlier. The spools as per their commonly functional use are as follows:

(i) Fully locking spool (i.e. closed centre valve)
(ii) Open center spool
(iii) Partial open center spool
(iv) Fully bypass spool
(v) Partial bypass spool

Some advantages of spool type valves are:

(i) Very flexible porting configuration
(ii) Relatively straight forward manufacturing up to 125 bar
(iii) Low inertia
(iv) Low operating load. (sometime as low as 10 gmf only)

Manufacturing of spool type valves needs lot of precision and fine machining. The most commonly used threads used are BSP or SAE O-ring boss with UNF thread except for larger sizes which have bolt SAE split flange type.

5.5 CENTER CONDITIONS OF SPOOL VALVES

While the two extreme positions of a 4-way direction control valve, controls the two extreme direction of motion of a cylinder or hydromotor, very often the valves may have an intermediate position or a center position designed to satisfy a specific need or condition of the system performance. Varieties of center positions are possible in direction control valve by suitably designing the spool. According to the center position, the valves could be termed;

1. Open center valve
2. Closed center valve
3. Tandem center valve
4. Float centered valve, etc.

Apart from the specified neutral position of a direction control valve, one can also have the following neutral conditions.

1. Tank port closed but P opened to A and B port.
2. P and B closed, A opened to T.
3. A closed, P, B and T opened.
4. P opened to A, B and T closed, etc.

Some of these are shown schematically in Fig. 5.8 indicating various center position of direction control valves.

5.5.1 Open Center Valve

In this valve (shown in Fig. 5.8) in the center position all ports, i.e. P, A, B and T are open to each other. This condition will allow free movement of an actuator when flow from the pump is returned to the tank at a low pressure.

Advantages of the open center valve: The advantage of the valve is that as soon as the cylinder completes its cycle, the spool of the open center direction control

Direction Control Valves 155

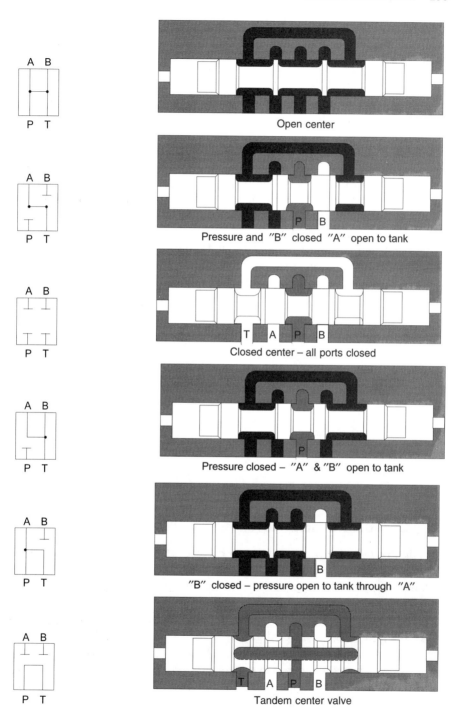

Fig. 5.8 *Center conditions for 4 way DC valves (Courtesy: M/s Vickers Sperry Rand, USA)*

valve is centered and the flow from the pump returns to the tank through the direction control valve at low pressure generating less heat and allowing the cylinder free movement to start its next cycle of motion. But the major disadvantage is that no other cylinder can operate when the valve is centered and hence, this valve is to be used mostly for a single cylinder or single motor circuit.

5.5.2 Closed Center Valve

The center position of such a valve, all the ports P, A, B and T are blocked to each other as shown in Fig. 5.8. In such a condition the motion of an actuator could be stopped as well as a number of individual cylinders can operate independently from a single power source. But the major disadvantage is that the pump flow cannot be unloaded to the tank through the valve when in center position. If the valve has to stay in the center position for a prolonged time, there is a possibility that leakage may take place through the land to the ports A and B and even port T, reducing the locking pressure significantly. In Fig. 5.9 a schematic circuit diagram indicating the use of a closed center valve is shown.

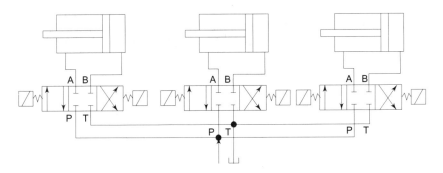

Fig. 5.9 *Closed center valve*

5.5.3 Tandem Center Valve

Here during the center position the ports P and T get connected and A and B ports get blocked. Therefore, while a tandem center valve stops the motion of the cylinder, the pump flow is unloaded to the tank through the direction control valve port T without passing through the pressure relief valve and thus generating less heat. But the disadvantage is that due to the cored passage inside the spool through which the port P gets connected to port T, for any normal tandem center valve there exists a pressure differential between P and T to the amount of 3 to 4 bar. In other words, when a number of cylinders are operated from a single source, the pressure differential for each tandem center valve will be 3 to 4 bar each while the valve is in its center position and the system pressure at this point may not be as low as a designer may like to have. This is shown in the circuit diagram in Fig. 5.10. Moreover the nominal flow rating of the valve is decreased.

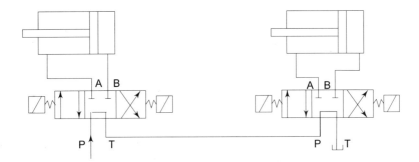

Fig. 5.10 *Tandem center valve*

5.5.4 Float Centered Valve

The use of such a port condition will allow independent operation of cylinders connected to the same power source and at the same time will facilitate free motion of each cylinder at start. No pressure will build up in the cylinder lines when P remains closed. Therefore there will be no drifting of cylinders during this condition. However the disadvantage is that the load cannot be locked in position during the neutral position. However a pilot operated check valve can be used in association with this valve to ensure positive locking as shown in Fig. 5.11.

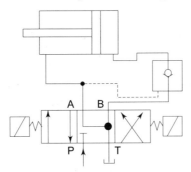

Fig. 5.11 *Float center valve*

In order to have a reasonably sized flow path from P to T during the neutral position, spool stem between the lands is much thicker is comparison to other spool type valves. This may result in a restricted flow path when the valve-spool is shifted to its extreme.

5.5.5 Other Center Conditions

As stated earlier, other center conditions are also designed apart from the four positions described above. This helps the hydraulic circuit designer to make a choice of the valve depending on the specific system requirement. In the symbols shown in Fig. 5.12 P is connected to A and B, T remaining closed and P, B,T are connected and A remains closed.

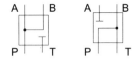

Fig. 5.12 *Other center position*

Valve materials:
(a) *Body*—Carbon steel, ductile CI, aluminium, stainless steel, etc. are used. Aluminium alloys are light and can be used for low pressure applications in general. But in aviation systems high strength Al alloys are also used. Stainless steel is used in an anti-corrosive atmosphere. Plastic could be used but they are in general temperature-biased. For CI anti-corrosive property is poor.
(b) *Spool*—Mostly case hardened steel, ground and polished, e.g. 15 Ni 2 Cr 1 Mo 15 is a common material having hardness of 60 ± 2 HRC—should be machined and polished up to 2-3 μm.

The normal valve spool/bore clearance is about 5-10 μm.

5.6 BALANCING GROOVE ON SPOOL LAND

Balancing grooves are very often out on valve-spool lands to eliminate hydraulic lock caused by machining inaccuracies or by silts. These grooves, mostly V-shaped, are circumferentially out on the spool lands with 1.5 to 3 mm spacing between the grooves and mostly with groove depth of 0.375 to 0.5 mm. The number of grooves to be cut depends on factors like frictional force, land length, etc. However with three grooves the frictional force due to hydraulic lock is reduced to 6% and with 5 to 7 grooves the reduction may be much higher, say up to 3%.

5.6.1 O-Rings in Valves

O-rings are very often used in the spool valves to stop leakage of oil. They are used both in the housing or even in the spool lands. The O-ring should be fitted in such a way so as to seal, support and float it in the valve spool isolating it from possible damage and distortion. Some stainless steel spools and sleeves eliminate O-ring—the lapped spool and sleeve each is precision finished and matched fitted to sub-micron tolerance—so the spool rides in the sleeve on a molecular film of air.

Some spools are anodized and are impregnated with Teflon and ride on specially designed Buna-N seals. The spool is cushioned at both ends. It must operate at zero pressure differential mid-position if the spool is driven by full line pressure and may require only 1-1.25 bar inlet pressure.

5.7 OVERLAP IN SLIDING SPOOL VALVE

The longitudinal difference between the length of the land and the port is called *overlap*. Therefore overlap in sliding spool valves is that length of valve housing, between the individual pressure chambers, which is blocked or sealed when the spool land comes in contact with it. One must here understand the principles between the overlap in unoperated as well as in operated condition of the valve. Underlap is opposite to overlap, i.e. the negative overlap. We can have the geometrical view from Figs 5.13 (a) and (b) of overlap and underlap in valve spool.

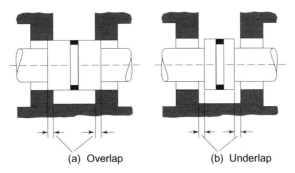

(a) Overlap   (b) Underlap

Fig. 5.13 *Overlap and underlap in spool valves*

(a) *Overlap in the unoperated condition:* The leakage quantity between the two pressure chambers is dependent on the geometrical accuracy of the spool fit in the spool bore and especially on the amount of overlap in the unoperated condition. The leakage is much higher if the clearance is more as may be noted is the graph in Fig. 5.14 (a).

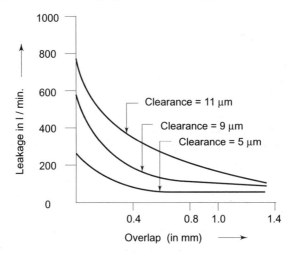

Fig. 5.14(a) *Relationship between overlap, clearance and leakage*

(b) *Overlap during operation*:
  (i) *For negative overlap*: During operation all passages are momentarily connected with each other.

Advantages
  (a) No pressure peaks, so no shock during start or stop
  (b) Soft switching
  (c) Less stress of the parts (Pistons, spools etc.)

Disadvantages
  (a) Certain loss of pressure liquid
  (b) Pressure collapses momentarily which may lead to load sinks, discharge of accumulators, etc.

(ii) For Positive overlap:
During operation all passages are momentarily closed.

Advantages
  (a) No collapse of pressure during operation
  (b) No loss of liquid under pressure
  (c) Ensured locking of connected load

Disadvantages
  (a) Pressure peaks and hits during start of switching

Loss of pressure
  Loss of pressure depends on the
  (a) nominal size of the valve.
  (b) rate of flow (Q in l/min.).
  (c) viscosity of the liquid.
  (d) construction of the valve.
  (e) circuit design.

5.7.1 Characteristics of Spool Overlap

With an overlapping spool the cylinder ports of the direction control valve get blocked in the neutral or null position resulting in locking of the load. For a manually operated valve the amount of overlap could be generally as high as 5.0 to 7.0 mm. This value however is much less in case of valves which are used for automatic actuation and in valves like electrohydraulic servo valves etc. Overlapping is normally desired to prevent leakage from port A or B to port T or pump Port P to cylinder port A or B during null position. However, the ideal fitting one likes to have is line-to-line fitting between the valve land and port.

In contrast, in an open center valve, the spool land is generally underlapped which implies that the load connected to cylinder ports A and B under null condition will be having equal pressure on both sides of the cylinder, i.e. when the spool is centered. This may be clear from here that even if the valve spool is slightly moved, the load will start moving as pressure differential will set in from one side to the other side of the spool. In open center valve, there will be a continuous oil flow which is desirable as the pump flow may be unloaded through this valve instead of through the pressure relief valve. However locking of the cylinder will not be possible at all.

In a close center valve full pump pressure is available when the valve starts operation (cracking). But in an open center valve full pump pressure will not be available until and unless the tank connection is fully cut-off. The same holds good for a tandem center valve too. It can be inferred that the pressure gain of overlapped valve is higher than the underlapped valve.

5.7.2 Influence of Spool Overlap on Pressure Peak Development

Figure 5.14 (b) illustrates the way of pressure peak development in 4-way direction control valve having 1 mm overlap as well as 1 mm underlap under no-load condition of the working cylinder. It is seen from the diagram that pressure peak in both overlap and underlapped valves are dependent on the switching time as well as flow rate increasing with increase in flow rate. In Fig. 5.15 (a) we observe the influence of positive overlap and the switching time on the rise and formation of peak pressure on a direction control valve having a spool with triangular notch cut on it. The pressure was set at three different pressures, i.e. at

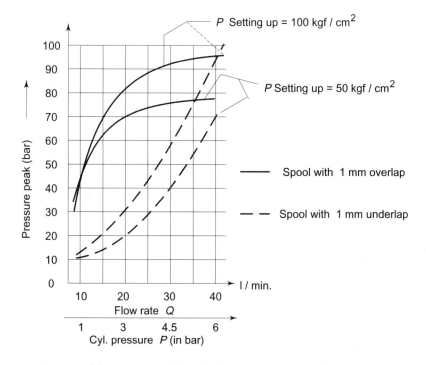

Fig. 5.14(b) *Pressure peak by switching of a 4/2 DCV on no load condition*
 [Courtesy: M/S Herion Taschenbuch]

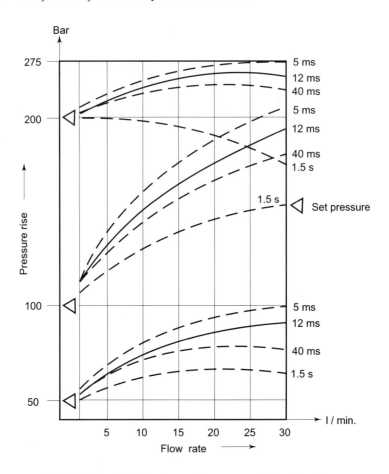

Fig. 5.15(a) *Switching time and pressure peak depend on flow rate*

50, 100 and 200 bar. From the diagram it is noted that the pressure peak depends not only on the flow rate capacity (Q) of the direction control valve but also on the duration of the spool actuation (T_s). From the figure it is noted that at $T_s = 5$ ms, and $T_s = 1.5$ s with $P = 100$ bar, the pressure peak is up to 250 bar that means 2.5 times the set pressure. At $T_s = 1.5$ s, the pressure peak is up to 136 bar, i.e. 1.36 times the set pressure. This indicates that with higher valve actuation time, the pressure relief valve can be set at lower pressure. By smaller actuation time, no reduction of the pressure peak is possible due to the inertia of the pressure relief valve.

The small on and off time of about 40 ms or less of a solenoid operated DC valve blocks/stops the pump flow to the system very quickly creating an undesirable impact on the flow lines and line fittings sufficient enough to damage them. In order to have a smooth change over of the on-off switch, triangular notches can be cut at the corner of the spool land as shown in Fig. 5.15 (b) through

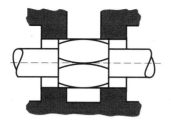

Fig. 5.15(b) *Triangular notch in spool land*

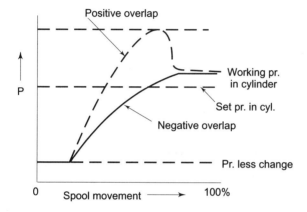

Fig. 5.15(c) *Pressure development in valves with positive and negative overlaps*

which at null position some oil may pass through but the pressure peak can be avoided significantly and pressure relief valve can be set at a much lesser pressure as shown in Fig. 5.15 (c).

5.7.3 Leakage through Spool and Housing Bore

Very often due to excessive clearance between spool land and spool bore, lot of leakage takes place. Depending on the nature of spool fitting the leakage path may be either (Figs 5.16 a and b) (i) concentric or (ii) eccentric. For a concentric spool with radial clearance (b) the maximum clearance passage is $2\ b$. The theoretical leakage quantity (Q_L) can be calculated from the formula given below.

$$Q_L = \frac{1.54 \cdot d_0 \cdot \Delta p \cdot b^3}{v \cdot L \cdot 1000} \text{ cm}^3/\text{min}$$

where
Q_L = quantity of leakage oil cm^3/min

b = radial clearance in μm

L = length of overlap leakage path in mm.

d_o = spool land dia, mm

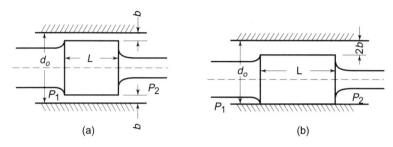

Fig. 5.16 *Concentric leakage path*   *Eccentric leakage path*

v = kinematic oil viscosity in cSt.

If the leakage path is eccentric, leakage area of a spool with a given radial clearance is almost 2.5 times larger than when it is concentric. The following formula can be used for calculating Q_L for eccentric leakage passage.

$$Q_L = \frac{\pi \cdot d_0 \cdot \Delta p \, b^3}{1.2 \, \mu \cdot L} \, (1 + 1.5 \, e^2)$$

where
- Q_L = leakage flow rate in cm^3/s
- d_o = dia of spool land m
- b = radial clearance in m
- ΔP = pressure loss in N/m^2
- μ = absolute viscosity in kgfs/m^2
- L = Passage length in m
- e = eccentricity = $\frac{x}{b}$,

where, X = difference between the spool and bore axis.

When $x = b$, $e = 1$ and hence the leakage loss is almost 2.5 times higher compared to leakage loss in a concentric leakage path. If $e = 0$; the leakage path will be equal to concentric path.

5.8 SPOOL POSITIONING, BACK PRESSURE AND FORCE ACTING ON SPOOL

Very often the following forces act on a spool and hence accurate spool positioning can only be achieved by taking into account their special features and influence on spool movement. The forces are:

(i) Friction
(ii) Flow force
(iii) Force due to hydraulic imbalances at end chambers.

5.8.1 Back Pressure

Owing to imbalances in the end chambers of a valve, a spool may experience certain forces due to back pressure even though they are connected to the tank. The back pressure in the tank connection can not be prevented as the hydraulic lines are not large. Hence a back pressure of 2 to 3 bar may very often act on the end faces of the spool stem. In case of a 15 mm diameter spool stem which passes through the wall on one end with the other end blind, the force due to back pressure may be as high as 35 to 50 N which may cause imbalance on the spool movement. This can be avoided if the stem is made to pass through the wall on both sides. This will balance the forces due to back pressure. But the construction cost of the valve may go up. Back pressure may also damage the seals causing undesirable leakage. This, however, can be prevented by separating the tank connections from the end connections by means of two additional lands on the spool and providing a drain connection from the end chambers. With this arrangement the seals may experience only drain pressure which will be quite negligible.

5.8.2 Flow Forces

While the fluid passes through a port in a DC valve, a spool experiences certain flow forces which are also called reaction forces. When the port B starts opening up to P, during spool movement, a metering restriction will be created before the port is fully opened. At this point of time, the flow velocity will tend to increase and it occurs while flow converges and diverges in passing through the restricted path of the port. Due to this there will be more kinetic pressure and less static pressure. Again as the static pressure P_1 at port A will be less than the static pressure P_2 at port B, the net flow force will tend to close the valve. One may note here that there will be two metering ports, one from the pressure port to cylinder load and other from the cylinder load to tank both being approximately of the same area ($\pi\, dx$) and pressure drop (ΔP) where d = spool dia., x = width of restriction, ΔP = pressure drop.

Therefore, the approximate magnitude of the combined axial flow forces can be assumed to be

$$F_{ax} = \frac{Q\sqrt{\Delta P}}{2g}$$

where F_{ax} = axial flow forces trying to close the valve

Q = flow rate in l/min

ΔP = pressure drop in bar.

5.8.3 Friction

Friction between the spool and the bore is another problem which needs to be tackled for easy and smooth spool movement. The spool should be assembled in the valve body with maximum care. The housing bore should be perfectly straight and fully circular in shape in order to reduce sliding friction and frictional resistance. The play between the spool and the bore is also equally important and should be around 5 µm. Both the spool and the bore should be fine-finished, honed, polished and chromeplated. The spool needs to be oiled before assembly so that the spool can reciprocate in the housing bore with the minimum actuating force.

5.8.4 Hydraulic Lock

A spool inserted in the valve body or inside the sleeve may show a tendency to generate a stick-slip motion when actuated. This is due to inaccuracies resulting from machining faults during fabrication and assembly of the valve which may result in hydraulic thrust that may force the spool against the valve wall. Due to inaccurate machining a spool may get tapered as shown in Fig. 5.17 (a). (The figure, however, has been exaggerated for easy understanding.) The actual difference in diameters at the two ends may be only a few microns.

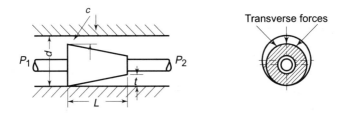

Fig. 5.17(a) *Spool land with taper*

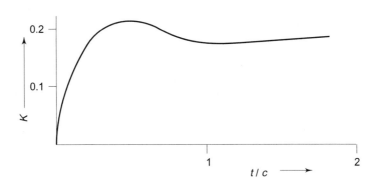

Fig. 5.17(b) *Value of k against t/c*

If the taper is diverging, in the direction of decreasing pressure, i.e. the small end of the taper is placed towards the higher pressure, then the spool is still self-centering. However if the taper converges in the direction of decreasing pressure, it results in a transverse force that may push the spool against the wall. This phenomenon is termed as *hydraulic lock*. Actually it may not necessarily result in a lock, but certainly it produces increased friction, noticeable particularly in case of servo-valves.

The transverse force = $F = k \cdot l \cdot d \cdot (P_1 - P_2)$,

where, k is dependent on t and c [(see, Fig. 5.17 (b)],

l = length of land,

d = diameter of land,

ΔP = pressure difference

For a force of imbalance of 50 N, with coefficient of friction between steel immersed in oil as 0.2, the resulting frictional force will be 10 N.

Even if machining is *perfect*, lands and bores are perfectly parallel, silting by minute contaminants on the spool surface may result in the same geometrical conditions as a tapered spool and this may result in sluggish movement of the spool causing problem to the system as explained above.

5.8.5 Centering Springs

It is generally desirable to have the spool center itself by means of a spring. Compression spring on both sides of the spool may be used, but very careful adjustment is required for perfect and true centering of the spool. Besides, the spring forces in the neutral position balance each other, and a small force of imbalance is sufficient to upset that balance. Sometimes this is desirable. But in many applications solid positive centering is needed for safety and security of the connected load. If the spool is moved to the right or left, the spring gets compressed. Without external force the spool will always tend to return to neutral position with the spring in the indicated condition. The spring must be slightly precompressed which means that a minimum external force is needed to start the spool moving. The position of the spring is also to be given due consideration. Centering springs do not necessarily have to be located inside the valve. They can be part of the actuating lever outside the valve body as shown in Figs 5.18 (a) and (b). The helical spring in Fig. 5.18 (b) is under pretension and its ends rest against two pins.

5.8.6 Detents

Detents are very often used to secure the desired valve position in a two or multi-position valve. The detent is a simple mechanical locking device which can detain the valve spool to its desired shifted position. A simple notch or groove is cut on the spool which acts as a receptacle for a spring loaded locking device like a ball. With the ball in a particular notch, the spool position is secured until and unless

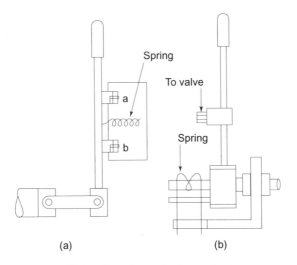

Fig. 5.18 *Centering spring in spool valves*

the ball is forced out of that notch and placed into another notch. A diagram of detent mechanism is shown in Fig. 5.19.

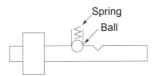

Fig. 5.19 *Detent mechanism*

5.8.7 Valve Size and Capacity Rating

Appropriate method of valve designation is a complex matter as a number of factors are involved in it. Apart from providing physical nomenclature of valves like constructional details, number of ports, etc., they are designated in varieties of other ways also to indicate their size and capacity. Port sizes of the valves are very often taken to specify the valve capacity, e.g. a direction control valve may be designated by its port size such as a DCV with a port size of 3/4". But this may be sometimes misleading as well as incomplete. Therefore apart from indicating the port dimensions including the base plate nomenclature of the valve one has to indicate the flow rate and range of flow rate of the valve. However flow rate capacity again may not carry the complete meaning if the pressure drop (ΔP) across the valve port is not mentioned. Even equally important is the nominal pressure which is to be mentioned along with this flow rate capacity. In order to avoid any misconception that a definite pressure drop is yielded at the maximum flow rate, it is better to have also the characteristic curve of pressure drop vs flow rate of the valve. Most valve manufacturers provide such graphs in their technical catalogue.

Standard valve sizes are also expressed in a dimensionless number, e.g. NS 4, 6, 10, 16, 20, 32 etc. For example, a dimensionless number NS 16 is taken as a rough indication of the nominal area of the port cross section. The important significance of such a system is standardisation of port configuration specially for plate mounted valves where otherwise standardization is difficult. Designation of valves using flow rate is widely used where there is no other standard designation possible or available, e.g. for modular valves.

5.9 FEW POINTS TO NOTE ABOUT VALVES

Solenoid controlled direction controlled valves can be either single solenoid or double solenoid operated. Most DC types are available with various power rating from 12 V onward and with solenoid coil indicator lights. These valves are available in sub-plate mounted type or modular/manifold type shown in Fig. 5.20. The spools are fine finished and very often chrome plated. For perfect sealing appropriate seals are used. It may be mentioned here that it is a common practice to lubricate seals and packings before installation as this makes the installation job easier. But one should ensure that the lubricant used is compatible with the seal material as well as the hydraulic fluid prescribed for the system. Ethylene propylene and butyl rubber swell significantly when in contact with petroleum based oils and greases which may cause premature seal wear and resultant seal failure. The spool bore in the valve housing or sleeve is also fine finished in order to minimize friction and increase overall life-cycle of the valve.

The valve-spools are to be cleaned occasionally with appropriate cleaning fluid like clean kerosene, CTC, etc. (keeping in mind the environmental obligation before using the cleaning solvent). Many a time the oxidized products (gummy materials) from the oil reservoir have a tendency to block small holes in valves or deposit on the spool land thereby making the spool movement sluggish resulting in loss of valve response. Proper care is therefore necessary to see that appropriate oil has been used and the gummy materials are regularly cleaned.

For manifold or modular valve blocks, appropriate precaution is to be taken to use adequate filtration to keep these free of all undesirable foreign materials, e.g. burrs, chips, etc. Manifolding provides flexibility, reliability and efficiency for mounting the valves and other components. They are easy to assemble, dismantle and repair or change. The other advantages of manifolding can be summarized as follows.

(a) They are service friendly
(b) Highly flexible
(c) It is possible to design a standardized system
(d) Greater pressure stability (due to shorter fluid column)
(e) Higher reliability
(f) Space saving
(g) Highly compact
(h) Less possibility of leakage
(i) Low or reasonable cost
(j) Elimination of many lines and fittings

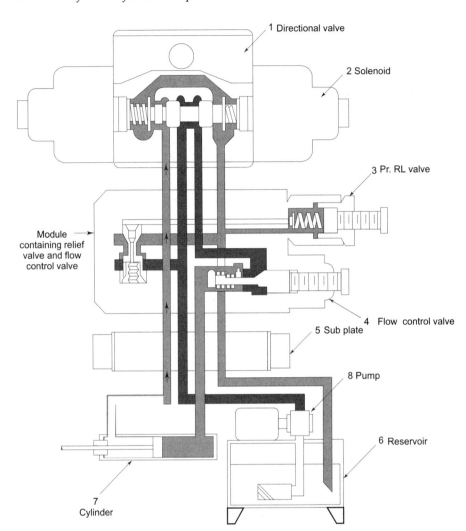

Representation of a circuitstack Modular Circuit. This includes a system RL valve in a flow control metering valve. Reverse flow in this line bypasses the flow control function.

Courtesy: Hydran

Fig. 5.20 *Modular valve construction*

5.10 DC VALVE SPECIFICATION

The following parameters are to be noted:
1. Rated flow
2. Material—for body, spool or valving element, seal
3. Rated pressure
4. Type of solenoids—AC or DC
5. Internal pilot supply

6. Spring centered or not
7. Outlet and inlet port size
8. Solenoid power, i.e. 12 VDC, 24 VDC, 120 VAC, etc.
9. Open or closed center application
10. 3-way or 4-way spool
11. Style of mounting
12. Sub-plate and modular construction and related details.

The pressure differential across the port is also an important factor. Generally the pressure differential (ΔP) across the port should be within 2 to 2.5 bar for normal valves used for general purpose.

Review Questions

1. State various types of hydraulic valves. What is a direction control valve? Why is it needed in a hydraulic system?
2. Differentiate between a seat type and a spool type DC valve. Which of these two types are mostly used in a hydraulic system? Why?
3. What is overlap in valves? What is the influence of overlap in the function of a DC valve?
4. What is meant by a 4/3 DC valve? State the art of actuation of direction control valves. What are the functions of a spring in valve actuation? Explain with examples.
5. Differentiate with sketches the function and characteristics of closed center and open center DC valves.
6. State the specification of a DC valve.

References

5.1. "New pumps and proportional valves at Mannesmann Rexroth", *Oel hydraulik u. pneumatics*, Aachen, Germany. April, 1987.
5.2. Backe, W. "Grundlagen u. Entwicklungstendenzen in der Ventil technik", Backe, *Oel hydraulik u. Pneumatik*, Aachen, Germany, 1990.
5.3. *The Power and Control Journal*, In house publication of Parker Hannifin Corporation, Ohio, USA.
5.4. *Industrial Hydraulic Technology*, Parker Hannifin Corp., USA, pp. 8-9.
5.5. *Industrial Hydraulic Manual*, Vickers Sperry Rand Corp., USA, pp. 7-16.
5.6. *Hydraulik Lehrgang fuer berufliche Bildung*, BBF, Berlin, Germany.
5.7. Ing. Thomas Krist and Vogel Verlag, Hydraulik Kurz u. Buending, 1973.
5.8. F. Und d. Findeisen, *Oel hydraulik in Theorie und Unwendung*, Schweizver lagshaus, Ag., Zurich, 1968, p. 165.
5.9. A. Duerr u. O. Wachter and Carl Hanser Verlag, *Hydraulik in Werkzeug maschinen*, Muenchen, 1968, p. 148.
5.10. M/s Herion Taschenbuch from Herion Werke, Stuttgart, Germany, 1969, p. 175.

6

Flow and Pressure Control

Apart from direction control valves as described in Chapter 5, many other types of valves are often used in a hydraulic system. They are:
1. Non-return valves or check valves
2. Flow control valves
3. Pressure control valves.

6.1 NON-RETURN VALVES

As mentioned earlier, non-return valves are also termed as check valves. The main function of a check valve is to block the reverse flow of oil or gas in a fluid power circuit (Fig. 6.1). Functionally a one-way traffic road sign may be appropriately compared to this valve. Additionally this valve may be used for pressure control or directional control. Although it functions like a two-port directional control valve, it does not always produce the same results; unlike a 2-position 2-port directional control valve, a check valve usually allows flow in one direction only.

Although a check valve normally controls the direction of fluid flow, sometimes its operation and application resemble those of a direct operated relief valve as we will see later. Thus a check valve with a heavy spring may act as a pressure control valve. Therefore it is more appropriate to allot check valves a

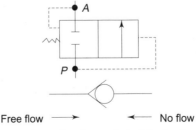

Fig. 6.1 *Graphic symbol of check valve*

separate identity between pressure control and direction control valves as they combine the functional features of both these valves.

The construction characteristics of a check valve are simple. It consists of a valve body with inlet and outlet ports and an internal moveable member biased by the spring. The moveable member can be a flapper or a plunger but most often in valves of hydraulic systems, it is a ball or poppet.

When system pressure at the check valve inlet port is high enough to overcome the biasing spring force on the poppet, the poppet is pushed off its seat allowing oil to flow through the valve. This is known as the check valve's free flow direction. When the fluid attempts reverse direction of flow, the spring pushes the poppet back on the seat to block flow through the valve.

Check valves can be direct acting or pilot operated. A pilot operated version is used where the no-flow characteristic of the valve is desired only for a portion of the system cycle. Figure 6.2 shows a pilot operated check valve.

It consists of a valve body (1) with inlet and outlet ports and a moveable poppet or ball (6) biased against a spring(7). Opposing the valve poppet is a plunger and plunger piston (2) biased by a light spring (3). When the pilot operated through port (5), this spring unseats the valve poppet permitting reverse flow. A drain (4) in the plunger spring chamber prevents any oil that passes the piston from being trapped in the spring cavity. A pilot operated check valve also allows free flow from its inlet port to its outlet port just as in an in-line check valve.

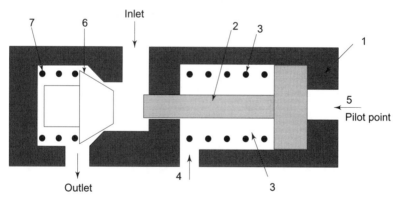

Fig. 6.2 *Pilot operated check valve*

The design and construction of a check valve makes it one of the simplest fluid power components. Though simple, its use can make the system sophisticated.

Application of check valves in various simple as well as sophisticated hydraulic systems has been depicted schematically in Figs 6.3 (a), (b), (c), (d) and (e).

While the circuit diagrams in Figs 6.3 (a) and (b) show the application of check valves in the pump suction and delivery line respectively, Fig. 6.3 (c) shows a check valve in the system return line. Check valves are also used in a hi-lo hydraulic circuit as shown in Fig. 6.3 (d) to isolate the high pressure from the low pressure pump. By application of check valves one can also design many sophisticated circuit diagrams as shown in Fig. 6.3 (e) where a constant flow can be maintained in both directions by using them in a bridge.

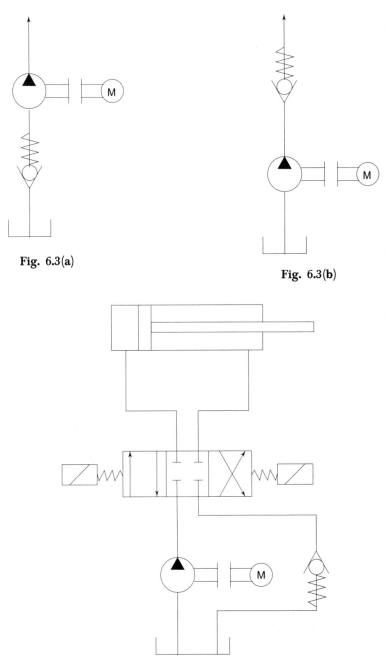

Fig. 6.3(a)

Fig. 6.3(b)

Fig. 6.3(c)

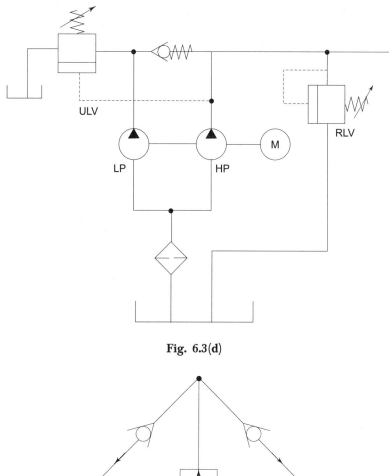

Fig. 6.3(d)

Fig. 6.3(e)

6.2 PRESSURE CONTROL VALVES

A pressure control valve performs the following functions:

1. Limiting maximum system pressure as a safety measure.

2. Regulating/reducing pressure in certain portions of the circuit.
3. Unloading system pressure.
4. Assisting sequential operation of actuators in a circuit with pressure control.
5. Any other pressure related function by virtue of pressure control.

The operation of a pressure control valve is based mainly on a balance between pressure and a mechanical load, e.g. a spring force biased against the oil pressure. The valve can assume various positions between fully closed and fully open conditions depending on the flow and pressure differential.

6.2.1 Types of Pressure Control Valves

They are classified according to their function, type of connection, size and pressure operating range. Relief valve, sequence valve, unloading valve, pressure reducing valve, etc are some typical examples of such valves.

6.2.2 Pressure Relief Valves

Pressure relief valves are found in every hydraulic system. A pressure relief valve is a normally closed valve connected between the pressure line and the oil reservoir. Its main purpose is to limit the pressure in a system to a prescribed maximum by diverting some or all of the pump output to the tank, when the designed set pressure is reached.

A simple sketch of a pressure relief valve is shown in Fig. 6.4 (a).

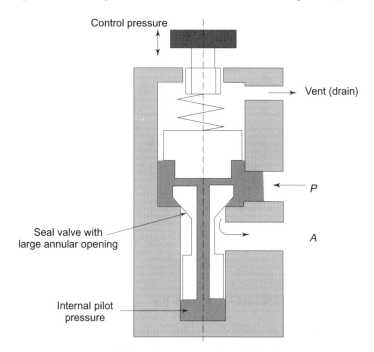

Fig. 6.4(a) *Direct operated pressure relief valve*

A simple relief valve may consist of nothing but a ball or poppet held seated in the valve body by the compressive force of a heavy spring. When the pressure at the inlet is insufficient to overcome the force of the spring the valve remains closed and hence it is very often referred as a normally closed valve. When the preset pressure is reached, the ball unseats and allows flow through the outlet to tank. In most of these valves, an adjusting screw is provided to vary the spring force. Thus the valve can be set to open at any pressure within the specified range. The pressure at which the valve first opens is called the *cracking pressure*. As the flow through the valve increases, the poppet is forced further, the resulting pressure increasing considerably thereby. The difference between the full flow pressure and the initial pressure may sometimes be objectionable to other system elements. In certain cases it can result in a considerable amount of wasted power due to the fluid lost through the valve before its maximum setting is reached.

From the graph in Fig. 6.4 (b) it may be seen that the cracking and closing pressure of the pressure relief valve is not same. Moreover in most cases, the valve poppet cracks at a pressure lower than the adjusted pressure but the valve closes at a lower pressure than at which it cracks. This condition is not favorable from the point of system safety and security. The favourable opening and closing characteristics of such a valve should be as shown in Fig. 6.4 (b) by the dotted curve. The closing pressure difference depends on flow-volume, valving art, spring characteristics, adjusted pressure, etc. as may be seen from the curves illustrated in Fig. 6.5. The opening (cracking) characteristics of spool type and seat type, pilot operated pressure control valves is comparatively depicted in the Figs 6.5 (b), 6.5 (c) and 6.5 (d). Figure 6.5 (a) shows the ideal opening characteristics.

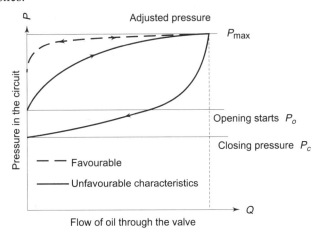

Fig. 6.4(b) *Opening and closing characteristics of pressure control valve*

6.2.3 Pilot Controlled Pressure Relief Valve

Direct controlled pressure relief valves are used where the flow rate and the system pressure are reasonably smaller or there is not much variation in system pressure or flow rate. When a valve has to maintain the poppet or spool seated in its place to contain a large pressure, one needs to provide a bigger spring to match the high system pressure. We know that a bigger spring will have a higher spring rate with its attendant problems and its cross section will also be larger requiring more space and sometimes being not feasible at all in a compact hydraulic system as are being used nowadays. Due to this inherent spring problem, to control high pressure a direct controlled pressure relief valve has been found quite unwelcome. We also know that the controlled pressure does not remain exactly constant but depends on the flow rate which is in turn influenced by the cylinder and valve position, shape and form of valve ports, spring characteristics, etc. Hence a direct controlled pressure relief valve is used where the hydraulic system does not need excellent P-Q characteristics. But for a large flow rate and higher pressure, the use of an indirectly operated valve, i.e. a pilot operated valve is most common. This valve is also called a compound relief valve. The great advantage of such a valve is that here the pilot valve can be kept spatially separated from the main valve such as on a control panel and one can introduce suitable d.c. valves in between to set different pressures by pilot valves. With reasonably good P-Q characteristics, the set pressure is almost constant with variable flow. A pilot operated pressure relief valve is shown in Fig. 6.5. Here from the main port P, oil is led through a throttle opening (5) in the main valve spool (1) to act against the valve element (2) which is held onto the valve seat by an adjustable relatively smaller spring (3) to hold it tight against the seat. The main spool blocks flow from P to A by means of another spring (4) plus the throttled oil pressure which is also acting against this spool.

Operating principle Let us assume that the system pressure is set at a pressure P and is acting against the main spool (1) as well as against the bottom of the adjustable valve element (2) through the in-built throttle. The pressure in this chamber will be reduced to P_1 due to the throttling effect. Therefore the adjustable valve is to be adjusted to a force just above or equal to the oil force commensurate

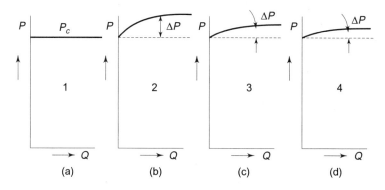

Fig. 6.5 *P-Q curve of various types of pressure relief valve*
(a) No pressure loss (b) seat valve (c) spool valve
(d) pilot operated pressure relief valve

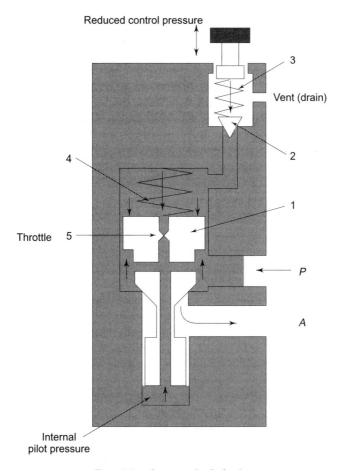

Fig. 6.6 *Compound relief valve*

with pressure P_1 and this equilibrium will continue until the main system pressure does not cross this set value. If this value however increases due to some reasons, the throttled value of pressure in the chamber above the main spool will also change to a higher value and thus act against the adjusted spring force with higher force thereby unseating it and opening the pressure in the pilot chamber to the drain line. This reduction of the pressure will also affect the main spool position which will crack immediately allowing system pressure to flow to the reservoir until the time the pressure equilibrium is restored.

6.3 SEQUENCE VALVE

Sequence valves extensively used in hydraulic systems are also pressure control valves. But unlike a pressure relief valve, a pressure sequence valve is used in a hydraulic system to cause various operations in a sequential order, i.e. one after another. For example a pressure sequence valve used in a clamping and machining circuit may permit the clamping operation to take place first and when the clamping cylinder is fully extended, the machining cylinder is actuated. This means this valve will cause actions to take place in a definite order and also to maintain a predetermined minimum pressure in the primary line when the secondary operation has to occur. Here fluid flows truly through the primary passage to operate the first phase. As soon as the spring setting of the valve is reached, the valve spool lifts and oil-flow is diverted to the secondary port to operate the next phase of the system.

A schematic diagram of a pilot controlled pressure sequence valve is shown in Fig. 6.7. Here the required sequential pressure can be adjusted manually. In this valve, the fluid flows freely through the primary passage to operate the first phase until the pressure setting of the sequence valve is reached. As the spool lifts, flow is diverted to the secondary port to operate the second phase. Application of a sequence valve is shown in a hydraulic circuit for a clamping and drilling sequential operation as illustrated in Fig. 6.8. The position-step diagram in Fig. 6.9 illustrates the sequential movement of the clamp and the drill cylinder.

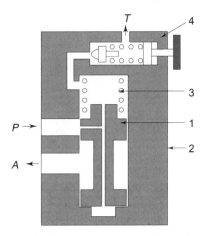

Fig. 6.7 *Pilot operated sequence valve*
[1-Main spool, 2-Valve body, 3-Spring, 4-Secondary adjusting screw]

6.4 COUNTER BALANCE VALVE

A counter balance valve is used to maintain control over a vertical cylinder so that it will not fall freely because of gravity. The primary port of the valve is connected to the lower cylinder port and the secondary port to the directional control valve. The pressure setting is slightly higher than is required to hold the load from falling.

When the pump delivery is directed to the top of the cylinder, the cylinder piston is forced down causing pressure at the primary port to increase and raise the spool, opening a flow path for discharge through the secondary port to the D.C. valve and subsequently to the tank. In cases where it is necessary to remove back pressure at the cylinder and increase the force potential at the bottom of the

Flow and Pressure Control **181**

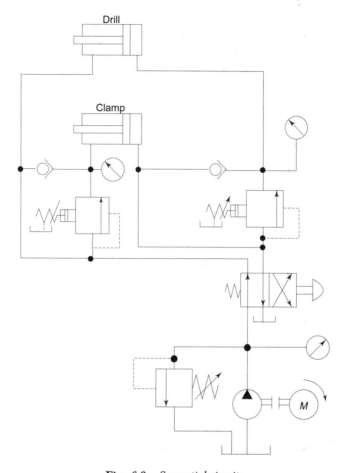

Fig. 6.8 *Sequential circuit*

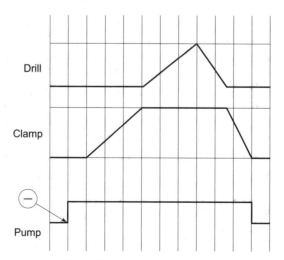

Fig. 6.9 *Position-step diagram*

stroke, this valve too can be operated remotely. When the cylinder is being raised, the integral check valve opens to permit force flow for returning the cylinder.

This valve is also called a back pressure valve. Application of this valve is shown in a schematic circuit diagram in Fig. 6.10.

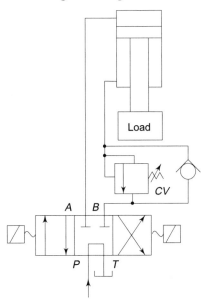

Fig. 6.10 *Counter balance valve in hydraulic circuit*
[CV-Counter balancing valve]

6.5 PRESSURE REDUCING VALVE

These valves are normally open pressure control valves used to maintain reduced pressures in certain portions of the hydraulic system. They are actuated by the pressure sensed in the branch circuit and tend to close as it reaches the pressure of the valve setting preventing further build-up of pressure.

A direct acting valve generally has a spring loaded spool to control the down stream pressure. If the main supply pressure is below the valve setting, fluid will flow freely from the inlet to the outlet. An internal connection from the outlet passage transmits the outlet pressure to the spool end opposite the spring.

When the outlet pressure rises to the valve setting, the spool moves to partly block the outlet port. Only enough flow is passed to the outlet to maintain the preset pressure. If the valve closes completely, leakage could cause the spool pressure to build up. Instead a continuous bleed to the tank is permitted to keep it slightly open and prevent down stream pressure from rising above the valve setting. A separate drain passage is provided to return this flow to tank. A schematic view of this valve is shown in Fig. 6.11.

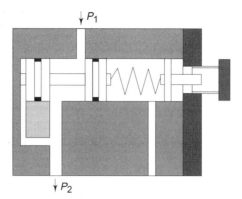

Fig. 6.11 *Pressure reducing valve*

6.6 UNLOADING VALVE

A pressure relief valve requires full system pressure to open thereby causing higher quantity of energy loss due to heat. Under certain conditions, the unloading valve may be used to unload the energy at a lower pressure. The primary port pressure in such a valve is independent of the spring force because the remote pressure operates the spool.

Uses These valves are used in systems where two pumps provide a large volume of oil at low pressure and one of them must be unloaded during a specific period requiring only a small volume of oil at high pressure. This will save the system from undesired heat energy to a great extent.

6.7 SPEED CONTROL DEVICE

The speed of hydraulic actuators is changed by varying the port opening of the flow control valve. This valve is basically a flow control valve which regulates the fluid flow by enlarging or reducing the port area while the oil is passing through the passage. Thus continuous step less control of speed of a cylinder or a hydraulic motor is possible with such a valve.

The flow control valves could be either:
(a) Throttle valves or flow restrictors which are pressure dependent.
(b) Flow control valves which are pressure independent. Before discussing further let us find out here how through various means the flow rate can be effected in a hydraulic system.

The fluid flowing to an actuator can be varied:
(i) By use of a variable delivery pump which is a relatively simple but quite an expensive method.
(ii) By using an orifice control, i.e. by using a simple flow restriction in the line of fluid flow. The principal disadvantage of a flow restrictor is that the flow rate is pressure dependent. The flow rate is influenced by the

pressure drop across the orifice, i.e. the rate of flow varies as the load changes.

(iii) By using a pressure compensated flow control valve, i.e. making suitable provision to maintain a constant flow rate regardless the pressure drop across the orifice.

6.8 PRINCIPLE OF FLOW CONTROL

The flow rate Q and linear cylinder velocity is related by the formula:

$$v = \frac{Q}{A}$$

where v = cylinder velocity cm/min
Q = flow rate cm^3/min
A = area of the cylinder piston, cm^2

Similarly rpm of a hydromotor is related by the formula.

$$n = \frac{Q}{V}$$

where V = volume capacity of the motor, cm^3/revolution

The above formula shows that for a constant speed we require a constant flow rate. In such a case, V and n are directly proportional to Q. For a simple regulation of speed, a throttle valve is used (depending on the pressure). For a control system where the speed of the cylinder or rpm of the motor is to be constant and independent of pressure drop, a pressure compensated flow control valve is to be used.

6.8.1 Physical Principles of Flow in a Throttling Aperture

It is known that at ventury throat, as shown in Figs 6.12 (a) and (b) the fluid speeds up, and loses some of its pressure. Two types of throats—one long and the other short, but both being fixed—are shown in the diagram. We know that laminar flow through small pipe-diameter is governed by the Hagen-Poiseuille's Law. The law states that pressure drop

$$P_A - P_B = \Delta P_{A-B} = \frac{8 \cdot \mu \cdot 1}{r^2} \cdot u$$

where μ = dynamic viscosity (kgf sec/m^2)

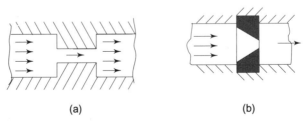

Fig. 6.12 *Throttle aperture (a) Fixed throttle (b) Diaphragm throttle*

$$r = \frac{d}{2} = \text{radius of flow path (cm)}$$

$u = $ Velocity of flow, cm/sec

or
$$\Delta P = \Delta P_{A-B} = \frac{32 \cdot l \cdot \mu \cdot u}{d^2}$$

where $d = $ diameter of flow path (cm) $= 2r$
$l = $ length of flow path (cm)

Now we know that
$$\mu = v \cdot \rho$$

where $v = $ kinematic viscosity in $\frac{m^2}{\sec}$

$\rho = $ density of oil in $\frac{kg \sec^2}{cm^4}$

Also

Reynold's No. $\text{Re} = \frac{\mu \cdot dw}{v}$ for a circular cross section.

where   Re = Reynold's No.
$dw = $ wetted flow diameter (cm)

and $\lambda = \frac{64}{\text{Re}}$

where $\lambda = $ laminar co-efficient of friction (isothermal)

$\therefore \quad \lambda = \frac{64}{\text{Re}} = \frac{64}{\frac{u \cdot dw}{v}} = \frac{64\, v}{u \cdot dw}$

$\lambda = \frac{75}{\text{Re}}$ for laminar non-isothermal condition

For a turbulent flow $\lambda = \frac{0.3164}{\text{Re}^{0.25}}$

$\therefore \quad \Delta P = \frac{32\, l\, \mu\, u}{d_w^2}$

$= 64 \cdot \frac{l\, v\, \rho \cdot u}{2\, d_w^2}$

$= \frac{\lambda \cdot u \cdot d_w}{v} \cdot \frac{l \cdot v \cdot \rho \cdot u}{2\, d_w^2} = \frac{\lambda \cdot l \cdot \rho \cdot u^2}{2\, d_w}$

$= \lambda \cdot \frac{l}{d_w} \cdot \frac{e}{2} \cdot u^2$

Now, Area $A = \frac{\pi d^2}{4}$ or $d^2 = \frac{4A}{\pi}$ or $d = \frac{4A}{\pi d} = \frac{4A}{C}$

where $C = $ circumference $= \pi d$

$$\therefore \qquad d_w = \frac{4 \cdot A}{\pi \cdot d} \text{ or } d_w = \frac{4A}{C}$$

where C = circumference = πd.

6.8.2 Coefficient of Resistance

This is also known as coefficient of local hydraulic loss in bends, tees, valves and areas of contraction or expansion.

$$\therefore \qquad \xi = \lambda \cdot \frac{l}{d_w}$$

$$\therefore \qquad \Delta P = \lambda \cdot \frac{l}{d_w} \cdot \frac{\rho}{2} \cdot u^2$$

$$= \xi \cdot \frac{\rho}{2} \cdot u^2 \quad \text{where } \xi = \text{coefficient of loss}$$

or flow vicocity $u = \sqrt{\dfrac{\Delta P}{\xi \cdot \dfrac{\rho}{2}}}$

From theory of continuity one can write flow rate = $Q = A.u = A_1.u_1$ where u and u_1 are velocities of oil flow at two different points of the pipe. A and A_1 = are of pipe cross section

$$\therefore \qquad Q = A \cdot u = A \cdot \sqrt{\frac{\Delta P}{\xi \cdot \rho/2}}$$

But $\qquad v$ = sp. wt. = $g \cdot \rho$ where g = acceleration due to gravity

$$\therefore \qquad \rho = \frac{v}{g} \text{ where, } \rho = \text{density of oil}$$

$$\therefore \qquad Q = A \cdot \sqrt{\frac{\Delta P \cdot 2g}{v}}$$

This is applied to laminar or turbulent flow up to a limited area of cross section. Whether a flow is laminar or turbulent is determined by the Reynold's Number. The flow character is determined by the coefficient of resistance and the area of cross section in wetted-diameter.

Flow through pipe is measured by the coefficient of friction (laminar) λ and flow through bends and tees is measured by coefficient of resistances (ξ).

$$\delta = \lambda \frac{l}{d_w} = \frac{64}{Re} \cdot \frac{l}{d_w} = \frac{64}{u} \cdot \frac{v}{d_w} \cdot \frac{l}{d_w} = \frac{64 \, vl}{u d_w^2}$$

The above formula shows that the coefficient of resistance ξ is dependent on:
(a) The velocity of flow u
(b) The kinematic viscosity v
(c) The throttle length l

(d) The wetted hydraulic diameter d_w
(e) The throttle form
(f) Reynold's Number Re

To calculate the flow rate through a diaphragm or an aperture, the following formula can be used:

$$Q = \alpha \cdot A \sqrt{\frac{\Delta P \cdot 2g}{v}}, \text{ or}$$

$$Q \approx \sqrt{\Delta P}$$

where Q = flowrate
 A = area of aperture
 ΔP = pressure loss
 g = acceleration due to gravity
 v = specific weight of oil
 α = dimensionless coefficient accounting for non uniform velocity distribution depending on flow chance form (form factor)

6.8.3 Calculation of Flow Rate

The following formula can be used to calculate Q.

$$Q = \alpha \cdot A \cdot \sqrt{\frac{2g \cdot \Delta E}{v}} \quad \text{where } Q = \text{flow rate in cm}^3 \text{ s}$$

α = form factor = 0.75 to 0.85, mostly.
v = specific weight of oil 0.855 at 20°C
   and 0.869 kgf/dm^3 for oil with a kinematic viscosity of 33 cS at 50°C.

From the above we also find that Q depends on
1. Pressure drop and area of cross section and throttle form
2. The specific weight and density of oil
3. The coefficient of resistance

If Q has to be constant, ΔP has to remain constant. It has been noted that if the oil gets heated up during operation from 20°C to 80°C, the density of oil ρ reduces by only 3.5%. Thus, it has only very limited or practically no influence on Q.

But the coefficient of resistance brings a lot of change in Q through change of temperature because the kinematic viscosity changes due to the change of temperature from 20°C to 80°C and is lowered from 115 to 10 cSt. This means a loss of kinematic viscosity by about $\angle 91\%$ during this range.

6.8.4 Throttle form and Influence of Viscosity

The throttling form of a valve is to be determined according to the influence of temperature or viscosity of oil on the flow rate Q. From the foregoing discussion it is clear that flow rate through a valve generally depends on
(a) Oil viscosity
(b) Friction

(c) Length of the throttle and diameter of throttling area

For designing a throttle, one must take into consideration the influence of the following factors:
- (a) Coefficient of frictional resistance
- (b) Viscosity
- (c) Wetted circumference of contraction
- (d) Flow rate, etc.

Selection of throttle form should be so that the largest hydraulic diameter can be achieved. The wetted diameter of the throttle form (d_w) can be estimated from the formula given below:

$$d_w = \frac{4 \cdot A}{C}$$

where A = area of orifice, i.e. throttle aperture,
C = circumference of orifice.

The ratio C/A is called the *form quotient* and should be low. The smaller the form quotient, the more favorable the form is. Table 6.1 shows a comparison of throttle form keeping almost the same value of area of cross section of certain probable throttle forms shown in Fig. 6.13 (a).

Table 6.1 Comparison of Throttle Forms

Sl No.	Throttle form	Area of cross section-A (mm^2)	Circumference C (mm)	C/A (1/mm)	dw (mm)
1.	Circular 5 mm dia.	19.62	15.70	0.80	4.99
2.	Square a = 4.43 mm	19.62	17.72	0.903	4.423
3.	Rectangular 3 × 6.54 mm	19.62	19.08	0.972	4.11
4.	Equilateral triangle of side a = 6.73 mm	19.62	20.19	1.029	3.887

Here we find the circular form is an ideal one. The next better form is the polygonal form. The decision to use a specific throttle form depends on whether the throttle shape is retained or not during the operational phase. When a circular form is used at the end one may get a blind point. By using a square form the shape becomes too complicated to maintain due to change in area. A triangular shape is more comfortable due to ease of manufacturing. In a triangular form it is possible to maintain the same shape if one cuts off an area parallel to the base of the triangle. Therefore an equilateral triangle form could also be selected as a desirable throttle form. The characteristics of some throttle forms are illustrated in the curves in Fig. 6.13 (b). It is seen that A/A_{max} is a function of h/h_{max} where curve (b) is better compared to curves at (a), (c) and (d).

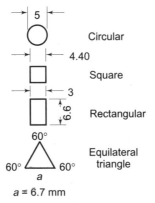

Fig. 6.13(a) *Four types of throttle forms commonly used and their form quotient comparison*

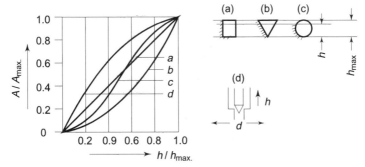

Fig. 6.13(b) *A/A max Vs. h/h max characteristics of few throttle form.*
[Courtesy: Herion Taschenbuch, m/s Herion Werke, KG, Stuttgart, Germany, 1969]

6.8.5 Throttle Form

(a) Fixed or constant throttle A simple throttle with fixed fluid flow control is shown in Fig. 6.12 (a). The area of flow here is $A = \dfrac{\pi d^2}{4}$ where d = diameter of the aperture. Here the following characteristics of the form are to be noted:
  (i) Throttle form is good.
  (ii) Wetted circumference is less.
  (iii) Due to the length of the throttle, Q depends on viscosity.

Diaphragm orifice: Shown in Fig. 6.12(b).

The flow area = $\dfrac{\pi d^2}{4}$

Characteristics The throttle form is good. The wetted circumference is less. Due to the small length of throttle, Q does not depend on the oil viscosity.

6.8.6 Adjustable Throttle

(a) Needle throttle From Fig. 6.14 (a), we see

$$\text{Area} = \left(d - \frac{h}{\sin \alpha}\right) \cdot \frac{\pi h}{\sin \alpha}$$

Characteristics
  (i) The length of throttle is small.
  (ii) The wetted circumference is bigger.
  (iii) Influence of viscosity is not much.
  (iv) By a little stroke of h, a relatively great area is opened to flow.
  (v) The range of flow path due to change in area is bad.

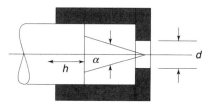

Fig. 6.14(a)   *Needle throttle*

(b) Triangular notch (longitudinal) [Fig. 6.14 (b)]

$$\text{Flow area} = \frac{h^2}{\sin^2 \alpha} \cdot \tan \beta/2$$

Characteristics
  (i) The length of throttle is relatively small.
  (ii) The wetted circumference is small.
  (iii) Influence of viscosity is limited.
  (iv) Range of possible flow path due to change in area is not so bad.
  (v) For a small flow rate it is good.

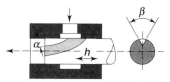

Fig. 6.14(b)   *Linear Triangular notch*

(c) Rectangular notch [Fig. 6.14 (c)]

$$\text{Flow area} = \frac{h}{\sin \alpha} \cdot b$$

Fig. 6.14(c)   *Rectangular notch*

Characteristics
  (i) The throttle length is relatively smaller.
  (ii) The wetted circumference is proportionally smaller.
  (iii) Influence of viscosity is less.
  (iv) Good form retaining capacity
  (v) For a small flow oil volume, a good form.

(d) Circumferential Notch (Triangular form)
The throttle form is shown in Fig. 6.14 (d).
The flow area = $a^2 \cdot \tan \beta/2$.

Characteristics
  (i) The throttle length is longer.
  (ii) The flowrate and pressure drop depends on oil viscosity.
  (iii) This form of throttle is not exceptionally good as the flow control is better only between 90° to 180° of the rotation of the notch.

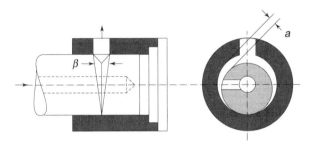

Fig. 6.14(d) *Throttle form*

6.8.7 Throttle Valve

A throttle valve controls the flow rate Q by increasing or decreasing the area of flow path. It can either be a fixed or an adjustable throttle. As already discussed Q depends on the pressure differential between two ports (ΔP). With changes in pressure, for the same opening the flow is different. As work resistance due to the load to be moved by the actuators has influence on the changes in pressure, the velocity of flow cannot be kept constant with a normal throttle valve which is graphically explained in the illustrations shown in Fig. 6.15. The flow rate Q changes with the change in work resistance due to the pressure drop in the throttle valve. Fixed throttles are generally inserted directly in the pipe lines. The l/d ratio of the throttle orifice may influence Q and pressure drop (ΔP) as stated below

  (i) l/d ratio is large – Q and ΔP depends on the viscosity of oil due to length of the throttle and resultant friction.
  (ii) l/d ratio is small – Q and ΔP independent of the viscosity of oil as throttle length is almost zero.

This means that the speed of the hydrocylinder and hydromotor will also change when the oil is passing through a throttle valve with larger l/d ratio.

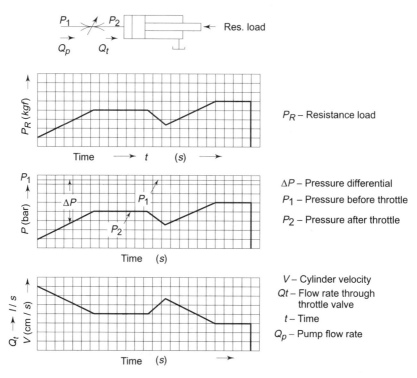

Fig. 6.15 *Velocity of cylinder changes with change in load resistance*

In a control technique system if the resistance due to work is kept constant, a constant speed is possible by a throttle valve only. Very often a non-return valve can be in-built with a throttle valve to provide regulated flow in one direction and full flow or free flow from the opposite direction. Figure 6.16 (a) shows a schematic diagram of such a valve where the flow from A to B is regulated and from B to A it is free.

6.8.8 Flow Regulation Valve (Pressure Drop Independent)

Precision machine tools and other equipment need a constant speed free from influence of external resistance and temperature. A pressure compensated flow control valve can meet such requirements of a hydraulic system which will provide a stepless adjustable speed control over a wide spectrum. The function of such a valve is to allow a constant predetermined amount of oil (Q l/min) independent of pressure drop and temperature across the valve, as demanded above. The above function can be achieved if the two conditions as given below, are fulfilled:
(a) Constant pressure drop on the adjustable throttle path;
(b) The form of the throttle should be such that the influence of viscosity is minimum or insignificant.

Flow and Pressure Control   **193**

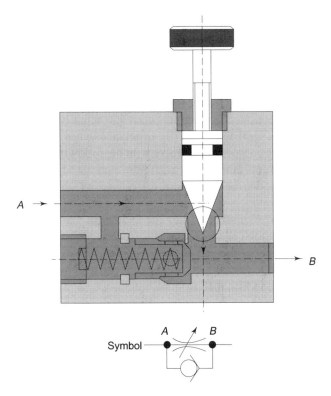

Fig. 6.16(a) *Non-return flow control valve*
A → B : Controlled flow
B → A : Free flow

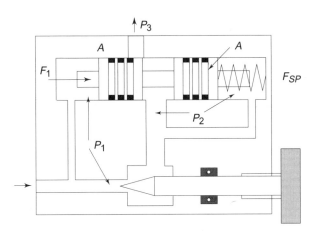

Fig. 6.16(b) *2 Way flow control valve (pressure compensated)*
$$P_1 \cdot A = A \cdot P_2 + F_{SP}$$
$$\therefore (P_1 - P_2) A = F_{SP}$$
$$\therefore \Delta P = P_1 - P_2 = F_{SP/A} = constant.$$

A schematic diagram of a pressure compensated flow regulating valve is shown in Fig. 6.16 (b).

1. How to achieve constant pressure drop

We have seen from Fig. 6.15 that due to the change in work resistance on a hydraulic cylinder, the pressure in the valve during oil entrance and exit also changes according to the position of the regulation valve.

The pressure compensated flow control valve has actually two valves arranged in series as shown in Figs 6.17 (a) and (b). In actual essence, these two valves are throttle and a pressure compensating valve with the following functions.

1. The inner and constant pressure difference on the adjustable throttle.
2. The outer and variable pressure difference which is fixed due to the regulator.

Such a valve consists essentially of the following main parts: (a) throttle, (b) spring loader of spool (compensating) and (c) the spool.

The pressure compensation can be positioned inside the valve before or after the throttler as shown in Figs 6.17 (a) and (b). It can also be arranged parallel to the throttler as shown in Fig. 6.17 (c).

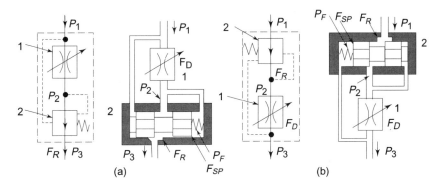

Fig. 6.17(a) & (b) *Pressure compensated flow control valves–Throttle before and after pressure compensation*
[Courtesy–Herion Taschenbuch, Herion Werke, KG, Germany, 1969]

In Fig. 6.16 (b) by equating the force against the spool, we get

$$P_1 \times A = A \cdot P_2 + F_{sp}$$

where P_1 = primary pressure before throttling.

∴ $A \cdot (P_1 - P_2) = F_{sp}$ A = Cross-sectional area of valve spool

or $$\Delta P = (P_1 - P_2) = \frac{F_{sp}}{A}$$ where F_{sp} = Spring force.

= Constant as F_{sp} and A are both constant.

The spring force determines the position of the spool and the amount of oil allowed through the valve.

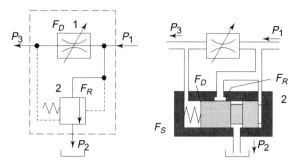

Fig. 6.17(c) *Pressure compensator in parallel*

If F_{sp} is constant, ΔP is also constant and hence the adjusted Q is constant. Therefore the cylinder speed will also be constant as shown in Fig. 6.18.

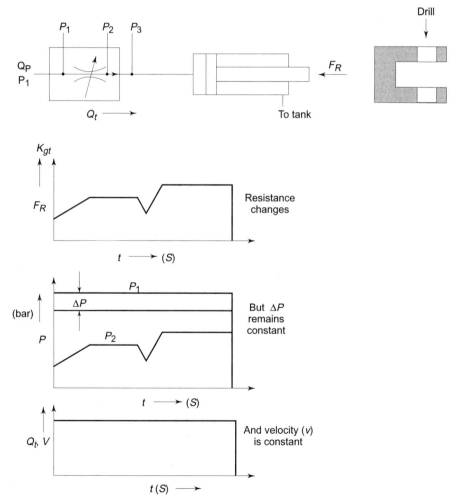

Fig. 6.18 *Application of pressure compensated flow control valve to provide constant velocity*

6.9 CALCULATION OF FLOW QUANTITY THROUGH AN ORIFICE

We know $\Delta P = \dfrac{\xi \cdot \rho_o}{2} \left(\dfrac{Q_o}{A_o}\right)^2$ kgf/m^2

$\therefore \quad Q_o = A_o \sqrt{\dfrac{2\Delta P}{\xi \cdot \rho}}$ m^3/s

where Q_o = flow through an orifice, A_o = area of cross section of orifice, m^2, ξ = flow resistance factor, ρ = oil density kgf/m^{-4}s^2.

For a given pressure drop

$$Q = \left(114 A_o \sqrt{\Delta P}\right) 10^{-3}$$

An useful method of calculating the area of the orifice required.
where A_o = area of orifice in mm^2
Q = flow rate in l/s
ΔP = differential pressure in bar.

One must keep in mind that the diameter of these orifices is usually small and often it is essential to provide for filtration on the up-stream side.

Conditions are worse for a needle valve where the orifice takes the form of an annular restriction which may be opened only a fraction of a millimeter and can be readily obstructed. Obstruction due to silting often occurs in restrictors which are used to damp out pulsations in the flow to a pressure gauge.

For precise speed control a pressure compensated flow control valve is to be used. A valve of this type maintains an almost constant pressure drop across a fixed orifice to produce a constant flow rate. For this purpose it incorporates an internal regulating system comprising of a variable orifice, with pressure acting on the ends of a spool against a biasing spring that has a low rate.

6.10 POSITIONING OF A FLOW CONTROL VALVE

When the flow from a fixed delivery pump is divided between particular sub-circuits, the problem of correct apportionment of flow to the orifice arises. There are three primary positions for the flow control valve.

1. Meter-in-to control   Here metered oil is sent to cylinder. Any excess fluid is forced through the relief valve to return to the tank at the maximum pressure P_s. Since excess flow passes through the relief valve a powerloss P_s equivalent to P multiplied by Q_r is encountered to reduce the efficiency. This method is useful for unidirectional opposing loads on rams that are frequently used, or when low power levels are involved. P_s is the system pressure and Q_r is the flow rate diverted to the tank from the relief valve.

2. Meter-out-flow   The oil from the actuator is passed through the flow control valve and with this position the loading conditions of all types can be controlled. The maximum design pressure for the ram must be greater than the system pressure as:

$P_2 = P_S \dfrac{A_1}{A_2}$ and A_1 is greater than A_2 resulting in pressure intensification.

Once again power loss is encountered given by $P_S \cdot Q_r$. The circuit is useful for loadings of all types, and particularly for infrequent applications at low power.

3. Meter-by-pass-often used in machine tools.

Here ram speed is effectively controlled by returning some of the operating fluid to the tank through a flow control valve. The remaining flow then determines the speed. The pump is connected directly to the actuator and the power loss is minimised at $P_1 (Q_P - Q_1)$ where P_1 is less than P_S.

6.11 PROPORTIONAL PRESSURE AND FLOW CONTROL VALVES

In the recent past, a new range of valves are being developed and used in hydraulic system. Proportional remote controlled electro-hydraulic pressure and flow controlled valves are such new addition in this respect. The main advantage of these valves is their ability to control the pressure and flow rate proportionately and provide remote function. The variation in pressure and flow is effected by varying the electrical input signal to a proportional solenoid which is used to actuate the valves. The input electrical current is derived from amplifying devices which could be located remotely from the valves.

Figure 6.19 shows a simple design of a proportional flow control valve and its principle of operation.

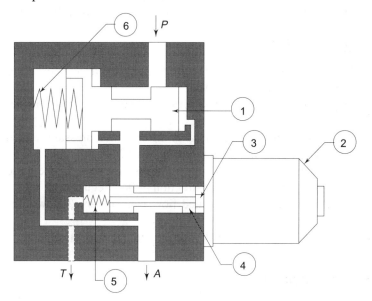

Fig. 6.19 *Proportional flow control valve*
[1-Compensator spool, 2-Proportional solenoid, 3-Push rod, 4-Moving spool; 5, 6-Spring]

The valve consists of a compensator spool (1) biased against a spring (6). A proportional solenoid (2) is fitted to the valve body such that the solenoid plunger when excited can push the smaller moving spool (4). An important feature of the valve is its ability to determine the neutral position of the smaller moving spool (4). A manual adjustment screw is provided in the solenoid cap to set the adjustment of the solenoid plunger manually if there is electrical failure and thus the valve can be operated as a conventional one to get the desired fluid flow. For initial flow rate adjustment there is a zero adjustment arrangement cap of the solenoid which is set by rotating the same. Very fine adjustment is possible due to the fine threads in the cap. The plunger friction is minimized significantly due to provision of very fine sliding bearings onto the plunger. This enhances the longevity of the valve to a great extent.

An infinitely fine flow variation within the valve's rated capacity is possible and the flow rate is directly proportional to the magnitude of the input electrical signal. The pressure compensation features minimize the flow variation against any pressure fluctuation that the valve may experience and thus avoid any system malfunction. These values are completely programmable and hence suit full automation.

In case of a proportional pressure control valve, the pilot control valve is actuated by a solenoid such that the solenoid plunger pushes the valve spool by means of the push rod. A remote controlled electrical input current results in movement of the plunger via the push rod. The remote controlled input signal current results in movement of the plunger which in turn makes the push rod to unseat the poppet in the pilot valve chamber and thus the main pressure from the pump is dumped to the tank when the system pressure is increased due to system malfunction. These valves have high response and are therefore very sensitive.

Review Questions

1. Why is a pressure relief valve used in a hydraulic system? State the basic types of pressure relief valves.
2. What is a check valve? Show various uses of a check valve in the hydraulic circuit.
3. How does a pressure relief valve differ from a pressure reducing valve? How does a pressure reducing valve work?
4. What is meant by "form quotient"? Describe some of the basic forms of valving element used in construction of flow regulators.
5. What is pressure compensated flow control valve? How does pressure compensation take place?
6. Describe a proportional solenoid operated flow control valve.

References

6.1 Herion Werke KG, Herion Taschenbuch Stuttgart, Germany, 1969. (p. 169).

6.2 White, F M, *Fluid Mechanics,* McGraw-Hill International Book Co., 348, Japan Boon Lay, Singapore.
6.3 Merrit, Herbert, *Hydraulic Control Systems,* John Wiley and Sons.
6.4 BoschGmbH, Robert, *Bosch Hydraulics,* Stuttgart, Germany, 1970.
6.5 Vickers Industrial Hydraulics Manual - 935100 A, USA.
6.6 Takeo Tanaka and Shichisaburo, *Proportional Electrohydraulic Remote Control Valves,* Yuken Kogyo Co, Tokyo, Japan.
6.7 "Check Valves—Simple designs provide sophisticated performance," a report published in *Hydraulics & Pneumation,* Penton IPC, USA information provided by S.R. Majumdar, 1980.
6.8 Findeisen F. und D., Oel hydraulik in Theorie und Anwendung, Schwetzer Verlagshaus AG., Zurich, Switzerland, pp. 165-166.

7
Hydraulic Servo Technique—Recent Trends

KEEPING pace with the spectacular growth in engineering technology, there has been tremendous development in the field of control technology. The recent trend in the field of sophisticated controls is to integrate various control media together so that the advantage of each technique could be gainfully utilized for the benefit of finer and more accurate control. In the case of hydraulic control, system engineers and system designers are continuously trying to provide accurate positioning of the load with improved control on speed, force and other physical parameters. Application of electronics has been a tremendous source of success in achieving higher accuracy in hydraulic system control. A new range of servo and proportional valves with electronic finer control element have been developed and are being increasingly used in various fields like production, material handling, aviation, shipping, robotics, etc. Let us take a look here on the evolution of servo valves from the earlier tracer control system and understand their respective field of application. By definition, a system in which a small input force is capable to control a larger output force can generally be called a servo control. Thus any basic hydraulic circuit is a servo system as a small actuating force on the valve spool provides a high output force from the hydraulic cylinder. In hydraulic systems, valves which are used to control cylinders and hydromotors are themselves regulated by various means like levers, plungers, air, oil, solenoids, etc. In modern automatic hydraulic systems, solenoids are more commonly used compared to other control medium. However the input power of a solenoid operated valve sometimes may be too small to operate a valve large enough to control the full power of the system, in which case it may be used to operate a low power valve which in turn, operates a second larger valve. The first valve in this case is called *pilot valve* and the second one as *slave valve*, although in practice in most cases the two are combined in a simple unit. The use of a control signal from a pilot valve to control a slave valve is common in electro-hydraulic circuits where the power level of the hydraulic circuit is greater than about 5 kW. However signal amplification or automatic correction of control parameters is not possible in an ordinary system using normal valves.

A servo system should provide both signal amplification and automatic correction of any deviation that may take place between the output quantity and quantity set by the command signal. This can be achieved by a "feedback" signal from the output stage by creating a "loop" between the actuator and the control valve and utilizing the error signal to reposition the output movement to compensate the error.

The control valve thus receiving both the command signal as well as the error signal (i.e. the signal due to error received from the system feedback) may be called a servo valve.

7.1 FUNCTION OF A HYDRAULIC SYSTEM

The major advantage of a hydraulic system is that it generates a very large force and at the same time it is quite compact with a very attractive power to weight ratio and it can be easily controlled. In spite of all the inherent advantages of hydraulics, to enhance the level of confidence commensurate with the current day technology, integration with other sophistication, hydraulic system needs control technology like electronics. It is, therefore, important to know the functional and accuracy limitations of a conventional hydraulic system and how both this technology and electronics can be bridged together for improved controllability and enhanced acceptability of the fluid power system.

The function of a basic hydraulic system can be described in a simple block diagram as shown below.

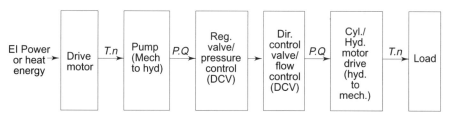

T-Torque; n-rpm; P-Pressure; Q-Flow rate

From the block diagram above it is quite clear that in order to perfectly position or control a load, the position control unit, i.e. the DCV should be fed with the correct command signal. The input command signal and the feedback signal from the actuator will be unbalanced as long as there remains a positive and negative error signal and until this error signal is eliminated, the control valve should continue to respond to both the input as well as the feedback signal.

Generally a conventional hydraulic system is not capable to estimate and quantify this positional and other error arising out of system malfunction.

Due to the mechanical nature of the system, a normal hydraulic circuit fails:

(a) To meet accurate cycle time thus affecting perfect sequencing of activities and loss of productivity which is demanded in sophisticated control fields.
(b) To protect system components against shock load due to fluid compressibility or other mechanical malfunctioning.
(c) To provide high efficiency of energy conversion and power transfer.
(d) To reduce noise level creating environmental hazards specially with certain types of hydraulic pumps.
(e) To sense system malfunction and take immediate and appropriate corrective action to stop it.
(f) To respond quickly to various system parameters.

Here use of a servo valve helps the system designer to design a fully controlled system combining the enormous hydraulic power with fine control techniques.

7.1.1 Open Loop and Closed Loop Circuit

Hydraulic circuits are designed with either an open or a closed loop. In a normal hydraulic circuit, the movement of an actuator or hydro-motor is commanded by the valve input signal but deviation, if any, can not be controlled as may be evident from Fig. 7.1. Such a circuit diagram is called an open loop system. In order to estimate and correct any deviation between the desired position and actual output position, a signal from the output is to be fed back to the control valve (servo valve) thus creating a closed loop between the valve and the actuator as shown in Fig. 7.2. If the feedback signal is very weak and difficult to apply, additional amplification arrangement is to be incorporated in the system to accept both the

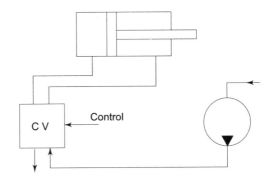

Fig. 7.1 *Basic open loop circuit*

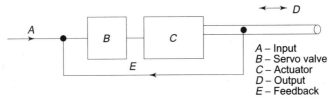

Fig. 7.2 *Closed loop circuit*

command as well as the error signal and provide automatic correction against any deviation.

This is shown as a block diagram in Fig. 7.3 (a) with two stage amplification and computation of linear or positional response. It gives a proportional response to the input signal regardless of effects exerted by the external load. A typical servo controlled cylinder circuit is schematically shown in Fig. 7.3 (b) to illustrate the mechanism of control of force feedback.

This type of system can fully guide the system load to its desired destination with the help of a controller. The current status is estimated with the help of feedback transducers. The controllers measure the output values, compare with the desired values and then issue the right command to the servo valve for repositioning of the load.

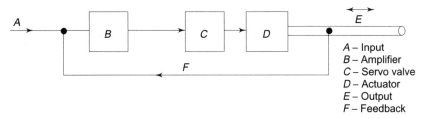

Fig. 7.3(a) *Closed loop circuit with amplifier*

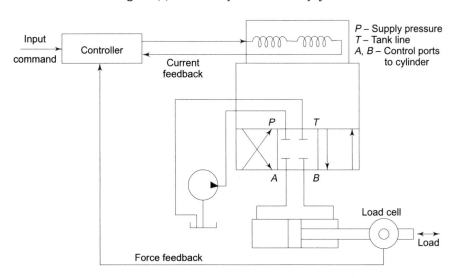

Fig. 7.3(b) *Closed loop control of a cylinder with force feedback*

7.1.2 Types of Feedback

A servo mechanism consists of a servo valve, servo amplifier and actuator connected together providing a closed loop feedback arrangement as stated earlier. The feedback can be either:

(a) Mechanical
(b) Hydraulic
(c) Electrical/Electronic

7.2 MECHANICAL FEEDBACK AND APPLICATION OF TRACER VALVE

Before the advent of electronics, tracer control was commonly used in a machine tool control system. A tracer control valve can be called a servo valve which is used to control the position of the cylinder as shown in the Fig. 7.4. The valve spool and the cylinder piston rod can be connected through a mechanical linkage—a differential lever as depicted in Fig. 7.5 which while it rotates about point 3, produces a displacement of the servo valve. The valve unit, the input lever and the linkage have all been arranged such that with actuation of input lever, the cylinder movement will be initiated and the valve position will continue to change until the ratio of velocities at points 1 and 3 corresponds with lever ratio a : b. Until the differential lever is actually stationery or the input and output members are moving simultaneously at velocities proportional to the lever ratio a : b, there will be an error signal which will displace the servo valve until the error becomes zero and the cylinder movement stabilises.

Practical examples of application of mechanical servo may be found in the power steering of automobiles and in hydro-copying attachment used in copy

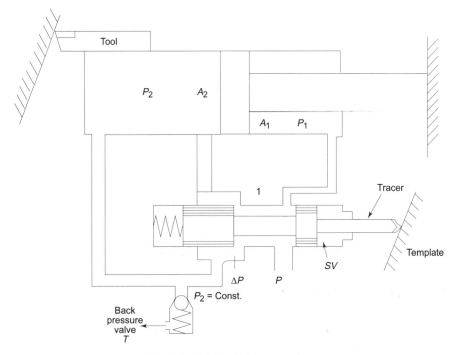

Fig. 7.4 *Mechanical copy turning*

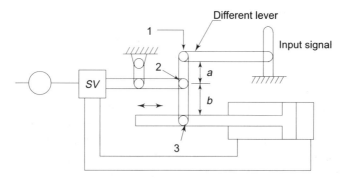

Fig. 7.5 *Functional links of a mechanical servo system*

turning, copy milling m/c, etc. The principle of a hydraulic copy turning system is explained here.

7.2.1 Hydraulic Copy Turning

As stated earlier, the hydro-copying device is a good example of a simple mechanical servo mechanism. The hydraulic piston is connected to the lathe saddle, (see Fig. 7.4), which has a rear slide way to support the hydraulic slide on which is mounted the tool post (the tool operates in this case from the rear of the m/c). Since the piston is attached to the saddle, its position is fixed and it is the cylinder which is displaced by the flow of hydraulic fluid, permitting tool movement accordingly.

The flow of fluid to the either side of hydraulic cylinder is controlled by the position of the spool relative to the ports in the spool valve. The spool maintains its relationship with the template through a stylus and is spring loaded to ensure constant contact. When the oil is applied to the system, the cylinder, slide, valve and stylus move until the stylus contacts the template and the spool is deflected to the null position. At this position, spool lands will cover the ports, and the oil supply to both sides of the hydraulic piston will be cut off. The radial displacement of the cutting tool, therefore, will cease, and if the lathe saddle is traversed along the lathe bed, the tool will produce a parallel work piece.

If we refer the diagram and consider that due to the saddle traverse, the template is moving downwards relative to the stylus; as soon as the vee-shaped position of the template is reached the spool valve will move to the left. Oil under pressure will then pass through the left side of cylinder, and oil from the right hand side of the hydraulic cylinder will return to tank. Consequently the cylinder (and cutting tool) will move to the left, and the shape of the work piece will follow the contour of the template. When the reverse slope of the vee is reached by the stylus, the spool will be moved to the right, oil will be ported to the right-hand side of the cylinder piston and the cutting tool will produce a taper on the work piece corresponding to the shape of template.

7.2.2 Hydraulic Gain

From the above description, the following points should be clear.
(a) At its null point, the spool cuts off the supply of oil to both ends of hydraulic cylinder.
(b) If the stylus and spool are displaced to the left of this center point, the cylinder, slide and cutting tool will be displaced to the left.
(c) If the spool is displaced to the right of null, the tool will move to the right.

Because the rate of flow of oil to the cylinder depends upon the position of the spool and on the conditions spelt above, the speed of the slide displacement is proportional to the distance of the spool from the null position.

By altering the design of the spool valve, greater or lesser amount of oil can be made to flow to the cylinder for a given displacement of the spool, and this would result in the speed of travel being increased or decreased, i.e. its characteristics could be changed. Therefore the sensitivity of the valve would increase or decrease and the gain of the system would be modified, i.e. the amplification of the system would have been changed. However with such a system the cutting tool may not respond instantaneously to the displacement of the stylus, and as a result, some error may occur between the shape of the template and the contour of the work piece produced if a low traverse rate is used. However, if the system is redesigned to give a greater gain (an increased flow of oil for a given spool displacement), the error will be reduced. Unfortunately, there is a limit to which the gain of a system can be increased with such a mechanical servo control. The obvious limitation is on the rate of oil flow through the valve; even where this consideration does not apply, e.g. in an electrically or electronically controlled circuit, a limitation to the increase in gain may arise from the point of system stability i.e.—if the gain is increased too much, uncontrolled oscillation may occur.

7.3 FEEDBACK IN THE SYSTEM

From the foregoing description of a hydraulic copying system, it can be seen that it is an error-actuated system. The radial position of the cutting tool remains unchanged unless the spool is displaced from the null position, i.e. when the tool and the stylus are not in their correct relative positions, the spool is displaced and oil flows to the hydraulic cylinder to correct the error. As soon as the error is corrected, the spool covers the ports and cuts off the oil supply to both sides of the piston. In valves with negative overlap, the land of the spool does not cover the entire area of the ports, but takes up a central position so that the flow of oil to the left of the piston is exactly balanced by the flow to the right hand side. Unless error or imbalance exists in the system, no action takes place.

7.3.1 Hydraulic Feedback—How it Happens

In a hydraulic feedback system, two cylinders of equal size are used—one being the actuator and the other an auxiliary cylinder which provides feedback. The body of the control valve is connected rigidly to the piston of the auxiliary cylinder. Thus any movement of the piston displaces the valve body in the same direction as the initial displacement of the valve plunger which is controlled by the input member. An illustration of such a system is shown is Fig. 7.6.

Initial movement of the input member displaces the valve plunger to direct fluid into one end of the main actuator and an equal amount of fluid is displaced from actuator into auxiliary cylinder. This results in movement of the auxiliary cylinder piston, maintaining flow to the main actual until a synchronized position is reached, when the servo valve is closed. Because of the two cylinders, the mechanism is bulkier and hence its application is limited. In spite of having a number of shortcomings this type of servo mechanism has certain advantages which are worth mentioning, e.g.

(a) It can be adapted to any size of actuator.
(b) It is completely free from backlash and mechanical wear
(c) Sensitivity and response is very high.
(d) Response lag can be reduced to as low as 1%.
(e) Though valve design is a critical factor due to size and capacity, this system could be gainfully applied where other forms of control medium do not suit.

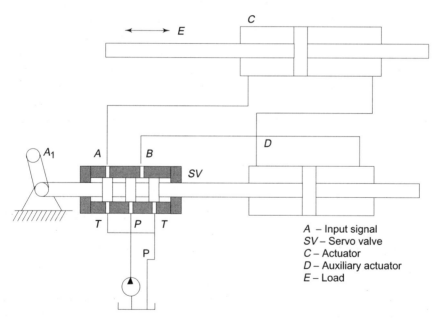

Fig. 7.6 *Hydraulic feedback*

7.4 ELECTRO-HYDRAULIC SERVO SYSTEM

The mechanical and hydraulic servo mechanisms suffer from various inherent and natural problems. They are inconsistent in their travel. Smooth and complete trouble free operation is difficult to achieve. They are to some extent sluggish and very often prone to stick slip motion. They are not suitable for application in remote control system. Hydraulic systems when controlled by solenoid operated valves may have certain added advantages over the other forms which come handy for better automation with an unlimited capability for remote controlling. Hence in critical applications with higher level of automation, electro-hydraulic controls are a better proposition. In this era of electronics many of the earlier control techniques have been replaced. It is now possible to easily integrate the muscle power of hydraulics with the precision of electronic control which is made possible due to the availability of electrical/electronic devices like torque motor, transducers, proportional solenoids and other sensors. Their easy adoption and integration in electro-hydraulic system has become quite common. The major advantages of an electro-hydraulic system combined with electronic control technique are given below:

(a) Precision remote control of position, force and speed of actuator easily achieved.
(b) Higher flexibility of operation is guaranteed.
(c) Better control of fluid-compressibility, system-stiffness and dynamic behavior of load is ensured.
(d) Enhanced capacity to perform multiple sequential functions with a high degree of positional accuracy is possible.
(e) Correct ratio of force and speed is ensured leading to higher energy conservation.
(f) Higher system performance and operating efficiency ensures less down time due to less mechanical failures.
(g) Easy operator control is possible with enhanced failsafe features and higher reliability.
(h) Less maintenance cost and higher machine utilization is possible.

Because of the above advantages, electro-hydraulic devices with electronic controls are very often used in modern engineering. Pressure, flow rate and flow-direction are mainly the three important parameters to be controlled by using sensors, transducers and other electronic devices which can be easily interfaced with the load and electrical (solenoid)/load variables (feedback) and commanded signal (i.e. input signal) and transmit information to the servo electronic or microprocessor control unit.

The introduction of electro-hydraulic devices has added new dimensions in hydraulic system controls like remote and multifunction controls. The block diagram in Fig. 7.7(a) illustrates how and where hydraulics can interact with electronics.

An electro hydraulic servo valve could be operated by an electrical signal to the torque motor or other electronic devices to directly or indirectly position a valve spool to its desired location. However in a closed loop electro-hydraulic servo system, the accuracy of response depends basically on the type and accuracy of the torque motors and transducers used. The various types of transducers mainly used are briefly described in subsequent sections. A block diagram of the feedback system of a position servo valve is shown in Fig. 7.7(b)

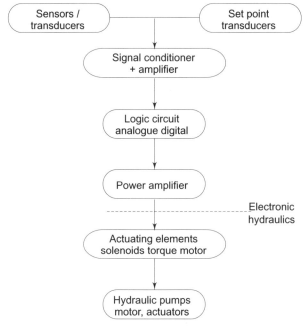

Fig. 7.7(a) *Hydraulic interfacing with electronics*

from which it is clear that the output signal is used to provide a easy feedback signal to the transducer, this signal being fed to the input side. The electronic regulator at this stage then compares the input signal with this feedback signal and varies its own output (driving or input current) to minimize or eliminate any discrepancy between input and output. The feedback signal thus received can indicate precisely all the physical parameters like position, velocity and force.

If necessary more than one control parameter may be fed back, e.g. position and velocity or position and force, etc.

Various sources of input and output signals commonly associated with such system are listed in Table 7.1.

Table 7.1 Sources of Input and Output Signals

Input	Output
(a) Reference voltage	(a) Controlled output
(b) Potentiometer	(b) Load feed valve
(c) Magnetic or punched tape control	(c) Spindle
(d) Tracer control	(d) Conveyer
(e) Programmer controller	(e) Test piece
(f) Function generator	

The main function of direction control valves in a hydraulic system is to start or stop, control and direct the flow in a system. In an electro-hydraulic circuit according to the function performed by a valve, the valves can be grouped as:

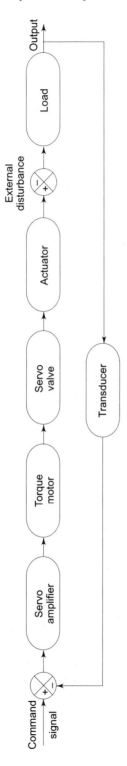

Fig. 7.7(b) *Block diagram of a position servo valve*

(a) On-off valves
(b) Proportional valves
(c) Servo valves

On-off valves when applied to control a cylinder provide the actuator two discrete positions as the valves mostly operated by fixed AC/DC solenoids either switches on the full flow or entirely shuts off the flow. These valves suffer from various problems like fatigue failure of the internal components, high and uncontrollable rate of acceleration or deceleration, burning of solenoids due to over voltage or sluggish stick slip motion of solenoid plunger due to insufficient force caused by under voltage resulting in poor spool movement, inaccurate starting or stopping behaviors, etc. The on-off valves are digital in nature and function.

Compared to an on-off valve, a servo valve is operated by a torque motor or transducer coupled with output sensor, command signal and comparator. In a normal electro-hydraulic servo mechanism, the output sensor is generally a rheostat or potentiometer, the command signal is an electrical voltage and the comparator is an amplifier that provides an error signal proportional to the difference between the input (command) voltage and rheostat voltage. This comparison is not digital in nature and a proportional analogue signal is provided to the servo valve though these valves are highly accurate and have wider application in hydraulic control systems.

A need was felt to design a control valve having performance characteristics between solenoid operated on-off valves and electro-hydraulic servo valves such that the valve element which controls the direction of flow can assume an infinite number of positions between the valves' minimum and maximum limits in which flow direction and control of pressure and flow can be combined in one component. The advent of proportional solenoids made it possible to develop proportional valves with mid-range performance characteristics. They are actually modified versions of on-off solenoid valves the solenoid parts being same but with higher spring rate and spool being precision finished to minimize overlap and solenoid plunger being more accurately sized to closely control spool travel.

A typical block diagram with a pulse width modulated proportional valve is shown in Fig. 7.8.

7.4.1 Transducers

A transducer is basically a sensor capable of providing quantitative information in the form of a signal proportional to the physical variable to be measured. However a distinction has to be drawn between a transducer and a sensor. While a sensor provides information about the physical operation, e.g. an electrical signal, a transducer delivers quantitative value of the physical variable being controlled. Transducers most commonly used for a hydraulic system are for measurement of motion like linear and angular/rotary displacement of the actuators and their velocity. Transducers are named according to the task they perform. They can be passive or active elements. Active elements are self-

212 *Oil Hydraulic Systems: Principles and Maintenance*

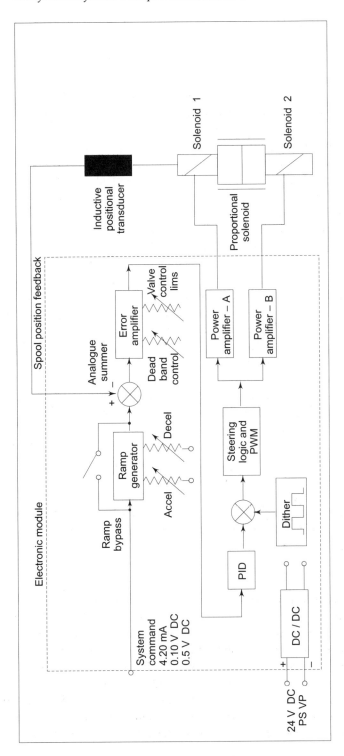

Fig. 7.8 *Block diagram of proportional valve with pulse width modulated amplifier*

generating components, e.g. thermocouples, which use the see-back effect for sensing the thermal condition of a device. Similarly a linear velocity transducer uses Faraday's laws of induction.

7.4.2 Transducers in Hydraulic Systems

Mostly passive transducers are used in fluid power control systems which need some external source of electrical power to function. In most cases, the physical input to the transducer is applied to a primary sensing element such as a diaphragm as in a pressure transducer. The changes in the physical characteristics of the sensing element (such as its change in dimension) are used to vary electrical parameters like resistance, inductance, capacitance, etc. Supporting electronics are added with the transducer to assist its functional parameters, e.g. it supplies the operating power, may amplify output signal, provides zero reference, filtering, analog to digital conversion, etc.

7.4.3 Types of Transducers

Common types of transducers used in hydraulic servo valves are listed below.

I. Linear Displacement Transducers They comprise one fixed and a moveable element connected by mechanical gearing or linkage to the output movement and the measurable value may range from few microns to a few meters. They are used in various fields like monitoring the position of machine tool elements, positioning the spool of a valve, etc.

The linear displacement transducers are available in the following types:
(a) Capacitive—mostly for small size applications
(b) Linear variable differential transformer (LVDT)
(c) Linear position potentiometer
(d) Pulse type

(i) Linear motion potentiometer A simple linear position transducer is a linear motion potentiometer as shown in Fig. 7.9. It consists of a DC power source with a metalized ceramic resistance element against which slides an axially electrical contact attached to an operating arm. When the slider is at one end of the resistance element, the output voltage is zero and the voltage increases in a linear relation, the maximum voltage being at the other end. The advantages of such a linear potentiometer are:

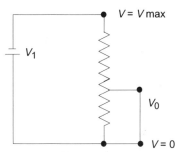

(a) They need simple and minimum electronic support
(b) They are easy to apply in the servo system
(c) They can easily interface with the analog-to-digital converter
(d) They can be interfaced with computers.

However the following disadvantages are also worth noting:

(a) They suffer from hysteresis effects from friction and electrical noise of slipring and are thus susceptible to quick wear out.
(b) Geometric misalignment of the slider to the resistance shaft poses accuracy problems of the output signal.

(ii) LVDT The problems associated with linear resistance potentiometers are overcome by linear variable differential transducers (LVDT). Compared to the linear position potentiometer, LVDT which is an electro-mechanical device, is found to be more attractive for fluid power systems and can be easily interfaced with the microcomputer system. A schematic diagram of such a device is shown in Fig. 7.10. In this device, one primary and two secondary coils are placed symmetrically on a cylinder inside which a free moving magnet is positioned to produce an electrical output proportional to the displacement of the core. Here the primary coil is fed with an external AC source which induces voltage of opposite polarity to the secondary coils. The magnetic core inside the coil assembly provides the path for the magnetic flux linking the coils which produces a phase related output voltage depending on the position of the core. The net output voltage of the transducer is the difference between these voltages which becomes zero when the core is at the center also called the 'null' position. The induced voltage increases in the coil towards which the core moves and decreases in the other coil which produces a differential voltage with linear motion of the core as depicted in Fig. 7.11. As may be seen from the diagram, this device indicates not only that the voltage is proportional to core position, but also the direction of

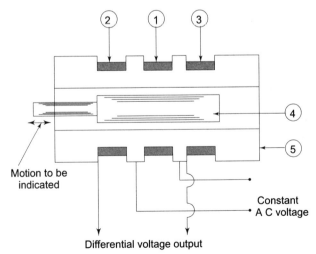

Fig. 7.10 *Construction of LVDT*
1. Primary coil 2. and 3. Secondary coil 4. Core 5. Insulating form

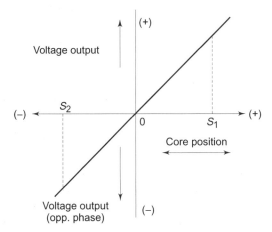

Fig. 7.11 *Output Voltage in opposite direction with change in core position*

phase change as the core moves from one side of the null to the other. This device is capable of indicating core movement from ± 125 μm to ± 50cm or more. A few notable advantages of LVDT are:
(a) Frictionless operation
(b) Long mechanical life
(c) Infinite resolution with its analog signal
(d) Extremely stable null position and high repeatability over a wide range of thermal conditions (from 150°C to 600°C)
(e) Capable of operating under hostile environmental condition
(f) Rugged construction
(g) Can withstand severe shocks and vibrations.

(iii) Linear Hall effect transducer   Here a small permanent magnet forms the non-contacting moving member inside a specially fabricated fixed member called a hall sensor, which generates a proportional linear output voltage by moving the constant flux magnet nearer to or farther from the hall sensor when activated by an appropriate DC voltage source. The output voltage is linearly proportional to flux but as flux is not usually linearly proportional to magnet's linear position, special arrangement is needed to develop a quasilinear output. They are sensitive to thermal effects but may provide low cost friction free measurement easy to operate from the DC power source.

(iv) Encoders   They are digital output position sensors which use an optical system and are capable of higher resolution than magnetic encoders which are also used in some cases. Encoders can be both absolute and incremental. Absolute encoders not only measure position change but also can have a memory about the starting position. Optical encoders are quite reliable and rugged in construction and are capable of providing high accuracy. But these sensors are quite costly and are used for sophisticated machine tools, robot controls, etc.

II. Pressure Transducers They comprise basically of one fixed element and a sensitive element which translates deformation of the element into an electrical signal.

The types of fixed elements used with pressure transducers are:

Flat or corrugated diaphragm, cap, bellows and common bourdon tube, whereas a potentiometer, differential transformer strain gauge, capacitive transducer, etc. are alternative types of sensitive elements.

Flat diaphragms and strain gauges are usually preferred because of their relative insensitivity to environmental noise as well as their higher accuracy and a wide range of response characteristics.

III. Angular Transducers They consist of a fixed unit or stator and a rotor, with provision for signalling the angular displacement between the two. The most commonly used angular transducers are:

(a) Rotary potentiometer available in single or multi-turn configuration. They have the same advantages and disadvantages as linear potentiometers described earlier.

(b) Rotary variable differential transformer (RVDT) is similar in function to a LVDT explained earlier. The output voltage varies linearly with the angular position of the shaft. The angular displacement is about 140° which maintains good linearity.

(c) Pulse type transducers have incremental or absolute, optical/magnetic encoders which are rotary devices and most suitable if the shaft angle measurement is desired in the digital form. In most common forms, the encoders and resolvers are coupled to ball screws or are used with a rack and pinion to indicate linear and angular position. An absolute encoder consists of a glass disc rotor supported on precision bearings and etched with a radial pattern of light and dark line segments, an optical system with a light source, aperture slits and photo sensitive devices.

IV. Speed Transducers They are of three main types:

(a) Linear Velocity Transducer (LVT) Here the feedback is obtained by transferring the linear motion into a rotary motion and measuring this motion, or by using the output electrical signal of a displacement transducer and expressing the same as a derivative of the displacement function.

They are basically a type of analog sensing devices capable of producing electrical output which is directly proportional to the linear displacement. A simple diagram of a LVT is shown in Fig. 7.12. It consists of:

- Double coil-assembly (1)
- Body casing (2)
- Stainless steel sleeve (3)
- A separable permanent Alnico magnet core (4).

LVT is an analog transducer capable of producing an output electrical signal proportional to the time rate of change of linear displacement of the Al-Ni-Co

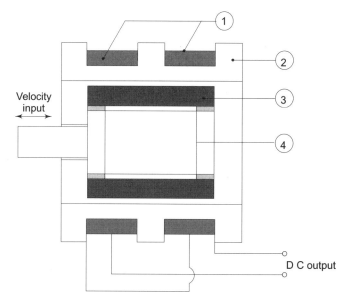

Fig. 7.12 *Cross-sectional view of a LVT*
1. Coils, 2. Body, 3. Sleeve, 4. Core

magnet core which is made to travel free axially inside the stainless steel sleeve around which the double coil assembly is fixed. The magnet core actually moves inside the sleeve with a distinct mechanical clearance thereby avoiding any contact with the coil assembly housing. As the pole or end of the Al-Ni-Co magnet travels axially inwards, a voltage is induced in each coil proportional to the rate of change of magnetic flux. As the flux of the permanent magnet is constant, the rate of change of flux in the coils is a linear function of velocity of the Al-Ni-Co magnet core. Therefore the magnitude of output voltage maintains a linear proportionality with the velocity of the magnet core.

The polarity of the output signal (i.e. the voltage generated) depends on the direction of travel of the magnet and orientation of its poles.

The advantages of such a device are:
1. Long mechanical life due to absence of friction.
2. Very high to low velocity input is possible (as high as 25 m/s). Minimum velocity is however very limited only by noise threshold of associated instrumentation.
3. Input velocity is limited only by winding insulation.
4. Infinite resolution of the device is possible.
5. They are highly sensitive.
6. They may not need on-site calibration.

(b) Pulse Speed Transducer This transducer used a pulse transducer when the frequency of the pulses is proportional to speed. This can be used to measure either linear or angular speed. This is a popular non-contacting rotary speed transducer and is known as variable reluctance (VR) tachometer generator. The constructional feature of such a transducer is shown in Fig. 7.13 from which it can be noted that it contains a coil surrounding a permanent magnet which is sensed by a rotating gear such that as each of the gear tooth approaches the magnet's pole, the reluctance of the flux path between the magnet and coil decreases inducing thereby an output voltage to the coil. Reluctance continues to build up as the tooth moves away from the pole and the voltage decreases generating a pulse. The pulses so generated are directly proportional to the gear rpm, which can be fed to a microprocessor unit for the feedback signal. The VR tachometer generator is cheap, reliable and easy to use.

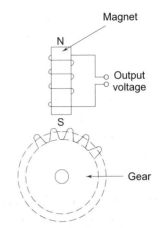

Fig. 7.13 *Variable reluctance tachometer generatar*

(c) Angular Speed Transducer Tachogenerator or alternator which provides an electrical signal and voltage proportional to the speed of rotation as described earlier is not direction sensitive. If one wishes to sense the direction or rotation also, a DC tachometer generator may be used.

7.4.4 Servo Amplifier

In most systems, a servo amplifier is required to drive the servo valve. An electrical feedback element senses the actual position of the load. Feedback is necessary in any control system that must provide accuracy and response. With the load at a given command position, the position servo will maintain the load position until the command signal changes.

As the command signal changes, the difference between the command and input signals increases. These two signals are compared, resulting in an error signal. The error signal is processed in the servo amplifier to drive the servo valve, displacing the load towards the new commanded position. As the load moves, its position is sensed by the feedback element, load motion reduces the difference between command and input.

When the error is close enough to zero, the load stops and the system reaches a condition of equilibrium. Any disturbance tending to move the load from this equilibrium position generates an error signal to restore the load; the system tends to remain at its commanded position until the input command is changed.

7.4.5 Types of Amplifiers

The signal output from most of the transducers is very weak. An amplifier is used to inject adequate energy to the signal to enable it to flow through to its desired

destination. Generally the following type of amplifiers are used in instrumentation systems:
(a) AC Amplifier
(b) DC Amplifier
(c) Chopper Amplifier
(d) Operational Amplifier
(e) Charge Amplifier
(f) Carrier Frequency Amplifier

Amplifiers suffer from two basic problems.
(a) Noise
(b) Drift

Noise is generated in almost all amplifiers due to random flow of electrons in tubes, transistors, resistors, etc. in the electronic devices. In order to ensure proper instrumentation, signal-noise ratio should be large. Drift is the main problem in DC amplifiers, and in other amplifiers this does not exist.

7.4.6 Sequence of Servo Control in Block Diagram

The sequence of control and feedback arrangement in a servo system is schematically illustrated in Fig. 7.14(a). In a closed loop position control system, the hydraulic flow from the valve is made to provide desired motion to a cylinder piston to move the load. The servo valve controls the flow in response to the electrical signal. The load position can be measured by use of an electric potentiometer as shown in Fig. 7.14 (b) or by use of electronic sensing devices as discussed earlier. The signal so obtained is fed back for comparison with a signal commensurate with the position desired. The difference between position command signal and piston feedback signal is the error signal. The error signal, if

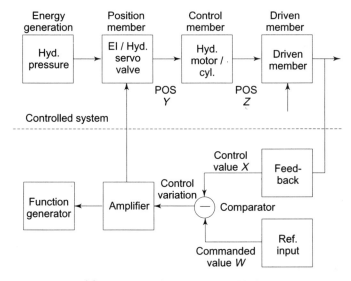

Fig. 7.14(a) *Sequence of servo control in block diagram*

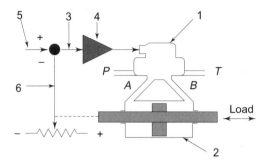

Fig. 7.14(b) *Use of potentiometer to detect load position*
[1-Servo valve, 2-Cylinder, 3-Error signal, 4-Servo amplifier,
5-Command signal, 6-Position feedback signal]

any, between the two is then amplified and fed to the valve actuating unit, e.g. torque motor, proportional solenoid, etc. as the case may be. The entire system contains generation of hydraulic energy, control of valves including the servo valve, control of the actuator (cylinder or motor), driven load, feedback sensing elements, amplifiers, etc. The block diagram indicating the sequence of logical operation of a position control unit with a hydraulic motor and its hydraulic circuit diagram is shown in Figs 7.15 (a) and (b) for easy understanding.

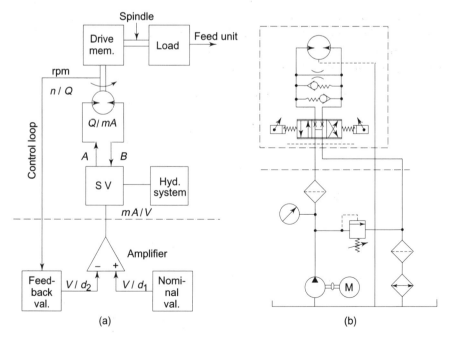

Fig. 7.15 *A position control system (a) block diagram (b) circuit diagram*

Hydraulic Servo Technique—Recent Trends 221

7.5 TORQUE MOTOR

A torque motor forms an integral part of a servo valve to produce the discrete positions of the valve. It is used to convert small electrical signals into equivalent limited mechanical motion. In simple terms it can be described as a magnetic bridge circuit as shown in Fig. 7.16 (a) and (b) where two permanent magnets form two arms providing static fluxes generated in the control coils. The armature takes up a normal position (shown in Fig. 7.16 (c)) which bridges the stator poles thus maintaining a state of magnetic balance in the four air gaps. If an electrical signal is now applied to the control coils, the armature will experience a magnetic imbalance providing a rotational motion to the armature as shown in Fig. 7.16(d). This will ensure a new position to the armature in order to rebalance the magnetic bridge circuit. The control coils are fed either from a single ended or differential

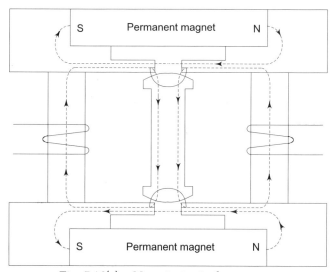

Fig. 7.16(a) *Magnetic circuit of a torque motor*

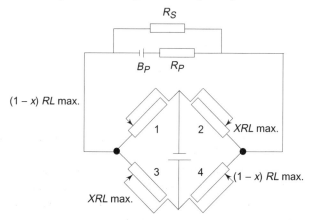

Fig. 7.16(b) *Torque motor bridge circuit*

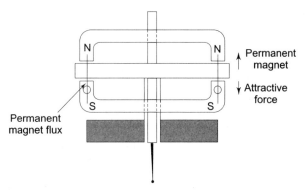

Fig. 7.16(c) *Magnetic balance maintained in armature of torque motor*

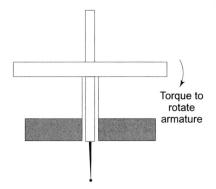

Fig. 7.16(d) *Armature starts rotation*

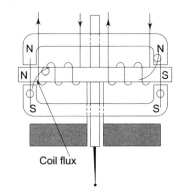

Fig. 7.16(e) *Torque motor internal*

source. The physical coil arrangement of the torque motor in a servo valve is shown schematically in Fig. 7.16 (e).

In a single ended system, the armature rotation is determined by the direction of current flow and its magnitude. In a differential application, the torque output and the deflection are approximately proportional to the difference between the currents flowing in each half of the control circuit.

In case of a single ended operation as the torque is directly proportional to the magnitude of current flowing, by regulation of the control current, the load can be positioned anywhere within the range of deflection of the armature. In a differential operation the torque motor can function as an error detector.

The output torque from a torque motor is a direct function of the control current. When the armature is deflected from its original position due to the applied external load, the unbalanced static flux created will produce a force opposing the applied load in order to restore the original state of magnetic balance till the armature takes up a new position depending upon the magnetic restraining force and applied load. The magnetic restraining force is measured as stiffness in terms of torque related to deflection. It is expressed in unit like gmcm/degree of deflection.

The coils on the torque motor are split into two sections correctly phasing and interconnecting them (as shown in Fig. 7.17) so that equal flux in each half of the control circuit can be ensured. For single ended applications terminal 1 and 3 are used and for differential applications terminal 2 is used as the center tap.

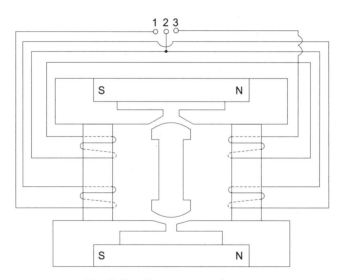

Fig. 7.17 *Torque motor coil arrangement*

7.6 TYPES OF SERVO VALVES

Electro-hydraulic servo systems were originally developed for aircrafts and missiles because of their excellent efficiency, compactness and high response capabilities.

Subsequently the use of servo systems has been extended to a great extent to include m/c tools and many other mobile equipments where a load needs accurate positioning. All electro-hydraulic servo systems include a servo valve which links the control electronics/electrical and hydraulic elements of a system as explained already. To provide correct specification and matching selection of a servo valve is one of the system designer's most important and possibly difficult task. The servo valve controls the: (a) dynamic response of the system, (b) the capability of rate of pressure, P flow, Q, (c) accuracy of the system operation. One can also say that the specific requirements of a system determine the configuration aspect of the servo valve. The system designer must recognize the relationship between the system requirement and the servo valve parameters that match such requirements.

7.6.1 Classification of Servo Valves

Servo valves may be: (I) single stage or (II) multistage. They may incorporate either torque motors or proportional solenoids, and may be spool valves, flapper valves, jet pipe valves or some other type of variable hydraulic control element. Let us study the schematic diagram of a few servo valves which are widely used in industrial hydraulic systems.

7.6.2 Single Stage Electro-hydraulic Servo Valve

A single stage spool type servo valve which is one of the most earlier developed servo valves is illustrated in Fig. 7.18. Here the torque motor directly actuates the sliding spool by means of a mechanical linkage opening the valve ports proportional to the electrical signal. Application of such single stage valves is limited due to force and travel limitation of the torque motor armature which results in lower flow capacity of the valve. However due to their simple and compact design, these valves were used with rotary actuators, e.g. hydraulic motors by means of manifold connection.

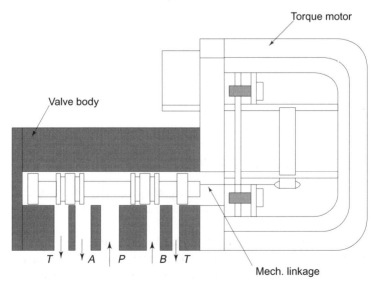

Fig. 7.18 *Single stage servo valve [Courtesy–Vickers Sperry, USA]*

7.6.3 Two-stage Electro-hydraulic Servo Valve

A schematic diagram of a two-stage torque motor actuated electro-hydraulic servo valve is shown in Fig. 7.19 (a). Here the first stage is a double nozzle flapper valve controlled by an electric torque motor and the second stage of the valve is a 4-way valve with a sliding spool whose output flow at a fixed valve pressure drop is proportional to the spool displacement from null. A cantilever feedback spring is attached to the armature flapper which controls two nozzles shown in Figs 7.19 (b), (c) and (d) illustrate how the feedback spring shifts the spool with the rotation of the armature which produces the torque if any

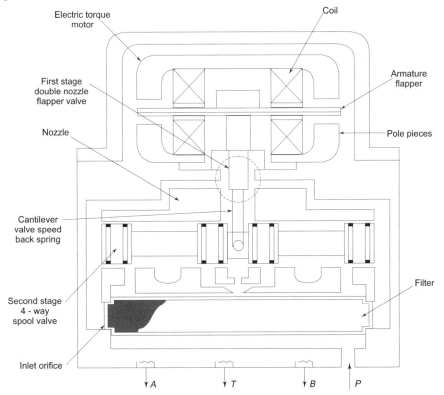

Fig. 7.19(a) *Two-stage electro-hydraulic servo valve*

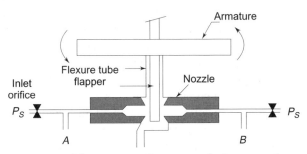

Fig. 7.19(b) *Armature flapper controls the two nozzles*

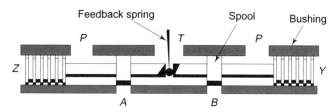

Fig. 7.19(c) *All ports closed*
[Courtesy-M/s Moog Inc., East Aurora, NY, USA]

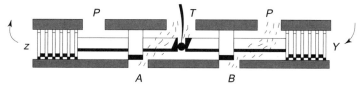

Fig. 7.19(d) *P opens to B and A opens to T*

unbalanced force acts in the torque motor and shifts the feedback spring to close either of the nozzles thus creating a variable force against the spool of the second stage valve.

The servo valve is connected to system supply and return by ports on the mounting face of the servo valve. The signal fluid to control the spool is filtered, and supplied through two fixed inlet throttles to the two variable orifices created by the nozzle-flapper arrangement. The fluid pressure developed between the fixed and variable orifices exerts a force on both ends of the second stage spool which shifts either way and the pressure fluid is ported to A or B to control the cylinder/hydraulic motor movement.

In some two-stage servo valves instead of a torque motor, the flapper is actuated by pneumatic energy as may be seen in Fig. 7.20. Apart from actuating the flapper by compressed air, the function of the two-stage valve is similar to that of a torque motor operated valve discussed above.

7.6.4 Operation of a 2-stage Servo Valve

Torque motor operated electro-hydraulic servo valve is particularly recommended in appliances where one intends to maintain two discrete extreme positions of the load. This valve consists of a polarized torque motor having restricted motion (mostly less than 10). If no input signal is fed to the torque motor, the sizes of the two variable orifices remain equal. Thus the pressure on both ends of the spool are also equal and the spool does not move preventing flow to either of the control ports A and B. As soon as a command signal is applied to the torque motor, it induces a magnetic charge resulting in depletion of the armature about its flexure tube support. The depletion of the armature-flapper unit increases or decreases the size of the orifice of the nozzle on one side and this motion creates variation of

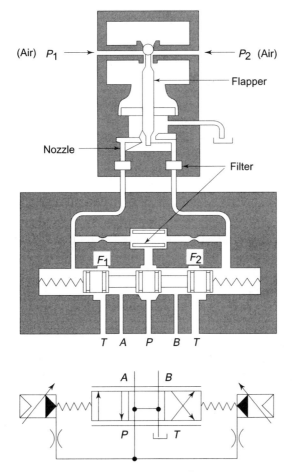

Fig. 7.20 *Servo valve with flapper actuation*

oil flow to the orifices resulting in a pressure imbalance and hence differential force acts on the two ends of the main valve spool shifting the spool which results the feedback spring to exert a torque on the armature/flapper proportional to the spool position. This torque opposes the torque exerted by the torque motor and increases if the spool shifts further until a position of equilibrium is reached when the two torques cancel each other. When the armature/flapper is centered, the variable orifices are again equal and there is no unbalanced force to move the spool from its new position. The spool is now positioned to supply controlled flow to one of the control ports, while the other control port is connected to the tank. The explanation above seems to be quite simple though a lot of complexity both in construction and operation are involved. In reality, servo valves are really complex and quite expensive and hence need utmost care in handling and servicing.

A servo amplifier is required to drive the servo valve. The actual position of the load is sensed by various feedback elements as discussed earlier. The feedback element must be accurate and should have high response so that the actual position of the load as per the input signal can be compared to the commanded position to have the correct error signal to reposition the cylinder.

7.6.5 Solenoid Control

A solenoid consists of a coil surrounding a moveable iron core which is pulled to the operating position with respect to the coil when the coil is energized. Thus a solenoid is an electromagnetic mechanical transducer which converts an electrical signal into a mechanical output force. In a normal pneumatic or hydraulic Direction Control Valve (DCV), the solenoid applies a mechanical force resulting from an electrical signal to move the spool to its position. A solenoid is either a push or pull type device. Basically it pulls the armature into the coil when the coil is energized, thus shifting the spool. If pushing motion is required a push rod can be added to the inner end of the armature. In both cases when the coil is deenergized a compression spring returns the spool and armature to their initial position. Figure 7.21(a) illustrates a normal solenoid. Figure 7.21(b) shows graphically the quantum of magnetic force a solenoid has to generate against the total resistances the spool experiences. As stated in earlier chapters, solenoids can be either AC or DC and their switching characteristics are graphically illustrated in Figs 7.21(c), (d) and (e).

Chief advantages of solenoid, (1) versatility, (2) convenience (3) remote controlling possibilities. The valve can be mounted in its most convenient and efficient place w.r.t. piping connection, power supply and output actuators while the input to the solenoid can be at a control station. Solenoid actuation eliminates mechanical cables or push pull rods which can be damaged or corroded creating mechanical troubles. Electrical signal may be given by pressing a button, flipping a switch or this can be fully automatic as a function of a pre-programmed device—offering many design options and wide design freedom. Two types of

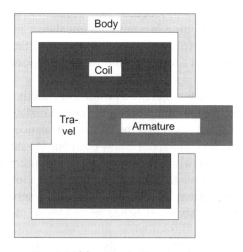

Fig. 7.21(a) *Sketch of a normal magnet*

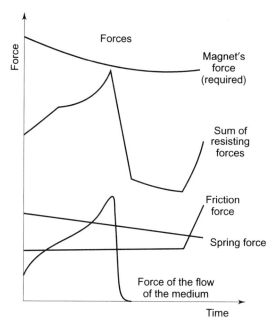

Fig. 7.21(b) *Magnetic force to overcome resistance*

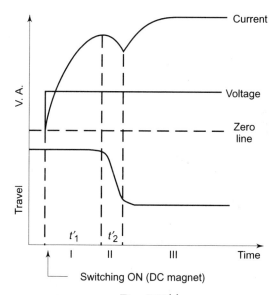

Fig. 7.21(c)

solenoids used are on-off and proportional. A vast magnitude of solenoids used are of on-off type. These in turn are divided into air gap and wet armature type. An air space separates the solenoid from the pneumatic or hydraulic circuit in the air-gap solenoid. They are specified for simple and complex circuits usually bolted into

230 *Oil Hydraulic Systems: Principles and Maintenance*

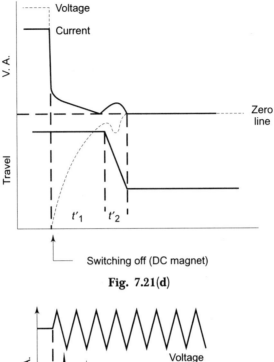

Fig. 7.21(d)

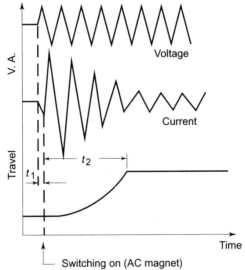

Fig. 7.21(e)   *AC magnet characteristics*

the valve body and are completely isolated by a dynamic seal from pneumatic or hydraulic circuit.

One drawback of air gap solenoids is the possibility that the dynamic seal may eventually leak and oil may seep into the solenoid cavity. When oil was relatively cheap and there was little concern about environmental pollution and dumping of waste fluid into sewers, users used to replace the seal when convenient. However, with present day concerns one has to be more careful about oil leakage.

In a wet armature solenoid, all moving parts of the solenoid operate in the hydraulic fluid enclosed in the solenoid chamber. An encapsulated coil and frame assembly form the outer part of the solenoid. Inside, a core tube acts as a container that retains the oil in the solenoid. The armature is enshimed inside the pin which shifts the spool. As the wet armature solenoid is an integral part of the fulid circuit, no dynamic seals are needed to seal the pushrod. A wet armature solenoid reduces the chance of oil leakage. In some m/cs there may be a good number of solenoid actuated valves. With a wet armature for such valves one can reduce at least one source of oil leakage and thus help reduce chance of pollution.

Because there is no dynamic seal to wear out, solenoids last longer. Because wet armature solenoids are cushioned by oil, armature deceleration at the end of each stroke is less abrupt compared to air gap armatures. They are more quick to react.

On the other hand air gap solenoids are more efficient electrically because less current is required to operate them. A wet armature solenoid valve requires 1.5 times or more electrical power to shift the valve spool than an air gap solenoid.

7.6.6 Proportional Solenoid

As the name suggests, a proportional solenoid allows movement of the spool in proportion to the input current fed to it.

In this device, armature position within the solenoid can be controlled by the input signal to the solenoid. The armature interacts with the hydraulic valve spool which can be shifted by the armature in a controllable manner allowing for accurate proportional control of the spool position providing precise control of fluid flow and appropriate load movement.

Proportional solenoids can control the main spool of power valves rated for flows to 30 lpm at 200 bar or even more. The hydraulic valve may have to be modified to accept the mounting of the proportional solenoid. This can be used in open loop or closed loop operation.

The simplest and least expensive proportional valve set-up is capable to control flow or pressure fluid through a spring return, 2-way valve with open loop electronics. This type of system could be used to control flow in a system, so as to, for example, control the rotating speed of a hydraulic motor.

After this solenoid is energized by the common signal, the armature moves within the solenoid which in turn shifts the spool inside the valve. The rated flow is a function of the distance the spool is shifted to. Increasing current and consequently armature spool movement increases the flow and vice versa. Further refinement of servo valve function has been made with addition of electronic control technique.

For good system performance, control flow should be proportional to the input current. If a flow curve with flow vs current is drawn (see Fig. 7.26 in Section 7.8) to show the effect of valve hysteresis during a complete cycle of the performance of the proportional valve, the locus of mid-point of the flow curve, called the normal flow curve, is used to evaluate servo valve behavior. The closer the curve approaches a straight line, the more desirable is the normal flow curve indicating better system performance.

The slope of the normal curve is defined as the normal flow gain of a servo valve. When examining servo valve flow gain, the system designer is primarily interested in the null region and the normal flow region. The null region is that area of low input current where flow is effected by the spool edges and sleeve openings (spool lap), beyond that is the normal flow region. Because of spool lap effects, flow gains in the two regions are originally not equal. Null flow gain can vary from 50% to 200% of normal flow gain.

Proportional solenoïd operated EHS valves are recommended to provide unidirectional control operational flow. In the case of torque motors the direction of rotation is effected by the rate of current through the coil. But in proportional solenoids the direction of force does not depend upon the direction of current. Hence they need two solenoids. The associated electronic control is however comparatively complex. Solenoid operated servo valves do not require any specific filtration but the power to operate this solenoid is 15 to 25 watts in general.

PW Modulator (PWM) The pulse width modulation technique is nowadays used to operate proportional valves. PWM allows a large amount of power to be delivered to the load (valve coil) while consuming the minimum amount of internal power in the final output amplifier stage. A judicious adjustment of on-time to off-time ratio of the valve with high PWM frequency makes the valve respond well to the average value of control current. A typical PWM circuit is shown in Fig. 7.22.

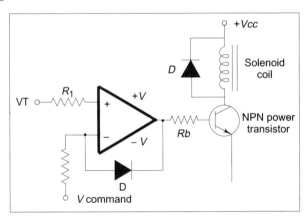

Fig. 7.22 *A PWM circuit*

7.6.7 Position, Velocity and Force Control

A load positioning servo can be created by using a servo valve, load actuator, position feedback transducer and a servo amplifier. A schematic circuit diagram is shown in Fig. 7.23 (a).

A similar circuit is illustrated in Fig. 7.23 (b) for a velocity control system and a force control system is illustrated in Fig. 7.23 (c).

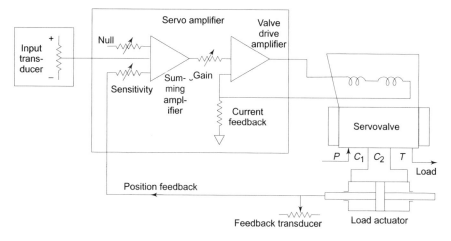

Fig. 7.23(a) *Position servo*

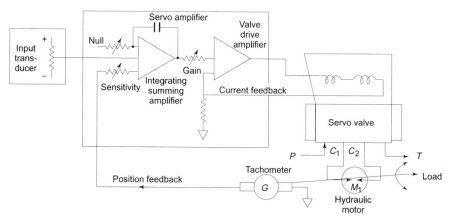

Fig. 7.23(b) *Schematic circuit diagram for a velocity servo*

7.6.8 Signal Accuracy and Dead Band Region

Integrity and accuracy of feedback signal in servo valves are very important especially when an electronic control unit is used in such systems. Signal integrity of feedback and command signals are maintained by signal conditioning. Special protection devices are used to make it immune to Radio Frequency Induction (RFI) and Electro Magnetic Induction (EMI). The error amplifier houses the dead band eliminator and maximum spool shift adjustment. The dead band is defined as that portion of the valve actuation in which the input current causes no output. Fig. 7.24 illustrates the dead band characteristics of proportional and servo valves. Large dead band is disadvantageous for smooth functioning of the valve and hence needs to be eliminated by means of a dead band eliminator which is an electronic means to eliminate the dead band of such valves.

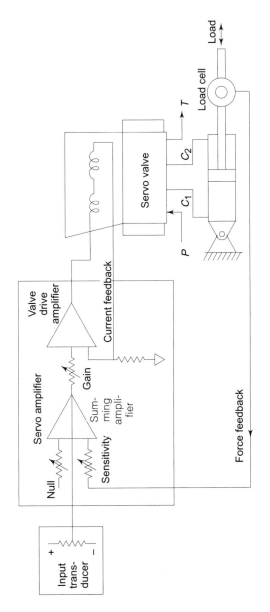

Fig. 7.23(c) *Force feedback circuit*

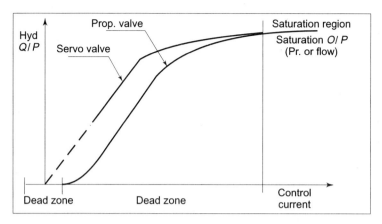

Fig. 7.24 *Dead band in proportional and servo valves*

7.7 SPECIAL SERVO VALVE FEATURES

Choice of servo valves is influenced by the degree of control required over direction, flow-rate, pressure and degree of accuracy required.

Torque motor operated electro-hydraulic servo valve is particularly recommended in appliances where it is necessary to maintain two discrete and accurate extreme positions of hydraulic actuator or motor. The valve has to be therefore very much responsive which is reflected in its mechanical and physical parameters and its manufacturing accuracy.

In nozzle flapper valves the diameter of the nozzle is about 500 µm and the diameter of the fixed orifice is about 140 µm. The clearance between the nozzle and the flapper is only 25 to 50 µm or even less. Particles larger than this will cause jamming of the flapper, silting and consequent clogging of the nozzle or a transient movement of the load until the particle is cleaned. To prevent this internal filters of very fine size are used immediately upstream of each orifice and nozzle.

These servo valves provide excellent performance and can meet the hardest requirements with regard to rapidity of response, excellent positional accuracy, good null point characteristics, etc. In order to avoid malfunction and to give a long life without wear, an efficient filtering system (5 to 10 µm) of the hydraulic fluid must be provided. The electric power required for operating the torque motor is very less. (0.1 to 0.2 W). For a proportional solenoid the power requirement is above 10 W.

Servo valves are very complex and their manufacturing needs lot of care and fine technique. Therefore they are quite expensive as they are to be made with fine machining so that accurate internal dimensions as given in Table 7.2 is posible to achieve.

Table 7.2 Typical Servo Valve Dimensions

Sl. no.	Dimension of Servo valve	Size in μm
1.	Air gap spacing (each gap)	250 to 375
2.	Maximum armature motion in air gap	±75
3.	Inlet orifice diameter	250 to 500
4.	Maximum opening of nozzle flapper	60 to 75
5.	Drain orifice diameter	250 to 375
6.	Spool stroke	250 to 500
7.	Radial clearance between spool and bushing	7.5 to 15
8.	Radial clearance between body and bushing	20

However, the above dimensions are not conclusive and may be referred to as an example to understand the complexity of the manufacturing process involved. Similarly special materials are used to manufacture various parts of a servo valve as shown in Table 7.3.

Table 7.3 Servo Valve Materials

Component part	Material
Body, end caps and accessories	17–4 PH stainless steel
spool and bushing	400–C stainless steel
Flexure tube	Beryllium copper
Pole pieces and armature	4750 Ni-iron steel
magnets	Alnico
Feedback wire	17–7 stainless steel
Torque motor cover	Anodized aluminium alloy.

Table 7.2 shows typical critical internal dimensions and clearances which give some idea of the precision machining required to produce a servo valve.

Table 7.3 lists various materials used in a servo valve. The materials were selected on the basis of wear, strength, magnetic characteristics and other requirements. Also the diverse types of materials used indicate the possibility of adverse thermal effects that may be caused by differential expansion of cache.

7.7.1 Design Features of Servo Valve

The following design features of a servo valve are of importance:
1. Should have a rugged, stainless steel body.
2. One-piece bushing with EDM prepared flow slots.
3. Spool and bushing diametral tolerances held within 20 μm.
4. 20 μm nominal filter (35 μm absolute) for pilot flow essential.
5. Symmetrical double nozzle hydraulic amplifier provides consistent performance over a wide temperature range.

6. Hydraulic amplifier integrated into the main valve eliminates several O-rings thus reducing seal friction and down time due to seal failure.
7. Torque motor is coupled in an environmentally sealed compartment.
8. Frictionless, flexure tube supported armature/flapper isolates hydraulic fluid from the torque motor.
9. Should be designed to minimise external magnetic field and to reduce sensitivity to external magnetic materials or fields.
10. Mechanical feed back with simple cantilever spring:
    (i) Rolling ball contact minimizes wear
    (ii) Feed back removable without damage to valve.

The spool is designed with certain special features such as:
1. 3-way spools having single control port.
2. Spool stops for limiting maximum flow-limit of flow is held to $\pm 5\%$.
3. Special spool null cuts:
    (i) Prescribed amounts of underlap or overlap, symmetrical or unsymmetrical.
    (ii) Graphite impregnation of spool for pneumatic and water application.
4. Non-linear slot width
    (i) Different flow gain for each valve polarity as used with some 3-way actuators.
    (ii) Stepped with slots for dual flow again.
    (iii) Spool driving force should be as minimum as possible.

7.7.2 Analog/Digital Control

In case of electronically controlled servo valves, the proportional signal may be analog or digital. A digital control is better because of the following reasons:
(a) Far greater resolution is possible.
(b) Utilize high threshold.
(c) Complete immunity from noise.
(d) Good response—of the order of 200 Hz or better.
(e) Current demand is 40–50 milli amp.
(f) Threshold sensitivity is 0.4 to 0.5 milli amp.

Servo valves and system requirements
While selecting a servo valve, one may take care of the following points
(a) Required positional accuracy
(b) System dynamic response
(c) Stability
(d) Rated capability
(e) Energy consumption
(f) Shock and vibration
(g) Thermal environment
(h) Oil contamination level
(i) Load acceleration rate

(j) Cost and life expectancy
(k) Physical size and weight

7.7.3 Control Flow vs Input Current in a Servo Valve

Looking at a servo valve as a component, its input is an electric current and its output is hydraulic fluid flow at the control ports, called control flow. Note that this is not the total flow supplied to the valve. For the example valve discussed earlier, some of the flow supplied to the valve goes to the first stage to position the second stage spool. The first stage flow is ported to system return and does not travel through the control ports. In addition, some fluid leaks through the metering edges of the second stage spool null and between the spool sleeve and body. Only the flow actually delivered to the control ports is defined as control flow. The relationship between control flow and spool uncovered slot area (proportional to spool position) is expressed as:

$$Q = CA\sqrt{2\Delta P/\rho} \qquad (1)$$

where,
Q = output flow
A = uncovered slot area
ρ = mass density
C = discharge coefficient
ΔP = valve pressure drop (supply pressure minus return pressure minus load pressure drop)

This indicates that for a given spool position, a servo valve represents a constant orifice opening. For a servovalve, the spool orifices can be considered to be sharp-edged. For sharp-edged orifices and turbulent flow, $C = 0.611$. For petroleum based fluids equation (1) can be written as:

$$Q = 100\, A\sqrt{\Delta P} \qquad (2)$$

Clearly servo valve control flow is proportional to the square root of the pressure drop across the valve. Since the first stage controls the second stage spool position, and since flapper position is proportional to input current, the relationship between input current and control flow is expressed as:

$$Q = Ki\sqrt{\Delta P} \qquad (3)$$

where
Q = control flow
K = valve sizing constant
i = input current

Control flow characteristics may change with fluid temperature $A + 30$ to $35°C$ temperature rise may increase control flow by as much as 3%.

7.8 TERMS USED IN SERVO TECHNOLOGY

Flow gain It is the nominal relationship of control flow to input current expressed as gm/milliamperes.

Linearity It is the maximum deviation of control flow from the best straight line of flow gain. It is expressed as % of rated current.

Threshold The increment of input current required to produce a change in valve output is termed threshold. Threshold is affected by
(a) Temperature
(b) Pressure
(c) Spool lap
(d) Oil viscosity
(e) Contamination level

It is actually to the measure of static friction present in the second stage of servo valve.

Lap It is the relative axial positional deviation in a sliding spool valve between the fixed and the movable flow making edges with the spool at null. Lap is measured as the total separation at zero flow of straight line extensions of the nearly straight portions of the flow curve, drawn separately for each polarity. It is expressed as a percentage of current. Underlap of spool increases flow gain at null, reduces valve pressure gain at null and increases valve null leakage. Overlap reduces flow gain at null, reduces null leakage flow and degrades pressure gain.

Null It is the relational condition between the spool and valve port where the valve supplies no control flow at zero load pressure drop.

Null shift The change in null bias resulting from changes in operating conditions or environment is called null shift.

Null bias In an ideal servo valve, control flow is directly proportional to input current. At $i = 0$, control flow is zero. But in practice input current required for zero control flow is not zero. The input current required to bring the valve to null excluding the effects of valve hysteresis is termed as null bias.

Frequency response The relationship of no-load control flow to input current when the current is made to vary sinusoidally at a constant amplitude over a range of frequency response is expressed by the amplitude rate (in decibels, or db), and phase angle over a specific frequency range.

Pressure gain Load fluctuation or disturbances may tend to alter the desired load position. Due to fluid compressibility, a valve with high pressure gain can cause the system to respond rapidly to such load disturbances and thus improve the positional accuracy under disturbances. Pressure gain is influenced by working temperature, contamination, spool lap and age of valve. Pressure gain is higher for valves with zero or over lapped valve compared to underlapped valve. Contaminated fluid may cause more erosion of the valve chamber, may cause rounding of metering edges and decrease the pressure gain. Spool lap plays an important role on the pressure gain characteristics. The underlapped valves produce an intermediate pressure to the actuator at null compared to overlapped valves which may be understood from the graphical illustrations shown in Fig. 7.25. A valve with underlap has high null-flow gain but avoids hydraulic shock.

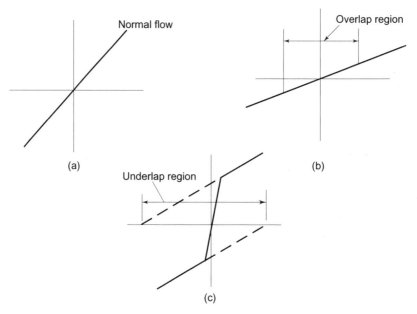

Fig. 7.25 *Flow characteristic of servo valves at null region*

Hysteresis We have already seen in Section 7.6.6 the effect of hysteresis on performance of a servo valve. Hysteresis is defined as the difference in valve input current which may be fed in order to produce the same output of the valve when it is slowly cycled between plus and minus rated current. It is expressed as a percentage of rated current as shown in Fig. 7.26 graphically. The rated current is the specified input current of either polarity to produce the specified flow rate. Maximum hysteresis under normal condition is > 3% and may increase to 4% at sub-zero temperatures. For special high temperature valves servo valve hysteresis

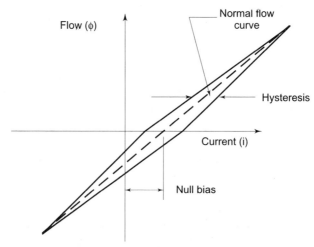

Fig. 7.26 *Servo valve hysteresis*

limit is > 4% (when temperature is 150°C). The most important factors which influence the servo valve hysteresis are:
(a) Temperature; (b) Electromagnetic effect; and (c) Contamination level.

Dither Dither is 'low amplitude-high frequency' signal superimposed on input signal to improve threshold. Servo valve performance normally measured without dither. Dither current may be applied to some types of servo valves. Dither usually improves both the servo valve and actuator threshold and increases spool null leakage.

Dither characteristics
   Usually 100 to 400 Hz; Peak-to-peak dither amplitude may be as high as (+) 20%.

Review Questions

1. Differentiate between open and closed loop circuits.
2. What is meant by feedback? State the methods of feedback in a hydraulic circuit.
3. How is feedback arranged in a copy turning lathe?
4. (a) What is a transducer? What is its function in a servo system?
   (b) Classify transducers.
5. What is a torque motor? Describe its function in a servo valve.
6. Differentiate between single-stage and multistage servo valves.
7. What is a proportional solenoid? How does it differ from an ordinary solenoid?
8. What specific points should be taken care of while selecting a servo valve.
9. Define: (a) Gain (b) Null bias (c) Lap (d) Null shift (d) Hysteresis

References

7.1 Publication of M/s Moog Inc., Industrial Division, East Aurora, New York 14052, USA.
7.2 Heyen, Gerhard "Position Control with Low Cost Proportional Valves", *O + P Oil Hydraulic & Pneumatics—European Journal for Fluid Power*, Feb. 1992.
7.3 Tanaka, Takao and Shichisaburo Shibuya, "Proportional Electrohydraulic Remote Control Valves", Yuken Kogyo Co., Japan.
7.4 Herceg, E E "Transducers", *Hydraulics & Pneumatics*, August, 1983.
7.5 *Oil-Hydraulics*, M/s. Sperry Vickers, USA.
7.6 Ing Feuser A., "System Applications in Servohydraulic Drive Technology", *Oel-Hydraulik und Pneumatik*, Sept. 1989, Germany.
7.7 Ing H. Hesse, Digital Electronic Hoist Control for Agricultural Tractors, Oel-Hydraulik und Pneumatik, Germany, Nov. 1991.
7.8 Findeisen F. und. D., *Oil Hydraulic in Theorie und Anwendungen*, Schweizer Verlagshaus, Zurich, Switzerland, 1968, p. 193.
7.9 *Taschen buch*, M/s Herion Werke KG, Stuttgart, Germany, 1969, p. 260-261.

8

Linear Actuators

RELIABILITY of a hydraulic system not only depends on the system design but also on factors such as component design and manufacture and their correct choice. This is true while selecting cylinders too. A good number of hydraulic system failures may be attributed to defects in cylinder design.

As the main source of transmission of power, a hydraulic cylinder should have optimum reliability. A correct cylinder in a hydraulic system contributes to:

1. Optimize system maintainability
2. Ensure minimum down time
3. Ease the process of repairing and trouble shooting
4. Maximize the rapidity of recommissioning of the machine and plant
5. Ensure maximum work accuracy
6. Maintain least economic liability and financial losses.

Though maintainability of a system is very much influenced by the end user, equipment reliability should essentially be built into by proper design and process of systematic manufacturing. It is important for designers and manufacturers of hydraulic systems and components to understand that a faulty system or a faulty component may cost him very dearly in the long run, by way of loss of reputation and image, loss of market and consumer and overall financial losses.

Today, in spite of rapid growth and development in metallurgy, production techniques and manufacturing methods enable cylinders to be designed and produced with a predictable performance, it is not uncommon that we come across numerous cases of failures due to faulty cylinders.

Recurrence of such problems could probably be reduced at the design stage. Special care and precaution is taken to specify explicitly the need and requirement of cylinder characteristics application-wise instead of a generalized view which may prove unrealistic in some cases. Hence a system designer should study case by case the technical compatibility of the system and the cylinder for use in the system. This will certainly pay him rich dividends at the end.

8.1 HYDRAULIC CYLINDERS

Various types of actuators are used in hydraulic systems, e.g. hydraulic cylinders, motors, etc. A cylinder is a device which converts fluid power into linear mechanical force and motion. It usually consists of a moveable element such as a piston and piston rod, plunger or ram operating within a cylinder bore. In contrast to a cylinder hydraulic motor provides rotational motion and is used in a hydraulic system for a variety of applications where rotary movement is the need.

A fluid power cylinder is a linear actuator which is most useful and effective in converting fluid energy to an output force in a linear direction for performing work such as pulling or pushing in a variety of engineering applications such as in machine tools and other industrial machinery, earth moving equipment, construction equipment and space applications. To gain a better understanding, a fluid power engineer needs to study the principle of selection, operation and maintenance of fluid power cylinders.

8.1.1 Types of Cylinders

Functionally cylinders are classified as:
1. Single acting cylinders
2. Double acting cylinders.

In single acting cylinders, the oil pressure is fed only on one side of the cylinder either during extension or retraction. When the oil pressure is cut off, these cylinders return to the normal position either by a spring or by an external load. The cylinders which are actuated by an external load are essentially mounted in the vertical position since the return is effected through the force of gravity.

In the case of spring return cylinders, the spring may either be external or internal. Spring size is a function of load and desired operating speed and hence needs careful and judicious selection for optimization.

Double acting cylinders are operated by applying oil pressure to the cylinder in both directions. Due to inherent mechanical problems associated with the spring, single acting cylinders with spring return are not used in applications using larger stroke lengths. As a result cylinders in various applications are mostly double acting type only.

Double acting cylinders may be either single rod ended or double rod ended type. The single rod end cylinders as shown in Fig. 8.1 have pistons connected to a smaller diameter piston rod. For a given pressure single rod end cylinders exert greater force when extending than when retracting. When it is required to exert equal forces in both directions, the double rod end cylinder is used. However, the maximum force of the cylinder for a given tube size is smaller than the single rod end type.

Apart from the above commonly used functional designs, there are special cylinders such as:
1. Plunger or ram cylinders
2. Telescoping cylinders

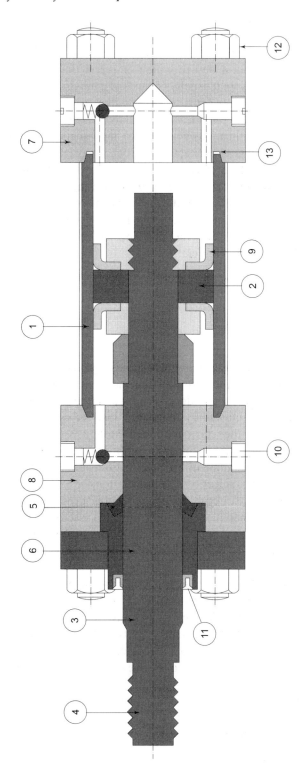

Fig. 8.1 Single ended double acting cylinder

3. Cable cylinders
4. Diaphragm cylinders
5. Bellows cylinders
6. Tandem cylinders
7. Duplex cylinders and
8. Rotary cylinders, etc.

The piston and piston rod of a ram cylinder are of almost equal size and therefore can be used as a single acting cylinder. Ram cylinders are used in a vertical position so that the load on the cylinder can retract when the oil supply is stopped. Cylinders commonly used as a lift in automobile service stations is a classic example of the application of such cylinders.

Telescoping provides long working strokes of cylinders in a short retracted envelope. Telescoping cylinders find use in mobile applications such as tilting of truck dump bodies and fork lift trucks, hydraulic cranes, etc.

Cable cylinders are double acting cylinders which can be powered either pneumatically or hydraulically. These are normally used in applications requiring relatively long strokes and moderate forces and can be operated in limited spaces. The advantage of these cylinders is that the load is attached to cables which run over pulleys at the end of the tube barrel, thereby the load can be moved almost over the entire length of the tube while the cylinder is only slightly longer than the tube. Diaphragm cylinders are usually used in pneumatic applications. They are either of the rolling diaphragm or flat diaphragm-type and are essentially single acting and spring return type with the additional advantage that they have very low break-out friction with absolute zero leak across the piston. For very low strokes and low force application in sensitive pneumatic control systems one can use bellows cylinder. The pressure and the spring rate of the bellows determine the amount of tension and contraction. Cylinders with metallic bellows may be used for basic servo-control systems since metal bellows have a linear spring rate. In tandem cylinders used commonly in hydraulic and pneumatic systems, two cylinders are mounted in line with the pistons connected to a common piston rod in order to multiply the force in a limited lateral space. However, they occupy more linear space. Another modification of tandem cylinders is the duplex cylinder which is used for multiple position operation.

8.1.2 Classification According to Construction

Depending on their use fluid power cylinders are available in a wide variety of styles and constructions. The following are the basic construction styles in industrial power cylinders:

(i) Tie rod cylinders.
(ii) Mill type cylinders.
(iii) One-piece welded cylinders.
(iv) Threaded head cylinders.

(i) Tie rod cylinders This is the most common style of construction and is extensively used in industrial applications. These cylinders find use in medium

and heavy duty applications where shock loads are experienced. The tie rods are capable of taking impact loads. The problem of elongation of tie rods under pressure in long stroke cylinders is overcome by pre-stressing the rods and tube be applying a higher torque depending on function. This eliminates leakage through the tube end joints under pressure due to elongation.

(ii) Mill type cylinders The construction of these heavy duty cylinders is similar to that in tie rod construction except that the length to diameter ratio of the fastening bolts seldom exceeds the ratio 2:1. Fixing flanges are provided at the two ends of the tube through which these short bolts are fixed. This type of construction is generally preferred in steel mills and metal processing industries where operating conditions are severe and hence the cylinder tube and bearings require extra ruggedness. The most reliable and strongest way of fixing the flanges to the tube is by welding prior to finishing. However circlips and other methods of retainers such as wires, wedges, etc. are also used in some designs.

(iii) One-piece welded construction In this type of construction, the body is either cast integrally or the ends are welded or crimped to the tube. These are simple and inexpensive in design and are non-serviceable. They are used in mobile and farm equipment where the duty cycle is moderate. The welding of the ends to the tube after finishing may result in distortion of the tube, though welding of the cap end may eliminate seals.

(iv) Threaded head cylinders In this type of construction, the ends are screwed on to the tube by threading outside or inside. However this weakens the tube and problems such as perfect concentricity of threads with the bore of the cylinder are inherent. Also if the threading is done after the bore is finished, it might result in ovalities and taper of the bore because of pressure exerted while clamping the job. In some modified designs retaining rings, wires or keys hold sliding ends in position. In these designs, leakage can be a severe problem unless proper care is taken.

8.2 CONSTRUCTION OF CYLINDERS

A double acting hydraulic cylinder consists of the following major parts as shown in Fig. 8.1.
1. Cylinder body or tube
2. Piston
3. Piston rod
4. Threaded end of piston rod (rolled finish).
5. Bush seal
6. Bush
7. Front end cover or head end
8. Rear end cover or cap end
9. Piston seal
10. Cushioning assembly
11. Rod wiper

12. Tie-rod fastener
13. Circular groove for wire to join tube and end cover.

For light duty cylinders, the tube or barrel is fabricated from hard drawn seamless tubes of Al-alloys or steel with pistons invariably made of aluminium alloy castings. However for a medium duty cylinder, one uses a stronger cylinder tube with end covers made from compatible materials and suitably secured by fasteners. The barrel or cylinder bore is chrome-plated to provide a smooth corrosion and scratch resistant surface. Light and medium duty cylinders may be thin-walled but heavy duty cylinders are thick-walled with hard drawn seamless tubes made of steel, stainless steel, titanium, etc. Figures 8.2 (a) to (d) illustrate various methods of fastening the tube or barrel with the end covers. The end covers can be joined with the tube by screw (if the tube wall thickness is structurally adequate) as shown in Fig. 8.2 (a). They can also be joined by welding or tie rod [(Figs 8.2 (b), (c) and 8.2(d)]. The tube and cover assembly can be strengthened by shrunk fitted stainless wire on machined semi-circular groove on the tube ring as illustrated in Fig. 8.2 (d). More details about different type of tube materials used in hydraulic cylinders are given in Section 8.2.1.

From the maintenance point of view, materials used for all the important parts of the cylinder and their features need very careful selection which is discussed here in subsequent paragraphs.

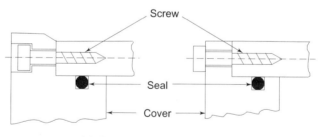

(a) Screwed to the tube wall

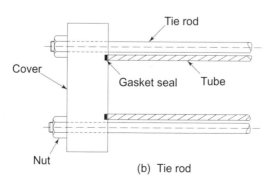

(b) Tie rod

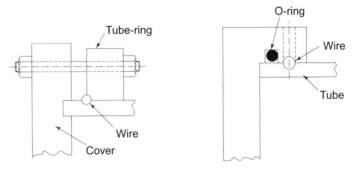

(c) Tube-ring with stainless steel wires on machined semi circular groove on tube-ring

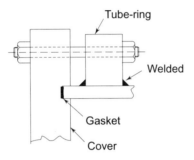

(d) Welded tube ring to cover

Fig. 8.2 *Assembly of cylinder body*

8.2.1 Cylinder Material

(i) Cylinder barrel Both cast and drawn materials are used for the cylinder tube. Material stress and manufacturing feasibility are two important criteria for material selection. Ordinary grey C.I. tubes are found to suit a stress value of 300 bar, but special heavy duty C.I. and mechanite castings may give higher stress values of 500 to 600 bar. Brass, bronze and aluminium alloy castings find their use in certain applications. While for brass and bronze the maximum permissible stress is 400 bar, for aluminium alloy the maximum permissible stress is 550 bar.

But the highest stress value is possible with cold drawn deep polished low carbon steel (up to 1250 bar).

Except metal castings, cold drawn deep polished low carbon steel or stainless steel, tungsten alloy, titanium and aluminium alloys finely ground and polished with tolerance range of H_7 or H_8 and sometimes H_9 are also used. Bore is carburized or induction hardened, finished with chrome plating and polished. The finishing tolerance for various parts of a cylinder and the desired fittings have been shown in Fig. 8.3.

Linear Actuators 249

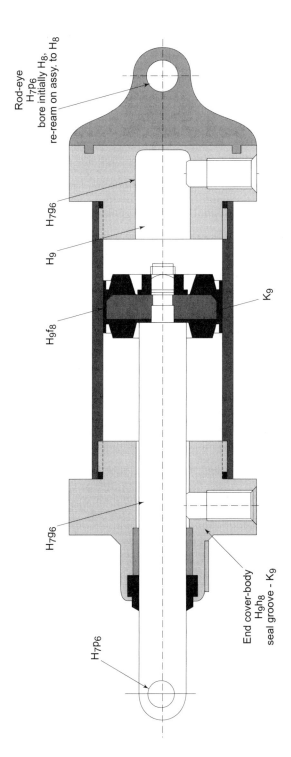

Fig. 8.3 *Assembly fitting of a double acting cylinder*

(a) Cold drawn tubes: promote better machinability but show increased tendency to distortion when machined.
(b) May develop ovality.

On the other hand stress relieved tubes avoid ovality, but may pose machining problems. Tube tolerance is an important manufacturing criterion. It is noticed that if diameter to wall thickness ratio is higher than 20 : 1 it may be difficult to generate the expected tolerance.

(ii) Piston rod May be made of steel bar (stainless steel for corrosive medium) carburised or induction hardened, finely ground, polished and finished by hard chrome plating with rod end finished with rolled threads of all types.

Hardness value may range from 60–65 HRC for the rod and 40–45 HRC at the rolled thread. The rod is generally finished to a tolerance of K_9. The clevis may have a fitting with the locking system with $H_7 P_6$ fitting.

(iii) Bush Made of bronze or other similar metal honed and polished with preloaded seal.

(iv) Cylinder head and cap Metals such as cast steel, spheroidal C.I. of minimum yield strength of 30 kgf/cm^2 or meehanite castings are used. It should be finely finished and welded or screwed together by tie rods made of heat treated steel of yield strength of above 70 kgf/cm^2 with rolled threads at their ends. The bush fitted to the cap end may have a fit of say $H_8 J_7$ type.

(v) Bush Made of brass, bronze, babbit metal or split nylon, finely ground, honed and polished with preloaded seal. Piston rod and guide fitting may be of the order of $H_7 g_6$.

(vi) Piston Should be of softer material than the tube and finely finished. The final fitting of piston bore with seals may be of the order of $H_9 f_8$.

8.2.2 Comparison of Common Tube Materials

A comparative picture of the main characteristics of various metals used for cylinder tubes is provided in Table 8.1. However, readers may note that the list is not conclusive.

Table 8.1 Characteristics of Various Metals used for Cylinder Tubes

Metal used	Characteristics
1. Aluminium alloys	Light weight, inertness to flow medium and water, ease of machining. But it has poor fatigue property. Unsuitable for systems with vibrating and pulsating pressures.
2. Cold drawn low carbon steel	Carbon percentage in general varies from 0.08 to 0.15%. Widely used. High strength as high as 6300 bar under drawn condition, easy to produce a surface finish of 0.8 to 1.2 μm or better after machining and finishing.

(Contd.)

Metal used	Characteristics
3. Stainless steel	Maintains strength at high temperature and has good anti-corrosive and scratch resistant properties. Higher strength to weight ratio next only to titanium. Higher strength than medium carbon steel (4 times) helps to reduce the wall thickness and weight for a given pressure rating and diameter. But high cost is prohibitive in certain applications.
4. Tungsten alloy	An Al– Si–Ni-brass alloy, high strength and good fatigue property and excellent corrosion resistance and high pressure enable it for use in certain aviation hydraulics. Free from cracks and brittle failure.
5. Cast iron	Used for bigger sizes and moderately high pressure. Stress relieving can be effected by providing generous radii or fillets at end covers. Easily machinable. Meehanite castings have good tenacity. Sand particles from finished body may cause premature seal failure.
6. Titanium's alloys	Very good strength to weight ratio, corrosion resistant and suitable for high tensile strength (14000 kgf/cm^2) and high temperature application. Marked increase in strength on cold working. Heat treatment, e.g. stress relieving should be done under controlled atmosphere but titanium is susceptible to contamination in hot atmosphere.

8.2.3 Types of Fit in Cylinder Assembly

Though it is very difficult to prescribe accurate manufacturing tolerances and type of fits for the assembly of the components of a cylinder, attempt has been made here to project probable types of fits between the mating components for the benefit of the component designer, production and machine shop engineers and maintenance mechanics. However this should not be taken as conclusive.

The following fits may be used for normal cylinders as shown in Fig. 8.3:

Piston and bore interface	– $H_9 \, f_8$
Piston rod guide and head end cover	– $H_7 \, g_6$
Cap end cover and bore	– $H_9 \, h_8$
Piston rod	– K_9
Cap end cover and bush	– $H_8 \, J_7$
Piston rod and bush	– $H_8 \, f_8$
Piston rod and eye bore	– H_7
Head end cover	– H_7

8.3 SEALS IN CYLINDER

Various types of seals that are used in cylinders, will be discussed subsequently. Piston seals are mostly of 'u' or 'v'-ring types used in molded form which may be loaded with an in-built spring for taking up wear. A wiper seal is used with the piston rod to prevent dust, dirt or other contaminants from entering the cylinder tube. Misalignment, shaft run-out, seal shrinkage, single pressure, mechanical friction, etc. may create problems for seals and ultimately lead to their failure. Hence selection of cylinder seals needs functional, environmental and economic consideration. A seal is the major friction element of the cylinder and friction related problems need careful study and consideration.

While selecting a seal, seal material compatibility with the oil to be used should be given utmost importance. Type of elastomer, its concentrate, solidification point, oil viscosity, shrinkage and swell characteristics of the seal are also equally important.

8.4 CYLINDER RELIABILITY

The major contribution to cylinder failure emanates from seal failure due to excessive friction and the amount of unpredictability connected with this. Seal surfaces are prone to deform introducing an additional factor of surface straining. Parameters which may reduce friction include high speed of cylinder, operation at higher temperatures, high surface finish of seal bore, seal rod, lower seal hardness, less cross-sectional area of seal, higher dimension of seal groove, small cylinder size, etc.

Many a times cylinder performance may pose a problem due to increased elastic deformation of the tube under pressure creating more seal bore gap and this can be tackled only by proper selection of both seal and tube materials so that the overall clearance between the seal and tube remains unchanged or at least minimally changed. Amongst other vital factors, seal performance is greatly influenced by working speed, working time, idle time, pressure and temperature. Lubrication of component parts and operation time play a role in enhancing reliability against failure. It is known that at speeds above 0.25 m/s, better hydro-dynamic lubrication prevails. Total cycle time is equally significant. Depending on enhanced or reduced seal reliability longer travelling time of piston or bigger cylinder stroke length may subject the seal to more stress and strain while longer idle time under pressure may result in dry friction condition in the seal bore gap due to rupture of lubricating film. As most of the elastomers used for the seal have 10 to 15 times higher coefficient of thermal expansion than metals, hydraulic systems working at high temperature—ambient or otherwise—may induce higher thermal stress on seals. Non-compatibility between seal hardness, friction and rate of wear may pose a serious reliability problem. Under very low pressure application lower value of hardness may be used, but in such cases the degree of surface finish of the mating components should be higher, otherwise it will result in excessive wear with frequent seal failure and increased downtime. Better wear resistance and abrasion characteristics are possible with shore hardness value between 70° A to 80° A reducing the running friction substantially, but above 85° A, the chances of brittle failure of seal is much higher thereby limiting the value. Every designer is already aware of various discussions and reports on the characteristics of various seal materials. Polytetrafluoeoethylene (PTFE) is widely used as seal material. Some of its characteristics are given below. The frictional characteristic of PTFE is quite remarkable compared to most other seal materials.

8.4.1 Characteristics of PTFE (TEFLON)

(i) PTFE has remarkable frictional characteristics. The coefficient of friction is mostly between 0.15 to 0.20, but could be as low as 0.05 in certain cases.
(ii) High coefficient of thermal expansion
(iii) PTFE in bulk is mechanically not strong
(iv) Poor conductor of heat
(v) Poor adhesibility—very difficult to stick other metals or solids on it
(vi) Amount of material transfer is not detectable
(vii) Time of loading and load play important role on friction characteristic of Teflon.

In comparison to PTFE elastomers have more friction. They deform elastically, but easily return to original shape when the force is removed. This is in contrast to polymers which deform elastically and also may have viscous characteristics. Friction of rubber depends on seal geometry or load which is in contrast to metals in which case shape or geometry may not play a significant role. Deformation of elastomers also depends on the temperature and the rate of deformation.

Frictional coefficient between PTFE and metal can be compared to the polyurethane. (see Table 8.2)

Table 8.2 Comparison of Friction of PTFE and Polyurethane on Metal

Seal Material	Friction on metal without lubrication	
	Static	Kinematic
PTFE	0.10	0.15
Polyurethane	0.30	0.40

However one must keep in mind that friction is comparatively less important than surface damage and wear whereas surface finish has a lot of influence.

8.4.2 How does Wear Affect Reliability of Cylinder?

Mutual wear possibility between the mating parts of the cylinder may be anticipated if one knows the following minimum parameters and their relationship.
(a) Pressure
(b) Material hardness

(c) Seal and groove geometry
(d) Seal-bore clearance
(e) Coefficient of friction—both static and sliding
(f) Working temperature
(g) Surface finish.

When the coefficient of friction for the most diverse system and material may not vary by more than a smaller factor, the difference between a commonly used seal material PTFE and steel may differ by a factor much higher.

8.4.3 Wear Rate

The knowledge about wear rate specially at the design stage may play an important role in predicting the life of cylinder parts.

Wear rate depends on:

(a) Geometric shape of contact surfaces and height of individual asperities
(b) Coefficient of mutual overlap
(c) Hardness
(d) Contact pressure and running load
(e) Possible nature of lubrication
(f) Working temperature
(g) Elastic property of the material
(h) Working speed
(i) Environment
(j) Running condition
(k) Design features, etc.

8.4.4 Surface Roughness

Time rate of wear is very much dependent on the nature of surface roughness of the components used to make a cylinder and the finishing process used. The following Table 8.3 provides a glimpse of possible range of roughness which may be generated in the various parts of the cylinder through capping and honing.

Table 8.3 Range of Roughness for Lapped or Honed Surfaces

Method of surface finish technique used	Possible range of roughness in μm	Optimum limit to which can be used in μm
1. Lapping of internal cylindrical surface		
(a) for steel	0.025 to 0.100	0.35
(b) for C.I.	0.050 to 0.160	0.40
2. Honing	0.04 to 0.63	
(a) for steel		
(b) for rubber seal tip to steel surface	0.16 to 0.32	
(c) C.I. piston ring	0.04 to 0.08	

8.5 PREDICTING WEAR

Nowadays designers are to be more alert about this vital sphere of machine operation. Though it is a difficult task, a theoretical exercise in this regard may be paying. Before selecting a particular kinematic pair, it is the duty of a designer to

(i) Select kinematic pairs after comparative study of expected wear resistance of various such pairs.
(ii) Collect the test results of the specimen of mating materials.
(iii) Specify working conditions and estimate the effect of this on the expected wear characteristics.
(iv) Study the feasibility of the specified conditions.
(v) Specify the expected rate of wear.
(vi) Incorporate necessary data on the wear characteristics of mating elements in the machine manual for the use of maintenance and operational personnel.
(vii) Specify the maximum and minimum levels of tolerance range to the component production and assembly shop.
(viii) Specify the component durability and system reliability with probable service life.
(ix) Incorporate a diagnostic warning system to monitor the working parameters.

Chrome plating helps to increase wear resistance of the metal. Wear life is prolonged by 5–15 times and parts made of steel, C.I. and non-ferrous alloys may be chrome plated. Chrome plated parts have least deformation and accuracy is retained from previous operation. But wear resistance of Cr-plated parts does not improve if the working temperature changes the hardness of chrome. In such cases one can go for hard chrome plating.

Roller burnishing is also possible for C.I., steel, non-ferrous alloys and titanium alloys. Wear resistance is increased by 20 to 30% but the surface thickness of the hardened layer is much higher than that possible with Cr-plating. A comparative table of dimensional accuracy of various surface treatment processes are given below in Table 8.4.

Table 8.4 Surface Treatment Process and Probable Accuracy and Thickness of Hardened Surface

Process	Workpiece material	Dimensional accuracy	Thickness of hardened layer
1. Carburizing	Low carbon steel	Warpage 50–150 μm	0.5–2.0 mm
2. Nitriding	Steel, C.I.	50–100 μm	0.05–0.60 mm
3. Induction hardening	Steel	30–70 μm	0.2–10.00 mm
4. Cr-plating	Steel, Ci, and non-ferrous alloys	No-deformation Accuracy retained from previous operation	0.005–1.0 mm
5. Roller burnishing	C.I., steel, Ti and non-ferrous alloys.	-do-	1.0–20 mm

8.6 CYLINDER FORCE, ACCELERATION AND LOSSES

It is the desire of any designer to get the maximum possible thrust from the cylinder. The system pressure is calculated from the total load to be overcome by the cylinder piston rod. The losses which one encounters with cylinders are mainly the resistances due to friction from piston as shown in Fig. 8.4 and other mechanical elements. To ensure normal reliability of cylinder function with adequate efficiency, the total frictional losses should not exceed 2 to 5% or at the most 8% of the theoretical force. For double-acting cylinders the effect of back pressure should also be considered. During the period of acceleration including the dynamic force on the piston, the total force to be developed in the cylinder is given by

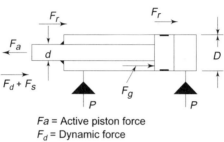

F_a = Active piston force
F_d = Dynamic force
F_s = Static force
F_r = Friction force
F_g = Back pressure force
P = Pressure of oil
D = Piston diameter
d = Piston rod diameter

Fig. 8.4 *Forces acting on a cylinder*

$$F_o = F_s + F_d$$

where F_a = maximum piston force to be generated
F_s = static piston force comprising of load, friction due to external mechanical elements F_r, rod seal and piston seal, and effect of back pressure (F_g).
F_d = dynamic force due to accelerating mass.

The system pressure should be calculated from F_a. During steady travel, F_a equals F_s.

An ancient Greek philosopher Themistius pointed out over 2000 years ago that "it is harder to start a body moving than to keep it moving". Hence the static friction (i.e. the break-out friction) of a cylinder is always more than of the running friction. Naturally F_s is always greater than F_d. During steady travel F_a = F_s. But during acceleration, the total force includes both F_s and F_d.

Frictional losses for an average cylinder is taken anywhere between 2–5% and coefficient of friction for a hydraulic cylinder μ is 0.15 to 0.20.

The cylinder piston is expected to overcome the resistance of the load W, sum of the friction forces ΣF_r and the back pressure Force F_g and the dynamic force F_d which is required to reach specified acceleration.

To get the maximum useful work from the very beginning of the piston stroke, it is necessary to design and manufacture the cylinder such that optimum piston speed is attained as quickly as possible. Cylinder size and rigidity, pressure oil compressibility, stick-slip effect between the piston seal and cylinder, bore, manufacturing and assembly defects, etc. are factors that may affect the process of acceleration. For cylinders with normal size and used at normal speed, it is desirable if the piston attains steady state velocity within 0.5 s to 1 second or 10 –15 mm of its travel from the start. This will ensure better service from the cylinder, but one should note that in comparison to the overall cycle time, this is quite insignificant in most cases and hence may not be given much importance in sizing the cylinder parameters. While extending, the load and the frictional resistances offered by mechanical elements act on the piston rod subjecting it to a buckling load, and while retracting, the rod is subjected to a tensile load. An element of back pressure also acts on the piston including the frictional resistance due to piston and rod seal. In general the back pressure lies mostly between 4 to 6 bar.

The effect of back pressure may be neglected if the ratio of the effective piston areas (i.e. piston area: annular area) is 4:3 or less.

8.6.1 Friction Force and Losses in Cylinder

A major portion of the cylinder losses comes from the friction due to seals and packings. Some of the main factors influencing frictional losses are:

1. Surface finish of the mating elements
2. Coefficient of friction between seal surfaces and their interface
3. Size of the sealing surface
4. Oil pressure
5. Pre-load of seal.

A broad estimate of the friction force may be given by

$$F_r = F_p + \mu P \cdot A_c$$

where, F_r = frictional resistance

μ = coefficient of friction (may vary from 0.1 to 0.2)

A_c = circumferential area of seal = $\pi l D$

l = length of seal surface

D = cylinder I.D.

F_p = Σ pre-load force

The value of pre-load force (F_p) per ring may be taken from 0.5 to 1.0 bar per ring for a majority of cylinders. A number of seal rings (mostly V or U-cup rings) used in a stack may vary widely according to:

(i) Sliding speed (ii) Pressure rating (iii) Seal material (iv) Seal size and seal geometry (v) Types of cylinders and the nature of their duty cycle and (vi) Frictional properties of all related materials.

The number of rings to be used may vary from two to four in general.

At high sliding speed not more than two rings are used to reduce the friction and wear rate. More rings are used for higher pressure and lower sliding speed.

Table 8.5 below illustrates estimation of number of seal rings to be used depending on the type of seal material and operating systems pressure.

Table 8.5 Seal Material, Number of Seal Rings and Possible Pressure Rating

No. of seal rings	Pressure (in bar) Elastomer material	Elastomer/fabric material	Leather material
2	Low pressure		
3	30 bar	30 bar	30 bar
4	100 bar	100 bar	150 bar
5	200 bar	350 bar	450 bar
6	350 bar	700 bar	1000 bar

In cases of stacked seal rings the first ring is subjected to 70 to 80% of the total pressure. Number of rings required depends on clearance, permissible leakage quantity per second and diameter as well as operating pressure. If frictional resistance varies from 3 to 8%, we may take

the total frictional force $F_r = 0.03 \times F_s$ as maximum pressure

and $F_r = 0.08 \times F_s$ as minimum pressure

where F_s = Static force encountered

8.6.2 Effective Cylinder Force

In this era of global energy crisis, every designer should endeavor to generate the highest possible force from the cylinder. Out of the maximum available theoretical force $F = 0.7854 \, D^2 P$, it is expected that after meeting the losses due to friction, etc, the force that may be available from the piston should equal at least $0.7 \, D^2 P$. A well designed hydraulic cylinder should not have losses more than 2 to 8% in general.

The effective force available from the cylinder has to overcome the sum total of resistance force due to friction, work load, and back pressure.

8.6.3 Pressure Drop

Total pressure drop in a double-acting cylinder may be taken as

$$\Delta P = \Delta P^1 \left(1 + \frac{1}{r}\right) \text{ where, } \Delta P = \text{Total pressure drop}$$

ΔP^1 = pressure drop in inlet line

$$r = \frac{A_1}{A_2} = \frac{\text{Piston area}}{\text{Annuler area}}$$

If the ratio $\frac{A_1}{A_2} = r$ is 4 : 3 or less, effect of back pressure may be neglected.

Pressure losses in pipes may be taken into consideration if the total line length $1 > 100\ d_i$ where d_i = inside diameter of the pipe. Pressure losses due to local resistance can be determined fairly accurately if the distance between two adjacent resistances is at least equal to 20 d_i. If it is less, the value is mostly maximum.

Ratio $\frac{A_1}{A_2}$ is dependent on the piston rod diameter d.

8.6.4 Piston, Piston Rod and Rod Length

If piston velocity for extending and retracting is almost equal, then we can assume that piston rod diameter d is

$d = 0.2$ to $0.3\ D$ where D is the piston diameter

In other cases, d is taken as $d = 0.5$ to $0.7\ D$

Factors such as system rigidity, ease of installation and application may limit the length to diameter ratio of cylinder to 20 or less for the most economical and maintenance advantage. High pressure cylinders should be of low stroke if possible. But if a longer cylinder is to be used in a high pressure system such cylinders with longer stroke should be tension proof.

If the piston diameter is given, the required pressure of the oil entering the cylinder is determined from the magnitude of force F for the acceleration period.

8.6.5 Buckling and Tensile Load on the Cylinder

For extending a cylinder, the total force including the load and frictional forces is applied to the rod subjecting it to a buckling load and the force of friction.

If the cylinder is retracting, the rod is subjected to a tensile load

$$F_t = W + \Sigma F_r + F_d$$

where F_t = tensile load on piston rod, ΣF_r = total frictional force F_d, dynamic piston force, W = Load to be moved. The back pressure, in this case, acts on the piston.

8.7 CALCULATION OF CYLINDER FORCES

To calculate the forces in a hydraulic cylinder, it is necessary to know the following:

(a) Total load to be moved

(b) Desired acceleration and deceleration of the load
(c) The coefficient of static (or breakaway) friction
(d) The coefficient of sliding friction.

Let us take two examples to calculate cylinder load.

Example 8.1 *Calculate the breakway force and the force required to bring a hydraulic cylinder to its operating force and speed when the table load is 1000 kgf and the operating acceleration is 20 m/s².*

Solution The table of the machine tool weighs 1000 kgf and assume static friction between the mating surfaces is 0.16.

Once the table starts moving, let the coefficient of friction drop to 0.14.

∴ Breakaway load = 1000 × 0.16 = 160 kgf

Load to keep it moving = 1000 × 0.14 = 140 kgf

$$\text{The accelerating force} = \frac{20}{100} \frac{m}{s^2} \times \frac{1000 \text{ kgf}}{9.81 \frac{m}{s^2}} = 20 \text{ kgf (approx.)}$$

(assume acceleration to be $20 \frac{m}{s^2}$)

Consequently load to take the table to speed

= 140 kgf + 20 kgf = 160 kgf.

or, we can say (approximately)

Breakaway force = $F = w \cdot f_s$

where w = load to push

f_s = coefficient of static friction

Again force required for horizontal acceleration

$$F = W \cdot f_s = \frac{W}{g} \cdot a + W \cdot f_d \text{ where } a = \text{acceleration}$$

g = acceleration due to gravity

f_d = coefficient of sliding friction

Example 8.2 *A cylinder has to operate a vertical load of 600 kgf. If the operating acceleration is 30 m/s² calculate its running force.*

Let a hydraulic cylinder lift a load of 600 kgf.

Here F = load which the cylinder encounters

$$= \frac{W}{g} \cdot a + W$$

$$F = \frac{600 \text{ kgf}}{9.81 \frac{m}{s^2}} \times \frac{30 \text{ m}}{100 \text{ s}^2} + 600 \text{ kgf when acceleration assumed}$$

is 30 cm/s²

$$= 18 \text{ kgf} + 600 \text{ kgf} = 618 \text{ kgf}.$$

Sometimes the specification may not give acceleration but may state that a speed of say, $30 \frac{m}{min}$ has to be reached in 1.50 cm of cylinder travel or within a certain time, say, within 0.05 sec from the start of the cylinder.

Since it is generally safe to assume linear acceleration we can find by the formulae:

$$a = \frac{v^2}{2S} \cdot a, \text{ where } a = \text{acceleration}$$

$$S = \text{distance travel}$$
$$v = \text{speed}$$

or

$$a = \frac{v}{t} \text{ where } t = \text{time}.$$

Thus to reach $30 \frac{m}{min} = \frac{30 \text{ m}}{60 \text{ sec}} = \frac{1}{2} \frac{m}{s}$ in 1.5 cm of stroke.

This means

$$a = \frac{\left(\frac{1}{2} \frac{m}{s}\right)^2}{2 \times \frac{1.5 \text{ m}}{100}} = \frac{\frac{1}{4} \cdot \frac{m^2}{s^2} \times 100}{2 \times 1.5 \text{ m}} = \frac{25}{3} \frac{m}{s^2} = \frac{8.3 \text{ m}}{s^2}$$

Force which the cylinder encounters in case of 500 kgf load.

$$F = \frac{500}{9.81} \times 8.3 \text{ kgf} + 500 \text{ kgf}$$

$$= 498 + 500 = 998 \text{ kgf}$$

Theoretically in determining the total load, the acceleration of the oil column should also be considered. This can be calculated by

$p = h\rho a$ where h = height of oil column in m

$$a = \text{acceleration}, \frac{m}{s^2}$$

ρ = density of oil, kg/cm^3 or kg/m^3

But from a practical point of view this is insignificant and hence it can be ignored for a low flow rate system.

Stoppage of load

Stopping a load may create large pressure. Here to calculate force F, we can use the formula $F = \frac{W \cdot a}{g} - W \cdot f_m$ where a = deceleration in $\frac{m}{s^2}$.

Example 8.3 *A cylinder has to carry a load of 3000 kgf and it is to move over a horizontal surface at a speed of 30 m/min. and has to be brought up to speed as well as stopped within 15 mm. Calculate the pressure.*

Solution Here, acceleration $a = 8.3 \dfrac{m}{s^2}$ [from previous example]

Let us take coefficient of friction $= 0.15$.

$$F = \dfrac{3000 \text{ kgf} \times 8.30 \text{ m/s}^2}{9.81 \dfrac{m}{s^2}} + 3000 \times 0.15 \text{ kg} = 2540 + 450$$

$$= 2990 \text{ kgf}$$

Let the cylinder diameter $= 40$ mm $= 4$ cm.
The effective area of the cylinder $= 12.56$ cm^2.

$$\text{Pressure} = \dfrac{\text{Force}}{\text{Area}} = \dfrac{F}{A} = \dfrac{2990 \text{ kgf}}{12.56 \text{ cm}^2} = 238 \dfrac{\text{kgf}}{\text{cm}^2}$$

Hence it is found that a pressure of $238 \dfrac{\text{kgf}}{\text{cm}^2}$ can provide the necessary force.
If the valve is closed to stop the motion while still under acceleration, the pressure $238 \dfrac{\text{kgf}}{\text{cm}^2}$ must be absorbed together with deceleration force.

$\therefore$ $F = 2990 - 450 = 2540$ kgf

$\therefore$ $\text{Pressure} = \dfrac{2540 \text{ kgf}}{12.56 \text{ cm}^2} = 202.23 \simeq 203$

A surge pressure of $238 + 203 = 441$ kg/cm^2 results for which the cylinder has to be designed.

8.8 FLOW VELOCITY

Oil flow velocity in the line of the hydraulic system is to be taken generally as 1 to 7 m/s. Flow velocity of oil depends on pressure side or suction side of pump, oil pipe-construction, smoothness of surface, oil viscosity, oil pressure, pipe diameter etc.

This is 6 to 7 m/s for higher viscosity oil and is between 3 to 4 m/s for all other cases depending on the pipe diameter.
when $d_1 = 12$ to 50 mm, flow velocity is 3 to 3.5 m/s. When $d_1 > 50$ mm flow velocity lies between 3.5 to 4 m/s.

Flow velocity through high pressure delivery line is 3 m/s average and flow velocity through intake or suction line is 1 m/s average.

The flow requirement of a cylinder can be calculated from the formula is given below.

Flow requirement for cylinder forward stroke $Q_f = \dfrac{Q_p}{r}$,

where, $Q_p =$ pump flow (without losses).

The velocity of the cylinder for its forward and return motion can be calculated from the following two formulae.

Forward velocity of piston $V_f = \dfrac{4Q_P}{\pi D^2}$ were Q_P = flow rate of pump

Return velocity of piston, $V_r = V_f \left(1 - \dfrac{d^2}{D^2}\right)$

At very low speed (i.e. $v = 0.01$ to 0.015 m/min) the option of the hydraulic cylinder is unstable and should be avoided. When piston speed varies within a wide range say 200:1 and cylinder stroke is very high, it is better to go in for rotary system rather than a linear (cylinder) actuator to improve system rigidity.

8.9 CYLINDER EFFICIENCY

The overall cylinder efficiency is mostly dependent on the frictional losses encountered by the piston and rod during its stroke. With close tolerances and absolute alignment of the mating parts, the total frictional resistances can be contained to 2 to 5% or 8% of the theoretical force, the value increasing with increase in pressure. Frictional losses may also increase if the cylinder design is poor and excessive losses take place. Mathematically, cylinder efficiency can be expressed as:

Cylinder efficiency $\eta_{cyl} = 1 - \dfrac{\Sigma F_r}{F_s}$ where (F_r = sum of frictional force F_s = static force.)

One may get cylinder efficiency as given below:

$\eta_{cyl} = 0.94$ to 0.95 for normal cylinders

$\eta_{cyl} = 0.88$ to 0.90 for telescopic cylinders

With increasing bore diameter, the cylinder efficiency has been found to increase. Due to involvement of a lot of factors a correct estimation is however very difficult. The normal cylinder efficiency with increasing piston diameter is shown in Table 8.6.

Table 8.6 Piston Diameter and Cylinder Efficiency

Piston diameter	Cylinder efficiency
20–50 mmφ	80–85
50–120 mmφ	85–90
120 mmφ	90–95

8.10 SIZING OF CYLINDER TUBES

A good design of a cylinder starts with proper selection of geometric size of piston and rod diameter, wall thickness, stroke length, etc. While calculating the size, a designer should not depend only on the calculated value but he has to consider the feasibility aspect of the values to be sure how much these dimensions may serve the specific need as well as if they are commercially available. For example, with a system pressure of around 75 bar, the theoretical wall thickness for normal cylinder material may prove unreliable in practice as it may be too thin for economic manufacture or handling. In such circumstances, the component dimensions may be selected bigger than needed.

8.10.1 Calculation of Tube Thickness

Tube thickness of a cylinder barrel is a very important factor.

The strength of the cylinder tube, is proportional to its wall thickness. A cylinder that is too thick or too thin may pose serious safety and operational problems and hence the tube thickness of the cylinder has to be carefully chosen. For thin cylinders (diameter to thickness ratio greater than 10:1), thickness t may be calculated from the normal Barlow formula as shown below. We may refer here Fig. 8.5.

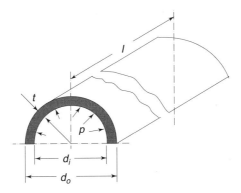

Fig. 8.5 *Cylinder tube thickness*

The cylinder must be of such material and the wall thickness to be such as to safely resist the hoop stress produced by the hydraulic pressure.

$$\sigma = \text{hoop stress} = p \frac{d_o^2 + d_i^2}{d_o^2 - d_i^2} \text{ where, } \sigma = \text{hoop stress}$$

p = oil pressure

d_o = outer diameter of cylinder

d_i = inner diameter of cylinder

When wall thickness is less than $\frac{1}{20}$ of the diameter, $d_o \simeq d_i$ and hence

we can write $\sigma = P \dfrac{d_o^2 + d_i^2}{d_o^2 - d_i^2} = \dfrac{P(d_o^2 + d_i^2)}{(d_o + d_i)(d_o - d_i)} = \dfrac{P(d_o^2 + d_o^2)}{(d_o + d_i)(d_o - d_i)}$

$$= P_2 \frac{2d_o^2}{d_o(d_o - d_i)} = P \frac{d_o}{d_o - d_i}$$

As t = thickness of wall = $\dfrac{d_o - d_i}{2}$ (Refer Fig. 8.5)

we can write $\sigma = P \dfrac{d_o}{2t}$

The maximum hoop stress depends on the tensile strength of the material and the safety factor used.

Cold drawn carbon steel used for the cylinder has a tensile strength of 7350 kgf/cm^2.

Tensile strength of C.I = 4000 to 5500 kgf/cm^2.

For continuous severe application safety factor = 4:1.

If there is no shock load and if maximum pressure is a relatively low occurrence, a factor of safety of = 2:1 or 3:1 is acceptable.

For critical and reliable selection of thickness of the tube some extra allowances may be necessary. These are:
1. allowance for material tolerance
2. allowance for thinning of the material in drawing
3. allowance for decarburisation
4. allowance for corrosion

If one chooses to use a generous factor of safety, the above factors may not be necessary to take note of. Factor of safety most commonly used are (as per JIC standard):

8:1 for pressure of 70 bar (1000 psi)
6:1 for pressure 70 to 110 bar (1000–2500 psi)
4:1 for pressure above 170 bar

But the actual value of the safety factor will depend on the physical and metallurgical merit of the cylinder tube and the overall system characteristics, e.g. for continuous servere operation one may use a factor of safety of 4:1 and if there is no shock load and the peak pressure occurrence is low, use a factor of safety 2:1 or 3:1.

For calculating tube wall thickness of thick-walled cylinders one can use the formula

$$t = \frac{d_o}{2}\left(\sqrt{\frac{\sigma + P}{\sigma - P}} - 1\right)$$

But for thick-walled homogeneous tubes as the stress produced may not be uniformly distributed, the thickness can be calculated from the modified formula given below:

$$\sigma = \frac{d_o^2 - 2t + 2t^2}{2t(d_o - t)} \times P,$$

where d_o = cylinder diameter, outer
t = tube thickness
P = working pressure
σ = hoop stress.

8.10.2 Sample Calculation

Example 8.4 *Calculate the tube thickness of a hydraulic cylinder having dimensions as given below:*
Tensile strength of cylinder material = 7300 kgf/cm^2
Cylinder bore = 50 mm
System pressure = 200 kgf/cm^2
Factor of safety = 4:1

Solution

$$\sigma = \text{Working stress} = \frac{7300}{4} = 1825 \text{ kgf/cm}^2.$$

From the formula $\sigma = p \cdot \dfrac{d_o^2 + d_i^2}{d_o^2 - d_i^2}$, we can write

$$1825 = 200 \cdot \frac{d_o^2 + 5^2}{d_o^2 - 5^2}, \text{ where } \sigma = \text{hoop stress} = \text{working stress}$$

or, $\quad d^2 = 31.15 \text{ cm}^2, \quad d_o$ and d_i = outer and inner cylinder diameters

or, $\quad d = \sqrt{31.15} = 5.6 \text{ cm} = 56 \text{ mm}.$

∴ tube thickness = $\dfrac{56 - 50}{2} = 3$ mm.

Tensile strength may differ from material to material as given in Table 8.7.

Table 8.7 Tensile Strength of Few Steel Material Used for Cylinder Manufacturing

Steel nomenclature	Tensile strength
1. St 35.29	2300 kgf/cm^2
2. St 45.29	2600 kgf/cm^2
3. St 55.29	3000 kgf/cm^2
4. St 65.29	3500 kgf/cm^2
5. CK 35	3500 kgf/cm^2

8.11 PISTON ROD DESIGN

The piston rod of a hydraulic cylinder is highly stressed and therefore it should be able to resist the bending, tensile and compressive forces that it may encounter during its operation without buckling. Mostly piston rods are made of high tensile material finished and hardened with chromium plating to provide resistance to corrosion. Stainless steel is also used as a rod material due to its excellent anti-corrosive property. The cross section of the rod should be calculated after considering whether it is a stressed rod or a column. As a thumb rule, if the length of the rod exceeds 10 times its diameter, one can consider it as a column under compressive load and may buckle under the pressure of the load. If the piston rod is stressed as a rod, the rod cross-sectional area can be calculated from the formula

$$F = a\,\sigma$$

where F = compressive or tensile load
σ = material stress
a = rod cross section

In case the rod behaves like a column and is subjected to buckling, the rod diameter can be related to the critical load as per Euler's formula given below:

$$F_c = \frac{\pi^2 \cdot E \cdot I}{l_k^2}$$

where F_c = critical buckling load, N
L_k = free buckling length, m
E = Modulus of elasticity
    ($2.1 \times 10^{10} \times 9.80665$ Pa for steel)
I = Moment of Inertia, m^4
S = factor of safety can be taken as 2–10

The moment of inertia and therefore the maximum permissible stress in order to avoid buckling is dependent on the type of end fixing of the cylinder.

The moment of inertia (I) can be found from the formula:

$$I = \frac{\pi d^4}{64}$$

The maximum load on piston rod is $F = \dfrac{F_c}{S}$

where F = maximum buckling load on the piston rod
F_c = critical buckling load
S = factor of safety = 5 for general cylinders
  = 3 for cylinders with guided trunion or center trunion.

Depending on the manner of end fixing, cylinder load can be calculated from the four cases of Euler's Rule. The relationship between the effective column length with the actual length and the respective formulae for crippling or buckling load are shown in Fig. 8.6.

∴ Maximum piston rod load $= F = \dfrac{\pi^2 \cdot E \cdot I}{S \cdot l_e^2}$ kgf

∴ Critical or equivalent cylinder column length $l_c = \pi \sqrt{\dfrac{EI}{F \cdot S}}$ cm.

Euler's cases and crippling load					
I	II	III	IV		
Load = $\dfrac{\pi^2 EI}{4l^2}$	$\dfrac{\pi^2 EI}{4l^2}$	$\dfrac{\pi^2 EI}{l^2}$	$\dfrac{4\pi^2 EI}{l^2}$		
				$l_c = 2l$	One end free one end fixed or trunnion at end
				$l_c = l$	Tow ends pivoted and guided
				$l_c = \dfrac{1}{\sqrt{2}}$	One end guided and pivoted, other end fixed
				$l_c = \dfrac{l_c}{2}$	Two ends fixed and guided
				$l_c = \dfrac{3}{2}l$	Trunnion center

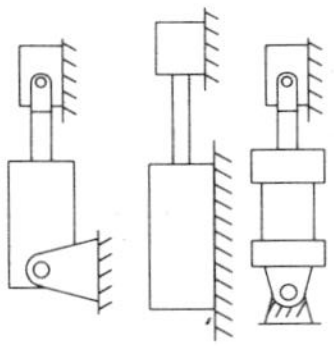

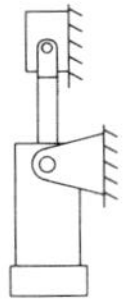

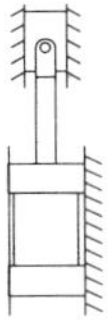

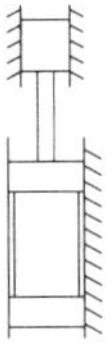

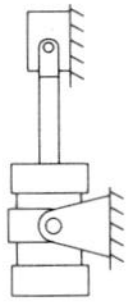

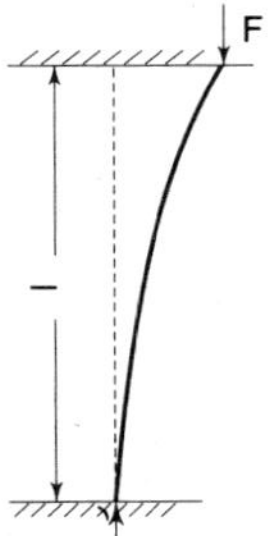

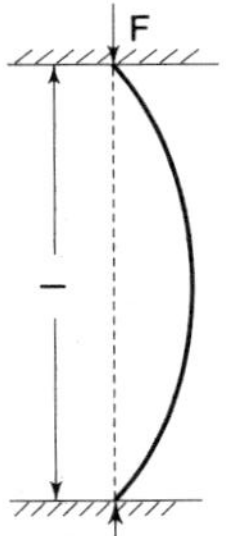

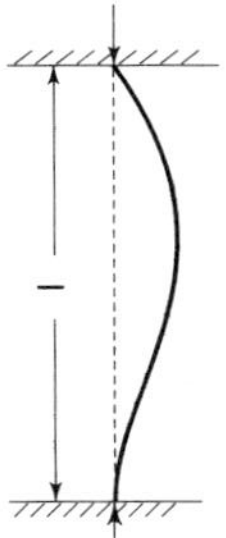

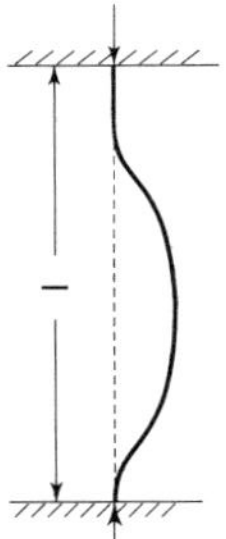

Fig. 8.6 *Crippling load as per Euler's rule*

where d = rod diameter in mm
 A = rod cross-section, cm^2
 l_e = equivalent length, cm

 I = moment of inertia, cm$^4 = \dfrac{\pi d^4}{64}$
 E = modulus of elasticity, kgf/cm^2 = 2.1×10^6 kgf/cm^2
 F = piston rod thrust, kgf
 F_c = buckling load, kgf
 S = factor of safety.

Semi-empirical data are mostly used to determine the rod size in relation to maximum thrust and rod length. In case one uses a bigger diameter rod, the speed of travel may get affected which needs careful consideration.

Let us take a sample calculation to determine the safe load for a double-acting hydraulic cylinder with two ends pivoted and guided and piston diameter = 120 mm, rod diameter = 70 mm, stroke length 900 mm, factor of safety = 3 and modulus of elasticity = 21×10^5 N/cm^2

The buckling load $\quad F_c = \dfrac{\pi^2 \cdot E \cdot I}{l_e^2} = \dfrac{\pi^2 \cdot 21 \times 10^5 \, \text{N/cm}^2 \times \pi d^4}{l_e^2}$

Here $l_e = l = 900$ mm = 90 cm.

$\therefore \quad F_e = \dfrac{\pi^2 \cdot 21 \times 10^5 \times \pi \cdot 7^4 \, \text{cm}^4}{(90)^2 \, \text{cm}^2} = 19271.494 \times 10^3 \, \text{N}$

$\therefore \quad$ Thrust load on rod $= \dfrac{F_e}{S} = \dfrac{19271.494 \times 10^3}{3}$ N

$= 6423.8 \times 10^3$ N.

8.12 MOUNTING STYLE OF CYLINDERS

The mounting style of hydraulic cylinders (shown in Fig. 8.7) is an important parameter both from design and maintenance consideration. The following forms of mounting arrangement are mostly used:

For the rod end of the cylinder
1. Plain
2. Threaded—mostly male threads
3. Clevis
4. Flange
5. Tongue or eye

For the cylinder body
1. Plain
2. Foot
3. Square bracket (single-ended) or pedestal

270 *Oil Hydraulic Systems: Principles and Maintenance*

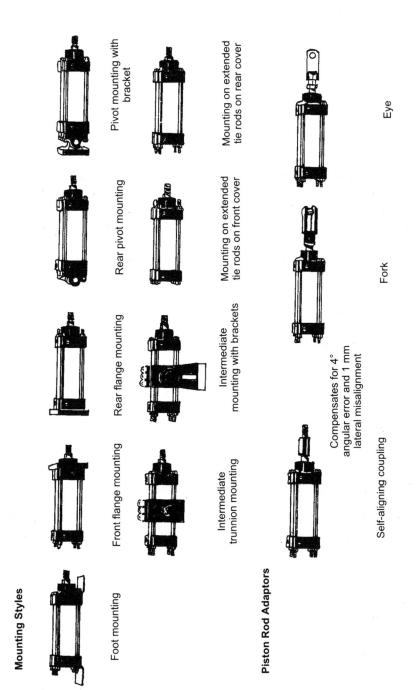

Fig. 8.7 *Mounting style of hydraulic cylinders*

4. Square bracket (double-ended) or pedestal
5. Trunnion–Cap or head-ended
6. Trunnion–Centre
7. Flange–rectangular or square
8. Clevis
9. Tongue and bracket
10. Tie rod extended

Mountings need not necessarily be same at both ends. Mounting styles influence the design life and frequency of maintenance of the hydraulic installation. All the above cylinder mountings can basically be categorized as:

1. Center line mounting.
2. Foot mounting
3. Pivot mounting

1. Center Line Mounting   The best way to support a cylinder is along its center line. The mounting bolts in this case will be under shear or simple stress. No compound forces act on the bolts.

However the alignment should be accurate and misalignment cannot be tolerated. The various styles in center line mounting area:

(i) Rectangular flange connected to cap end.
(ii) Rectangular flange connected to head end.
(iii) Square flange connected to cap end.
(iv) Square flange connected to head end.
(v) Tie rod mounting.
(vi) Lugs attached to the sides of both ends of the cylinder centre line. Flange mounting at the head end is ideal if the cylinder accounting tensile loads whereas flange mounting at the cap end is mostly suitable for compressive loads. Tie rod mountings are easy but less stronger than the flange mounting.

2. Foot Mounting   Foot mounted cylinders are subjected to a turning moment when they are loaded. This mount tends to rotate or bend the cylinder about its mounting bolts. These mountings are used where cylinders are to be mounted on to surfaces parallel to the axis of cylinder. Compared to center line mounted cylinders, foot mount cylinders are subjected to higher stresses but a certain amount of foot mount cylinders can tolerate a certain amount of misalignment. To avoid bending loads on mountings one can use shear pins or keys which enable mounting bolts to remain in simple tension. Shear keys should be located on the head end side for tensile loads and on the cap end side for compressive loads. The various styles in foot mounting are:

(i) Side end lug mounting
(ii) Side end angle mounting
(iii) Side lug mounting
(iv) Flush mounting

3. Pivot Mounting In many industrial applications a cylinder may have rotational freedom while it reciprocates. There are two basic methods of mounting so that the cylinder will pivot during a work cycle: Clevis and trunnion. These mountings offer rotational freedom in one plane. If freedom is required in all planes universal joints should be used. The different styles in pivot mounting are:
  (i) Clevis Mounting as shown in Fig. 8.8
  (ii) Trunnion Mounting—Head end
  (iii) Trunnion Mounting—Cap end
  (iv) Trunnion Mounting—Intermediate

The clevis is almost always part of the cap end and can be either fixed or universal. In the case of pivot mounting, the pivot centre line normally intersects the cylinder center line.

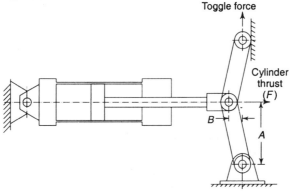

Fig. 8.8 *Clevis mount cylinder*

8.12.1 How to Achieve Perfect Alignment

Misalignment is a regular problem with heavy duty and medium duty tie-rod type cylinders. To secure good alignment* with moderate to high precision, some cylinder manufacturers provide machined groove at each end of the cylinder body as shown in Fig. 8.9 which is made concentric with the internal diameter of the

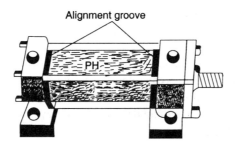

Fig. 8.9 *Groove cut on cylinder body for alignment*

---

* Beldon Rich, *Some Tips on Cylinder Alignment*, Parker Hannifin Corporation, USA

body tube and piston rod center line. The groove is machined 3/16" (4.763 mm) wide at each end of the cylinder body just in board of the head and cap. To check both horizontal and vertical alignment of the cylinder one can use a dial test indicator with a base to trace over the machined groove alongwith a straight edge.

8.12.2 Few Points to Consider for Correct Mounting

The following points may be noted as a checklist for securing correct mounting of cylinder:

1. Use of shear keys and dowel pins for fixed mount cylinders A cylinder which may be subjected to high shock loads is to be mounted in such a way so that it can take full advantage of its elasticity. Shear keys or dowel pins are never to be mounted at both ends of the cylinder otherwise the shock absorbing capacity of cylinder elasticity may be lost. Problems may also be faced if the cylinder length is changed due to temperature and pressure effects when both sides are firmly fixed by shear keys and pins. Shear keys or dowel pins are used with fixed-mount cylinders.

Cylinders with integral key mounts may be used if it is possible to cut key ways on the machine member on which the cylinder is to be mounted. Apart from providing accurate alignment and absorbing the shear force, both with the thrust load or the tension load, this will also help in simplifying the installation and provide easy servicing. The keys may be located as shown in Fig. 8.10 (a) and not at both ends as illustrated in Fig. 8.10 (b). When dowel pins are used to mount the cylinder, they should be located at one end only as shown in Fig. 8.11 (a) and not at diagonally opposite corners as shown in Fig. 8.11 (b) in order to avoid warping.

Shear keys can be used to absorb shear forces developed at cylinder mounting surfaces. Power location of keys depends upon direction of major shock loads; See Fig. 8.10 (a). Never mount keys at both ends of cylinder, as in Fig. 8.10(b), or shock-absorbing capability of cylinder elasticity can be lost. Also, changes in cylinder length, caused by temperature and pressure effects, can cause trouble, and component replacement can be difficult.

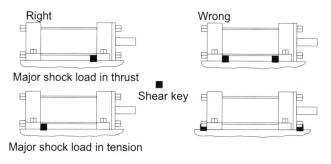

Fig. 8.10(a)   Fig. 8.10(b)

2. *Pivot mounted cylinders* In many industrial situations cylinders are required to be pivot mounted. For using pivoted mounts one should ensure that the mounts on the cylinder body and on the rod end are similar. It is to ensure that the pin in the rod is parallel to the pin on the cylinder. Trunnions for the cylinder pivot mounting are usually designed to resist shear load only and standard bearings are to be used as in Fig. 8.12 (a). Self-aligning bearings as in Fig. 8.12 (b) are not be used to support the trunnion mounts. Their small bearing areas act at a considerable distance from the trunnion junction subjecting it to excessive bending force and thus overstress the trunnion.

If dowel pins are used to help secure alignment and resist shock loads, both pins should be located at one end of the cylinder, as in Fig. 8.11(a). Choice of end depends upon direction of major shock load. Never locate pins at corners diagonally opposite each other, as Fig. 8.11(b), or temperature, pressure and shock loads may cause warping.

Trunnions for cylinder pivot-mounting are usually designed to resist shear loads only; see Fig. 8.12 (a). Self-aligning bearings, Fig. 8.12(b), should not be used for pivot-mounting. Their small bearing areas act at a considerable distance from the trunnion junctions, and bending forces may overstress the trunnions.

3. *Self-alignment possibility* With regard to aligning the cylinder with the machine member, it is true that a cylinder can tolerate a certain amount of misalignment as shown in Fig. 8.13 (a) and (b) but if the line of travel of the cylinder is in constant misalignment with that of the machine member, the cylinder will fail. For a long stroke cylinder where misalignment cannot be removed one can provide degree of self-aligning capability by making the head end to float to some extent on non-fixed dowel pins in the side legs of the cylinder as shown in Fig. 8.13 (c).

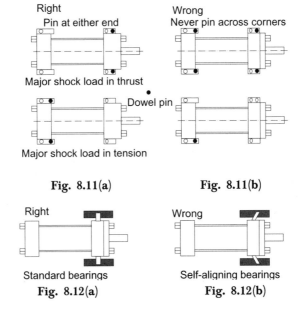

Fig. 8.11(a)   Fig. 8.11(b)

Fig. 8.12(a)   Fig. 8.12(b)

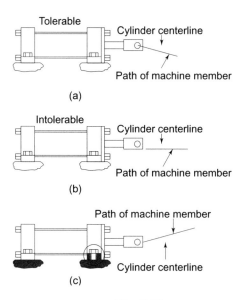

Fig. 8.13
(Courtesy–M/S Parker Hannifin Corporation, USA)

Misalignment of cylinder with work side may be of two types. A fixed mount cylinder can tolerate slight angular misalignment, shown much exaggerated in Fig. 8.13 (a), but it cannot tolerate constant axial misalignment, as shown in Fig. 8.13 (b).

A relatively long-stroke cylinder can be made self-aligning, to a degree, by allowing the head end to float on dowel pins. As shown in Fig. 8.13 (c), the hole in the side lugs at the head end to permit some vertical travel of the cylinder with respect to the pins. The cylinder body flexes slightly about its fixed cap end.

8.13 CUSHIONING OF THE HYDRAULIC CYLINDER

Cushioning of cylinders means gradual deceleration of the piston near the end of its stroke. It is very helpful to reduce shock or impact of load on the cylinder end covers especially when a heavy load is connected to the rod or the cylinder is working at very high speed. A schematic view of a cushion assembly is shown in Fig. 8.14 (a). A cushioning screw is used to control the cushion bore as shown in Fig. 8.14 (b). One must keep in mind that the cushion is not a speed controlling device but a shock alleviator. The cushion assembly is around 25 mm long for a standard cylinder. If the cylinder stroke is not completely used, cushions will be of little value. The cushion assembly consists basically of a small passage to allow entrapped oil to the port with a cushion needle with an integral ball check valve to allow free flow of oil during reverse start of piston travel. The end of the cushion nose is tapered, chamfered or rounded in order to allow it to enter more easily into the cushion chamber.

276 Oil Hydraulic Systems: Principles and Maintenance

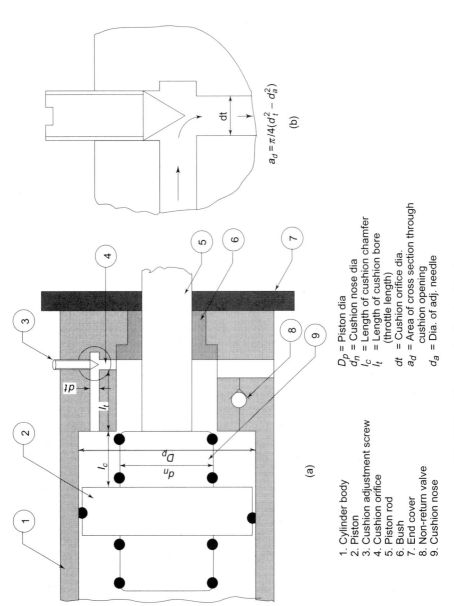

Fig. 8.14 Schematic details of cushion bore

Linear Actuators 277

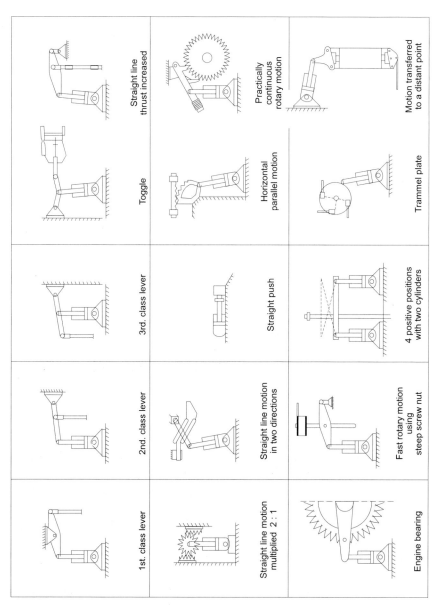

Fig. 8.15 *Typical cylinder installation procedure*

8.13.1 Typical Installation Methods

Figure 8.15 illustrates some suggested methods of use of hydraulic cylinders so that one can decide different types of operation, movement or installation. Cylinders can be installed to provide lever and toggle clamping, direct push or thrust, straight line, parallel and continuous rotary motion, multi-positioning of load, etc. as depicted here. The illustration is not conclusive and there are many other methods of cylinder installation which could be adopted for specific applications. For high force at a short distance and exact depth control, the toggle lever system is very useful. Referring to Fig. 8.8, we can find the toggle force as

$$\text{toggle force} = \frac{F \times A}{2B}$$

where F = cylinder thrust.

8.14 MAINTENANCE TIPS ON CYLINDER MOUNTING

It has been noted that for any operating problem with hydraulic cylinders, the major reason could be traced to mostly poor installation and misalignment. Misalignment of the cylinder could be defined as non-coincidence of axis of cylinder rod with the desired line or arc of motion. A foot mounted cylinder if misaligned may create the following problems.

1. Misalignment may cause wear on gland bearing.
2. Debris from the worn bearing may cut the seal.
3. Due to this an extrusion gap may form in the seal.
4. The seal may give in due to extrusion which may cause the cylinder to fail.

In order to have best possible alignment the following precautions may be taken:

1. Study the installation drawing carefully.
2. Before making any dowel pin holes, place the cylinder in the desired position on the machine, check for initial alignment and align it by means of shims to correct it if there is misalignment.
3. Recheck the alignment by hand stroking, if possible. Take measurement if necessary.
4. Now the cylinder can be mounted with appropriate tightening.
5. Apply power to the cylinder and run it to check up its alignment. Take appropriate measurements of any misalignment, if exists.
6. Turn or shim the cylinder as necessary.
7. For a heavy duty cylinder involving high shock loads, use shear keys as they will help to absorb the shock and maintain alignment.
8. Order special bearings for the clevis pin for high pressure continuous duty cylinder with clevis mounting. The clevis bearing needs continuous lubrication.
9. Angle iron mounts and lug mounts are vulnerable to misalignment and it is better to use them only for modular duty applications.

8.15 HYDRAULIC CYLINDERS AND THEIR CHARACTERISTIC APPLICATION

Even though there may be numerous applications of hydraulic cylinders, they could be grouped broadly into a few categories. Categorywise specific system requirements and matching cylinder characteristics are given in Table 8.7. (Please note that the categorization is not fully exhaustive.)

Table 8.7 Specific System Requirements and Matching Characteristics of Hydraulic Cylinders

Application	Specific system requirement	Cylinder characteristics required
1. Mobile applications and machineries e.g. excavators, dumpers and other earth-moving implements.	Need extreme flexibility of application and operation, high power to weight ratio, and high power conversion efficiency. Minimum nominal pressure rating is above 150 bar and minimum power requirements may vary from 70 kW onwards. Working environment may be hostile.	Required cylinder size to produce the expected output force will have to be optimized. Strong shock proof piston rods are essential. An optimal size of cylinder is to be chosen to strike a balance of all related factors of cylinder and pumps.
2. Shop floor appliances, e.g. fork lift, trolley, stackers, etc.	Mostly used for jacking, pushing, pulling, etc. Power requirement is moderate to high. Working environment may not be always quite satisfactory. Maximum power requirement may be around 50 kW of less. Higher load capacity may be essential.	Needs robust cylinder in design and construction with sufficiently stronger piston rods. Special self-aligning bearings are needed to manipulate eccentric loads and to avoid bending or buckling. Safety valves may be integrated in the cylinder body. Cylinder with gravity or externally loaded return may need special features.
3. Material handling appliances, e.g. mobile crane, hoist, lifters, etc.	Sensitivity and accuracy in load spotting is essential. Special safety arrangement (in the form of friction brake or otherwise) is needed to avoid accidents due to leakage or environmental problems. Load capacity may be as	Cylinder stroke length may be higher necessitating application of telescopic cylinder with 3 or 4 stages necessitating. Sizing of the cylinder is very important as lifting capacity is reduced in the proportion of rod to piston area and inverse increase of lifting speed.

(Contd.)

Application	Specific system requirement	Cylinder characteristics required
	high as 15 T or above. High lifting speed and less lowering speed are desired. Modern diagnostic warning systems may be of advantage.	Tie rod type cylinders are most popular.
4. Metal cutting machine tools and robotics	System requirement varies application and machine wise e.g. (a) High Speed (> 0.5 m/s) high power (> 50 kW) as needed in planner, or (b) High speed (1 to 2 m/s) Low power (< 5 kW) as in grinders or (c) Low speed (as low as 0.5 m/min) and low power (5–10 kW) e.g. needed in normal milling or lathes, etc. High sensitivity of response, low cycle time and low noise and high reliability, improved controll ability, energy efficiency, higher productivity, higher power to weight ratio are some important requirements of highly sophisticated machine tools, e.g. CNC machining centers, robotics, etc. System pressure may vary from very low to a very high value depending on application.	Cylinder characteristics may vary. Extra high precision in cylinder positioning for feed movement is essential. Size of cylinder bore and rod and their manufacturing characteristics may depend on the accuracy of the end product to suitably match the machine response and repeatability of the cylinder motion characteristics are to be adoptive to the modern instrumentation comprising of sophisticated electronic controls and microprocessors, position/speed sensing devices, etc. Electro-hydraulic servo actuations, smooth precise motions of heavy loads with zero backlash, ability to tolerate heavy shock loads, are of utmost importance in maintaining high performance rate of the machine. This may be achieved by appropriate material, higher rigidity coupled with improved manufacturing processes to maintain a very close tolerance and concentricity, stick slip effect should be fully eliminated in cylinder's mating components.
5. Press tools and other forming machines.	Generally very high pressure system in some cases above 400 bar. Speed and power to the machines are naturally dependent on the end	Cylinder geometry and sizes vary enormously depending on the type of machine and work capacity. Stroke length may vary from 10 mm upwards. Pressure and ram speed required to be under absolute

(*Contd.*)

Application	Specific system requirement	Cylinder characteristics required
	products being formed and their specific technological criteria and on types of machines. Hydraulic systems of rolling, die costing machine, etc. should have higher thermal stability. Fine blanking presses require accurate speed and feed. Compressibility effect should be taken into consideration.	positive and accurate control. Rams to be strong enough to withstand impact and shock load. Superior surface finish, close OD & ID tolerances and uniformity in wall thickness and material properties should be matching to high pressure and for high temperature applications. Metallic piston rings are often used in such cases.
6. Aviation hydraulics	System characteristics require "cent percent reliability" System pressure of a modern aircraft may range from 200 to 350 bar in general. Precise movements of all control elements are essential. High power to weight ratio and zero leakage system to be designed.	Cylinder response to command signal should be sharp and quick. High strength, low weight and corrosion resistance is needed. Compressibility effects are to be considered to increase cylinder response and to avoid failure. Cylinders to be designed to eliminate and minimise the undesirable effects of wind shear, cross wind, oscillation and mechanical malfunctioning.

Apart from what has been mentioned above, we find extensive use of fluid power systems in other fields, e.g. in mining, construction, agriculture, marine, oil rigs, automotives, etc. Many of the design criteria may be over-lapping but so far reliability and maintainability is concerned, the environmental condition plays an important role in the selection of cylinders, specially for their application in the open field such as for mobile equipment. Force, size, materials, manufacturing methods and assembly technique and mounting of cylinders will have direct bearing on the overall system working and efficiency. Let us identify the complexities of parameters which ultimately determine the reliability of cylinder operation.

8.16 CHECKLIST FOR CYLINDER DESIGN

A system designer or a maintenance mechanic need not design a hydraulic cylinder. However from the above discussion it is clear that designing a hydraulic cylinder for a specific purpose is a very complex task. Hence it is imperative that designing should be done only after taking into consideration all the relevant

factors given below which ultimately determine its reliability. A system disigner or repair mechanic should note the same:
1. Metallurgical and physical parameters
   (i) Selection of correct material
   (ii) Selection of the proper stress value
   (iii) Use of suitable safety factor
   (iv) Take into account the stress concentration factor
   (v) Consider the stresses due to impact load
   (vi) Analyze the load distribution
   (vii) Consider the material fatigue properties
   (viii) Consider the effect of compressibility on cylinder performance
2. Manufacturing parameters
   (i) Mechanical fitting of the mating parts
   (ii) Machining methods applied
   (iii) Superfinishing technique
   (iv) Hardening and heat treatment methods
   (v) Plating and coating of the parts.
3. Frictional parameters
   (i) Estimate the possible frictional coefficient
   (ii) Correctly position the seal and provide correct seal material and seal geometry
   (iii) Mutual wear rate of the mating materials, heat generation etc.
   (iv) Select the appropriate cylinder velocity, acceleration.
4. Pressure parameter
   (i) Estimate the total load
   (ii) Calculate the proper pressure
   (iii) Predict the total losses
   (iv) Find the possibility of surge pressure and shock and impact load
5. Environmental conditions
   (i) Type of equipment where the cylinder is used
   (ii) Type of cylinder
   (iii) Work place
   (iv) Ambient and working temperature
   (v) Thermal stresses
   (vi) Corrosion possibility of cylinder
   (vii) Contamination possibilities of the cylinder
6. Structural consideration
   (i) Cylinder mountings
   (ii) Link mechanism of the cylinders and installation
   (iii) Cylinder size
   (iv) Cylinder alignment arrangement
7. Operating conditions
   (i) Nature of working and area of application
   (ii) Cycle time

(iii) Possibility of contamination
(iv) Type of oil to be used
(v) Maintenance and servicing requirement, etc.

It is important for a designer and manufacturer of hydraulic components and system that a faulty system or a faulty component may cost him very dearly in the long run in the way of financial loss, low reputation and loss of customer.

It may not seem irrational if one attributes hydraulic cylinder failure as the sole cause for 10 to 15% of the overall failures in a hydraulic system. Or in other words we can say that the reliability index of any hydraulic system could be improved by that amount if a good cylinder is designed. The main impetus behind this is the need to make the system operate with optimum reliability. As with reliability, maintainability of the cylinder should also be built-in at the design stage itself. The component design of any hydraulic system should provide features and function that will contribute to the:

1. optimum maintainability with
2. minimum downtime ensuring
3. ease of repairing and
4. rapidity of recommissioning to serve at
5. best possible accuracy at
6. nominal economic liability.

One must keep in mind that maintainability is very much influenced by consumer but the equipment reliability is essentially built into by proper designing and systematic manufacturing. Component designer of the hydraulic system is responsible for the safety, durability and reliability of the entire system.

8.17 SOME COMMON CYLINDER PROBLEMS

Though cylinder maintenance problems and their solution will be discussed in detail later on a few physical problems are given below:

1. Sticky, slow start-up cushion
2. Scoring of piston and tube
3. Leakage of oil at rod end and at other parts
4. Premature seal wear out
5. Difficult servicing situation
6. Rod stripping and/or breakage
7. Tubing end leakage
8. Problems associated with link mechanism and machine member etc.

Increased cylinder speed leads to unacceptable hydraulic shock load. Hence a cushion is designed to allow kinetic energy to be absorbed gradually and smoothly over the entire cushioning stroke. A 90% shock reduction reduces machine noise, downtime and operating cost. Threaded tube and head eliminate bulky tie rod.

Speed Limitation At low speed (v = 12 to 15 mm/min), the operation of the hydraulic drives becomes unstable.

Leakage Though no leakage is allowed normally, up to 30 bar leakage of oil may be acceptable between the cylinder wall and piston surface at a rate very low. The clearance should be within 0.03 to 0.05 mm. Leakage introduces loss of pressure in cylinder. Unwanted pressure losses may be there due to line losses or friction losses also. Pressure drop due to line losses can be determined fairly accurately when $l > 20\, dp$ where dp = diameter of pipe (inside) and l is length. If it is less, the losses can be calculated in approximation only for which empirical formulae are available.

Fixing Possibility A cylinder which has got link movement, a compensating or correcting factor for side-wise swivel of $\pm\, 3°$ to $5°$ is allowed. When the link motion is in all sides, ball and socket joint is to be used.

Review Questions

1. State various types of linear actuators used in hydraulic systems. What is a telescopic cylinder? State at least three applications of such a cylinder.
2. (a) State the names of materials generally used for cylinder pistons and piston rods with regards to hardness, shock allevtation, corrosion etc.
   (b) What type of assembly fittings one should recommend for the following joints (i) piston rod and bush (ii) piston rod and eye bore (iii) piston and bore interface?
3. State the basic reason for cylinder component wear. What are the factors on which the wear rate depends.
4. What is PTFE? What is its common commercial name? State its characteristics.
5. The active force of a piston rod should be more than the sum total of friction. Dynamic force, load to carry and back pressure force. State the relationship of the active force with the static and dynamic forces:
   (i) during the steady state travel of the rod, and,
   (ii) during cylinder acceleration period.
6. Write down various styles of cylinder mounting. What is a clevis joint? Why is it used?
7. What is the effect of cylinder cushioning on the cylinder motion?

References

8.1 Publicity literature, Veljan Hydair Pvt. Ltd., Balangar, Hyderabad, India.
8.2 *The Power & Controls Journal*, Issue No. 2, Parker Hannifin Corporation, USA.
8.3 Rich Beldon, *Some Tips on Cylinder Alignment*, Parker Hannifin's Cylinder Div., USA.
8.4 Duerr A and O Wachter, *Hydraulik in Werkzeugmaschinen*, Carl Hanser Verlag, Munich, Germany, 1968.

8.5 Framklin D Yeaple, *Hydraulic and Pneumatic Power and Control* McGraw-Hill Book Co., New York.
8.6 Leiber, Wolfgang, *Hydrauliche Anlagen*, Carl Hanser Verlag, Munich, 1966.
8.7 Dipl.-Ing. Kurt Rauch, Fachbuchverlag Schaarschmidt, Einfuerung in die Hydraulik und Pneumatic—ihre Elemente und Symbole, 782 Goeppingen, Germany.
8.8 Bosch Hydraulics—Information and Data 1971/72, Part 1, Robert Bosch GmbH, Stuttgart, Germany.
8.9 M/s Seidel Stahlrohrhandel GmbH, Manheim, Germany.

9
Rotary Actuators and Hydrostatic Transmission

INTRODUCTION

IN hydraulic systems generally hydraulic motor provides rotational movement. A hydraulic motor transforms hydraulic energy into rotary mechanical energy which is applied to a resisting object by means of a shaft connected with the motor.

9.1 HYDRAULIC MOTORS

Hydraulic motors closely resemble hydraulic pumps in construction and size. The only difference is that instead of pushing the fluid as the pump does, in a hydraulic motor the rotating elements (i.e. vanes, gears, pistons, etc.) are pushed by the oil pressure to enable the motor shaft to rotate and thus develop the necessary turning torque and continuous rotating motion.

9.1.1 Motor Ratings

Hydraulic motors are rated according to:
  (i) Displacement (size)
  (ii) Torque capacity and
  (iii) Maximum pressure limitation.

(i) *Displacement*   Displacement can be defined as the amount of fluid which the motor may accept in turning one revolution or in other words, the capacity of one chamber multiplied by the number of chambers the motor internal mechanism contains. Motor displacement is expressed in cm^3/revolution.

(ii) *Torque*   Torque is the force component of the output of the motor. Mathematically, it is defined as the turning or twisting effort of the rotating body. Torque indicates a force present at a distance from the centre of the motor shaft. It may be mentioned here that motion is not required for a torque, but motion will result if the torque is sufficient to overcome friction and resistance of the load. The unit of torque in the ISO System is Nm. From Fig. 9.1, we can define torque as

$$\text{Torque} = \text{force} \times \text{distance from the centre of the shaft.}$$

The unit of torque mostly used is kgf.m or, Nm.
The further the force is from the shaft, the larger the torque at the shaft is.

Rotary power = torque × angular velocity.

$$= T \times \text{rev./s} \times 2\pi \text{ where } T = \text{Torque.}$$

The input hydraulic power $= \dfrac{T \cdot \text{rate of flow of oil to motor, l/s}}{\text{Motor capacity, cm}^3 \times 10^{-3}} \times 2\pi$

i.e.
$$P \cdot Q = \dfrac{T \cdot Q}{10^{-3} \times C} \cdot 2\pi$$

where P = pressure in bar,
Q = rate of flow of oil, l/s
T = torque in kN.cm,
C = motor capacity, i.e. displacement, cm^3

1 kN = 1 kilo Newton
    = 10^3 Newton.

∴ Motor torque $T = \dfrac{P \cdot C \cdot 10^{-3}}{2\pi}$ kN.cm

Starting torque is usually lower than the running torque because of breakaway friction characteristics. Gear and vane type motors will have an efficiency of 65–85% and piston type 75–95%.

Flow rate Q = rpm × displacement cm^3/rev.

$$\text{rpm (Motor shaft speed)} = \dfrac{Q\, l/s}{\text{displacement, cm}^3/\text{rev}}$$

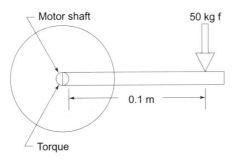

Fig. 9.1 *Definition of torque*

(iii) *Pressure* Pressure required in a hydraulic motor depends on the torque and displacement. A large displacement motor will develop a given torque with less pressure than a smaller unit.

9.1.2 Classification of Motors

Hydraulic motors generally may be:
 1. Uni-directional, or
 2. Bi-directional.

So far as construction is concerned, they are mostly:
 1. Gear type
 2. Vane type
 3. Piston type

However, other types of motors are also available, e.g. screw, gerotor, etc. as in the case of pumps. Let us discuss them in brief.

9.2 VANE MOTORS

A vane motor is a positive displacement motor which develops an output torque at its shaft by allowing hydraulic pressure to act in the vanes which are extended.

Basically a vane motor consists of vanes, rotor, cam-ring, shaft and port plates with kidney shaped inlet and outlet ports as we have seen in the case of pumps.

All hydraulic motors operate by causing an imbalance which results in rotation of the shaft. In vane motors the imbalance is caused by the difference in vane area exposed to hydraulic energy which is due to the eccentric position of the rotor to the housing. When pressure oil enters the inlet port, the unequal area of the vanes results in development of a torque in the motor shaft. The larger the exposed area of the vane, or higher the pressure, more is the torque that is developed. This makes the shaft rotate.

In a hydraulic motor two different pressures are involved
 1. System working pressure
 2. Pressure in tank outlet.

This results in side loading in the motor shaft.

To avoid shaft side loading the inner contour of the ring is changed from circular to cam-shaped. With this arrangement, the two pressure quadrants oppose each other and the forces acting on the shaft are *balanced*. Shaft side loading is therefore eliminated.

9.2.1 Construction of Vane Motors

Cartridge assembly A simple vane motor is shown in Fig. 9.2.

The rotating part of industrial vane motors is usually an integral cartridge assembly which consists of: (a) Vanes, (b) rotor, (c) cam-ring sandwiched between two port plates. The advantage of such a cartridge is easy motor servicing. After a time when motor parts naturally wear, the rotating group can be easily removed and replaced with a new cartridge assembly. Such a cartridge

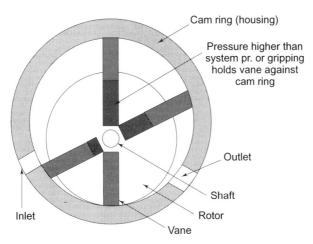

Fig. 9.2 *Basic design of a vane type motor*

assembly may help the vane motors develop more torque. When the same motor is required to develop more torque at the same system pressure, a cartridge assembly with the same outside diameter but larger exposed vane area can be quickly substituted for the original.

9.2.2 Extending Vanes

Unlike a vane pump, centrifugal force cannot be depended on to throw out the vanes and create a positive seal between the cam-ring and vane tip. Some other methods are used.

1. One method is spring loading the vanes so that they are extended continuously.
2. The other method is directing hydraulic pressure to the underside of the vanes.

Spring loading, shown in Fig. 9.3, is done either by a coil spring in the vane chamber or by using a small wire spring. The wire spring is attached to a post and moves with the vanes as it travels in and out of the slot.

In both types of spring loading, fluid pressure is directed to the underside of the vane as soon as the torque is developed.

Another means of extending the vanes is use of fluid pressure. In this method, fluid is not allowed to enter the vane chamber area until the vane is fully extended and a positive seal exists at the vane tip when the oil pressure is high enough to overcome the spring force biasing the internal check valve.

A balanced vane motor is shown in Fig. 9.4. As in the case of the vane pump, here also the rotor containing the vanes is concentrically placed creating two inlet and two outlet openings opposite each other thus maintaining the balance. However, here the fluid chambers are of fixed capacity contrary to the eccentrically positioned rotor where the chamber size can be varied by changing the amount of eccentricity.

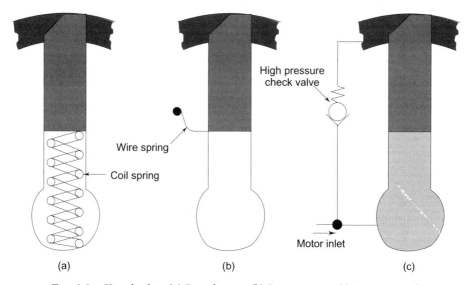

Fig. 9.3 *Vane loading (a) By coil spring (b) By wire spring (c) By pressure oil*

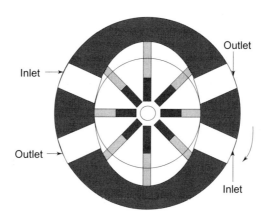

Fig. 9.4 *Balanced vane motor*

Pressure compensated vane motors are also used in hydraulic systems.

9.3 GEAR MOTORS

A gear motor is generally a fixed capacity motor. Both external and internal gears are used as discussed in the case of pumps. Gear motors are imbalanced motors. The hydraulic imbalance in a gear motor is caused by gear teeth unmeshing. As gear teeth unmesh it can be seen that all teeth subjected to system pressure are hydraulically balanced except for one side of one tooth on one gear. This is the point where the torque is developed. The larger the gear tooth or higher the pressure, more is the torque produced.

9.4 PISTON MOTORS

Generally three types of piston motors are used:
(i) In-line piston
(ii) Axial piston
(iii) Radial piston.

They are positive displacement motors which can develop an output torque at the shaft by allowing oil pressure to act on the pistons. However axial and radial piston motors are most common out of which the axial piston motors are used from medium to high pressure and moderately high power ranges whereas, radial piston motors are used for very high pressure ranges. Truly speaking, radial piston motors are high torque high power systems.

9.4.1 Construction of an Axial Piston Motor

Here again most of the motor designs are swash plate type. A schematic diagram of a swash plate piston motor is shown in Fig. 9.5 (a) and (b). The basic parts and flow of oil are marked in the figure.
1. Flow of oil from valve side
2. Motor shaft
3. Bearing
4. Return oil from motor
5. Control disc or plate
6. Housing
7. Cylinder block
8. Pistons
9. Swash plate
10. Cover
11. Pressure zone
12. Oil delivery zone
13. Maximum travel zone of piston.

The swash plate is positioned at an angle and acts as a surface on which the shoe side of the piston travels. The piston shoes are held in contact with the swash plate by the shoe plate and the bias spring. A port plate separates incoming fluid from the discharge fluid. The rotating shaft is connected to the cylinder barrel. An exploded view of an axial piston swash plate motor is shown in Fig. 9.6.

9.4.2 Working Principle

Let us understand the working principle of a swash plate axial piston hydromotor. As explained earlier, a hydromotor should produce a turning moment on the shaft so that hydraulic energy can be converted to mechanical work. In a swash plate motor the swash plate is held fixed at an angle and the cylinder block containing the axially placed pistons in the moveable block is made to rotate by admitting the pressure oil. From Fig. 9.7 we can see that a force F acts on the moveable piston—due to the oil pressure as well as the weight of the piston. This force F can be resolved as:

(a) F_N—normal to the plane
(b) F_T—tangential to the plane.

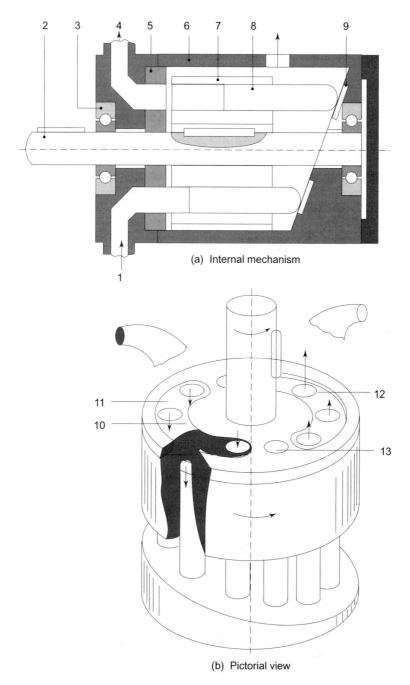

Fig. 9.5 *Swash plate piston motor*

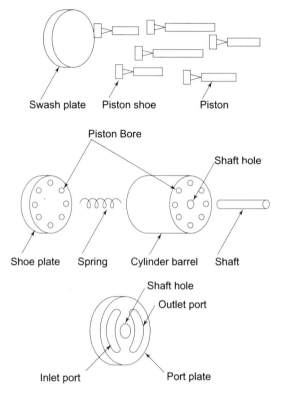

Fig. 9.6 *Parts of axial piston motor in an exploded view*

Let the distance of the force F_T from the centre of the shaft be r.

Due to this force the turning moment acting against the centre of the cylinder block can be written as:

$$\text{Turning moment (Torque)} = F_T \times r.$$

Hence due to this torque the cylinder block carrying the pistons is glided over the slant plate called the swash plate, on a circular path and rotation of shaft is created. Thus a torque is produced.

The torque continues to be developed by the piston as long as it is pushed out of the cylinder barrel by oil pressure. Once the piston passes over the center of the circle it is pushed back into the piston bore of the cylinder barrel by the swash plate. At this point the piston bore will be open to the outlet port of the port plate. It is to be noted that a single piston develops only 1/2 rotation of the full circle of rotation of the cylinder barrel. In practice a number of pistons are there attached to the barrel, which allow the barrel and shaft to rotate continuously.

The number of pistons vary, but to get a reasonable torque at least three pistons are a must. The more the number of pistons, the higher is the torque. It is better to have odd number of pistons.

As per the diagram in Fig. 9.5 (b) we find five pistons are receiving oil and five are delivering oil and one is at the dead point (maximum travel point).

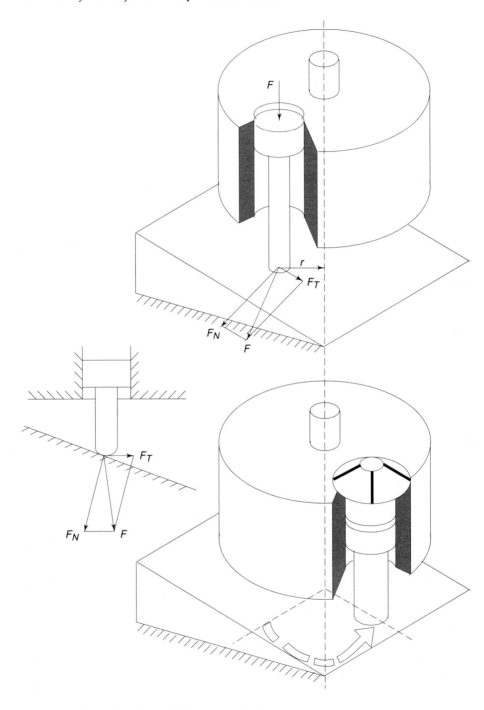

Fig. 9.7 *Principle of operation of a swash plate axial piston motor*

With the swash plate positioned at an angle, the piston shoe does not have a very stable surface on which to position itself. When fluid pressure acts on the piston pushing the piston out of the slot, it causes the piston shoe to slide across the swash plate surface. As the piston shoe slides, it is able to develop a torque at the motor shaft attached to the barrel. The amount of torque depends on the angle of slide caused by the swash plate and the oil pressure in the system.

For uniform rotation without much fluctuation, the number of pistons should be above seven. If the cylinder block contains more number of pistons, a high, continuous and stable torque is produced. Due to the slant of the swash plate, when one piston reaches its end, it starts withdrawing and the next takes its position producing the moment. Hence a continuous torque is ensured. This is shown schematically in Fig. 9.8.

The direction of rotation of the cylinder block and the shaft is made possible by changing the direction of flow of oil.

Axial piston motors are available in both fixed as well as variable capacity modes. The swash plate angle θ determines the amount of displacement and therefore the flow quantity. The maximum piston stroke is diagrammatically shown in Fig. 9.9(a). In the variable model the swash plate is mounted on a swinging yoke. The angle can be changed by various means, e.g. a simple lever, a hand wheel or a servo-controlled hydraulic cylinder. If the swash plate angle θ is increased, the torque capability of the motor will also increase but it may reduce the rotational speed of the shaft.

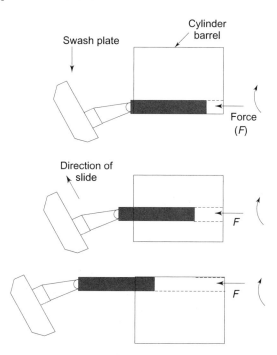

Fig. 9.8 *Sliding of piston on swash plate surface*

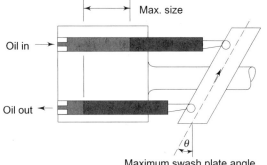

Fig. 9.9 *Variable displacement of piston motors—maximum and minimum torque against swash plate angle*

9.4.3 Bent Axis Piston Motor

Just like a bent axis pump, here also the axis of the cylinder block and the shaft are mounted at an angle to each other and the reaction is against the drive shaft flange. A schematic diagram of a bent axis axial piston motor is shown in Fig. 9.10.

These motors are used for machine tools like table movement of die sinking m/c for their smooth working.

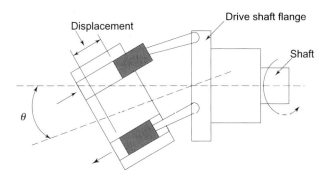

Fig. 9.10 *Bent axis motor*

9.4.4 Radial Piston Motor

The basic principle of operation of a radial piston motor is shown in Fig. 9.11(a). Design engineers in more and more industries are interested in *low speed high torque* motors to address multifarious problems in diverse power transfer applications. Radial piston motors have been found to provide solutions to such low speed high torque applications without any side loading or minimal side loading. Radial piston motors are designed with capacities from 200 cm^3/rev. to 10,000 cm^3/rev. at very high pressures upto 450 bar or even

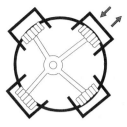

Fig. 9.11 *(a) Basic principle of operation of radial piston motor*

higher. In order to reduce or minimise the effect of side loading, robust roller bearings are used with such high capacity motors. They have found application in marine winches and rudders, construction equipment, high power industrial machines, etc. A pictorial view of a radial motor is shown in Fig. 9.11(b) and constructional view is shown in Fig. 9.11(c).

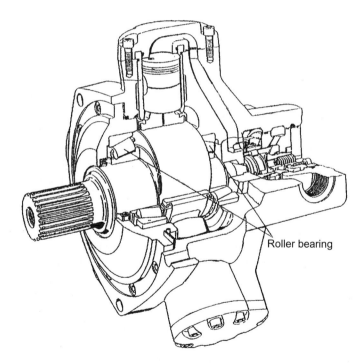

Fig. 9.11 *(b) Radial piston motor with two taper roller bearing*

9.5 SELECTION OF HYDRO-MOTORS

The following points are noteworthy while selecting hydro-motors for a specific application.

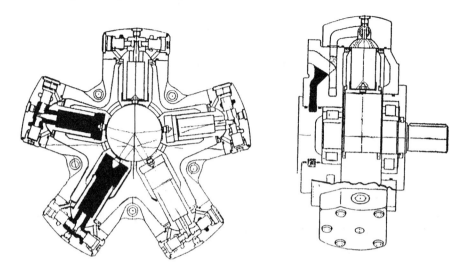

Fig. 9.11 *(c) Cross-sectional view of radial piston motor*

1. The amount of torque needed to move the work (breakway torque) has to be calculated.
   Breakway torque is the amount of torque required to initially start a load moving. This is much higher than running torque.
   Running torque is the amount of torque required to keep a load moving after it has been started.
2. System pressure (working)—After consulting the manufacturer's manual a suitable pressure is selected to produce the required torque. Depending on the system, it is necessary to oversize the motor as much as 40% or more to allow for the "breakway torque".
3. When hydraulic motors are used at a high rpm (such as flywheel conveyor system) a cross-over relief or breaking relief valve should be utilized. This valve will absorb most of the shock created by the centering of the DC valve and inertia of the rotating mass.
4. Hydro-motors should be operated at the highest possible speed. Generally 100 rpm is the lowest practical speed unless the motor is designed for low rpm high torque application. Gear reduction is sometimes used to obtain lower output.
5. Flow control valves may be used to control motor speed. In some systems, a meter-out application may create enough back pressure on the motor to cancel out some of the torque which the motor is otherwise able to produce. On meter-in application care should be taken on some lower rpm applications, where there may be binding or change in friction, this will affect the constant rpm.

6. With 'by-pass' application a pressure compensated FCV should be used. Some critical requirements may also require the pressure compensated FCV to be temperature compensated. This creates less heat in the system than in meter-out or meter-in HP savings.

9.6 HYDRAULIC OR ELECTRICAL MOTORS

Sometimes one may get confused in deciding whether to go for a hydraulic motor or an electrical motor for the solution of one's drive problems. Though the electrical motor has its universal appeal due to its versatile use, hydraulic motors have certain specific advantages over electrical motors which may decide the application of a hydraulic motor in certain specific fields. The advantages can be listed as:
1. Easy, smooth and instant reversing of motor's shaft without any shock.
2. Possibility of controlling the shaft torque throughout the operating speed.
3. Possibility of braking action at ease without any shock.
4. Most comfortable power-to-weight ratio of the hydraulic motor over the electrical motor.
5. Hydraulic motor can be stalled for an indefinite period without damage.
6. Very low starting torque of the hydro-motor.

9.7 HYDRAULIC MOTORS IN CIRCUITS

As has been explained earlier, hydro-motors are designed both in the uni-directional as well as bi-directional mode. Their symbols are shown in Figs 9.12 (a) and (b). In Fig. 9.13 (a), we find a hydraulic circuit used to control a uni-directional motor. Fig. 9.13 (b) depicts another circuit with a uni-directional motor having a pressure compensated flow control valve to control shaft speed. Readers may note here that most spring loaded check valves are having a spring rating equivalent to 4.5 bar to enable the motor shaft free-wheeling.

A simple circuit of a bi-directional motor is shown in Fig. 9.14 (a). Here we may note the use of four check valves along with a pressure relief valve for simple braking (at same level of braking) as well as the use of a 4.5 bar capacity non-return check valve positioned in the return line of the system for free wheeling. Readers may also notice the use of a normally open center or tandem center

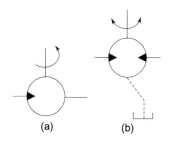

Fig. 9.12 *Motor Symbol (a) uni-direction (b) bi-direction*

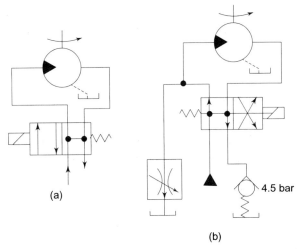

Fig. 9.13 *Uni-direction motor control (a) without speed and (b) with speed control*

direction control valve in the motor circuits. Contrary to Fig. 9.14 (a), in Fig. 9.14 (b), if level of braking is different, one may use two pressure relief valves for braking action. This is called dual relief braking.

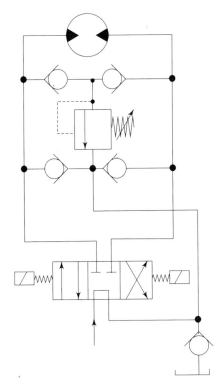

Fig. 9.14 *(a) Hydro-motor circuit with braking at same level*

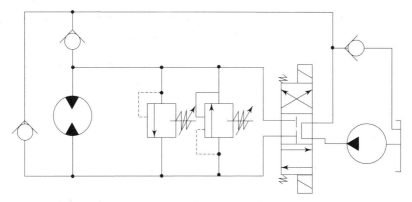

Fig. 9.14 *(b) Hydro-motor circuit with different braking*

Free wheeling When the load attached to a motor's shaft is allowed to free wheel, the load is allowed to coast to a stop. A motor, which uses hydraulic pressure to extend its vanes, requires 4 to 8 bar capacity check-valve in the tank line if the load is allowed to free wheel. The back pressure, which is generated because of the tank line check valve, keeps the vanes from retracting. This slows the load down more quickly.

9.8 TYPES OF HYDRAULIC TRANSMISSION

Most common types of hydraulic transmission drives may be categorized as shown below with their schematic view in Fig. 9.15.

(a) Rotodynamic or hydrokinetic drive to which fluid couplings and torque converters belong are schematically shown in Fig. 9.15 (a).
(b) Hydroviscous drive which is available in three types such as magnetic particle clutch, electroviscous slip clutch and silicone slip clutch.
(c) Hydrostatic drive which is a purely hydraulic drive with a pump and motor combined together shown in Fig. 9.15 (b).
(d) Hydromechanical drive shown in Fig. 9.15(c), which is again a hydrostatic drive combined with a gear box having three typical forms e.g.:

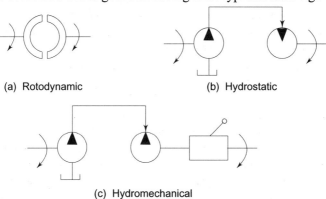

Fig. 9.15 *Types of drives*

(i) Split torque drive in which a variable displacement motor adjusts the output speed of a mechanically driven differential planetary system by varying the speed of one gear.
(ii) Torque drive mechanically shiftable but separate gear box driven by a hydrostatic or hydrodynamic drive.
(iii) Hydraulic planetary gear box.

9.9 PUMP-MOTOR COMBINATION

Various hydraulic pumps and motors are combined together by designers to achieve specific system requirements such as to develop smooth stepless change in transmission ratio and torque conversion. For example when a fixed displacement hydraulic pump is combined with a fixed displacement hydraulic motor, the system provides a constant output power. The motor generates constant torque and constant speed. A constant displacement pump when combined with an adjustable capacity motor, delivers constant power to the motor but the motor shaft speed and torque will be variable. In contrast if a variable displacement pump is combined with a fixed displacement motor, the latter develops a constant torque but may develop variable speed. When a combined pump-motor system uses both the pump and motor with variable capacity, the torque, speed and power developed by the motor will be variable.

9.9.1 Basic Pump-motor Combination

1. Fixed capacity pump and hydraulic motor with variable prime mover   Here both the hydraulic pump and motor are fixed displacement type but motor speed can be adjusted by varying the speed of the prime mover [shown in Fig 9.16(a)] which can be either an electric motor or an internal combustion engine. This arrangement is to control the pump flow rates.

2. Fixed capacity pump and motor with flow control valve   Here also both the hydraulic pump and the motor are fixed displacement type shown in Fig. 9.16(b). The speed of the motor can be controlled by a flow control valve by throttling oil which however may limit the overall efficiency of the system.

3. Fixed displacement pump with variable hydraulic motor   Here one uses a fixed capacity pump with a variable motor. The speed of the hydro-motor is adjusted by means of adjustment of pistons of motor. In this drive the torque decreases as speed increases or vice versa. The performance characteristics of such a PF-MV combination is shown in Fig. 9.16 (c) along with the layout arrangement.

4. Adjustable pump and fixed displacement motor   Here the motor speed can be adjusted by varying pump displacement to vary the pump flow [shown in Fig. 9.16(d)]. As the motor displacement is fixed and the pump is able to maintain full pressure, (neglecting the loss in efficiency) the motor torque remains constant.

At zero displacement, the pump acts as a clutch. A stepless increase in the displacement of the pump causes the motor to rotate at constant torque up to the

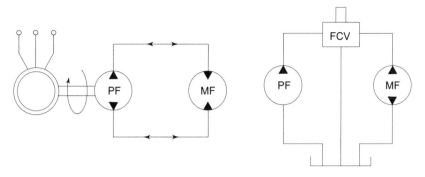

Fig. 9.16 *(a) Fixed capacity pump-motor variable prime mover*

Fig. 9.16 *(b) Fixed capacity pump-motor with flow control valve (FCV)*

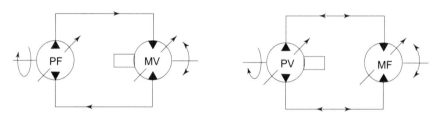

Fig. 9.16 *(c) Variable pump-variable motor*

Fig. 9.16 *(d) Variable pump-fixed motor*

maximum designed speed. The overcenter reversal control of the pump displacement will reverse the motor rotation. Such an ability of a variable pump and fixed motor combination may enable the system to have at ease variable speed, constant torque, reversing of motion and clutching action over a wide range of speed and power.

5. *Variable pump and variable motor*   As shown in Fig. 9.16 (e) both the pump and motor are variable. The motor speed can be controlled by manual adjustment of both pump and motor displacement. With such a design the speed range is quite wide such as 70 : 1. The performance curve for a variable pump-variable motor combination is shown in Fig. 9.16 (f).

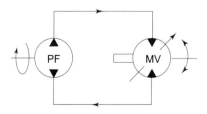

Fig. 9.16 *(e) Variable pump-variable motor*

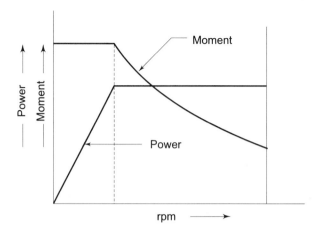

Fig. 9.16 *(f) Power and moment curve for a variable pump motor combination*

9.9.2 Hydrostatic or Hydrokinetic

In common terminology whenever a variable displacement pump or motor is used in a pump-motor circuit, it is labelled as hydrostatic transmission. But any hydraulic drive that can translate its pressure energy into a mechanical output may be termed as a hydrostatic machine. From this definition one can term a hydraulic pump as a hydrostatic device as oil under pressure is supplied to the system to do mechanical work. Similar is the case with a hydraulic cylinder. A simple hydraulic press is a good practical example of a hydrostatic device.

However, in specific terms a hydrostatic transmission is a case of energy transmission which consists of a system where the oil input device is a positive displacement pump and the output element is a positive displacement hydraulic motor both of which are designed with matching physical and operational characteristics with optimisation of energy transmission. The principle of hydrostatic transmission comprising a pump and motor involves in oil trapping moving the trapped oil and utilizing the trapped oil pressure with practically no change in the oil velocity. That is, the oil maintains a passive character which is in direct contrast to any normal hydrokinetic device. In the case of a hydrokinetic device it is change in the velocity of the fluid which provides the transmission of energy. In the case of a hydrostatic device a small flow of fluid at high pressure is responsible for the energy transmission.

But in a hydrokinetic system one uses a large volume of oil at a considerably low pressure. More precisely one can define a hydrostatic drive as a transmission system in which the circulated oil acts as the energy transporter between a primary unit, i.e. a pump and a secondary unit, i.e. a hydraulic motor and thus transmits rotary motion and torque. The oil transmission cycle comprises of the mechanical energy of the drive motor rotation being transformed into hydraulic energy by the pump to which oil flows from the motor. This means the oil is under pressure which from the pump again moves under pressure to the hydro-motor and the hydro-motor once again transforms the energy into mechanical energy (i.e. its shaft rotation).

9.9.3 Why Hydrostatic Transmission?

For applications where efficiency, wide adjustability and ability to hold constant speed at all loads is the main concern, a hydrostatic drive is most suitable. Let us see here what is the primary objective of any energy transmission system. The main objectives of a hydrostatic drive are

1. To accept energy input from a source
2. To transmit and modulate the energy within the energy transmission system
3. To deliver an energy output to the load.

Compared to other forms of energy transmission system, a hydrostatic device has certain advantages.

1. It provides stepless speed which ensures smooth and jerk free shock proof starting and stopping of the load.
2. In such a system speed changing is free of any shock.
3. Such a system transmits high power per cm^3 displacement with low inertia.
4. Continuous transmission between drive motor and wheel drive is possible in a hydrostatic drive.
5. This mode of energy transmission holds the current speed accurately against driving or braking load.
6. It can operate effectively in reverse at controlled speeds unaffected by output loads within design limit.
7. Dynamic braking by reduction of pump delivery rate is possible in hydrostatic drive.
8. Ample freedom of layout of pumps and motors is possible in such a drive.
9. Acceleration and deceleration can be limited to any value.
10. Faster response is achieved here than in any other type of transmission.
11. No creep at zero speed.
12. Simple and reliable overload protection is an added advantage.
13. Prime mover can be operated within its most economic speed range and power range.

9.10 OPEN LOOP OR CLOSED LOOP SYSTEM

Hydrastatic transmission is one of the latest introductions in the field of transmission technology. This facilitates infinite speed control without shocks. Because of its advantages, hydrostatic transmission has gained popularity for various applications in construction and earthmoving equipment as well as heavy vehicles including armoured vehicles used for combat application.

Hydrostatic transmission systems can be classified as:

(a) open circuit and (b) closed circuit.

Chart 9.1 *Classification of hydrostatic transmission systems*

```
                    Hydrostatic transmission
                    ┌──────────┴──────────┐
            Open circuit,            Closed circuit,
            discontinuous            continuous or
          or interrupted flow      uninterrupted flow
           ┌──────┴──────┐           ┌──────┴──────┐
      Open loop     Closed loop   Open loop     Closed loop
      shown in      shown in      shown in      show in
      Fig. 9.17 (a) Fig. 9.17(b)  Fig. 9.17 (c) Fig. 9.17(d)
```

The open and closed circuits depict the way the hydraulic lines are connected. In the closed circuit the flow path is continuous without interruption whereas the open circuit the flow path is not continuous being interrupted by the oil reservoir. In case of the closed circuit the pump flow moves to the motor inlet from the pump's discharge

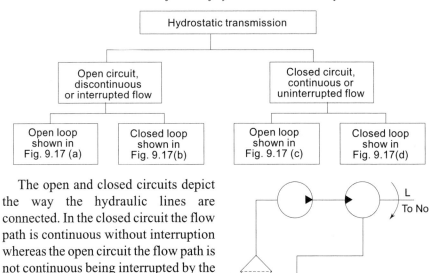

Fig. 9.17 *(a) Open loop open circuit HST*

Tr = Transducer
To No = Torque, power
C = Command
HST = Hydrostatic transmission
L = Load

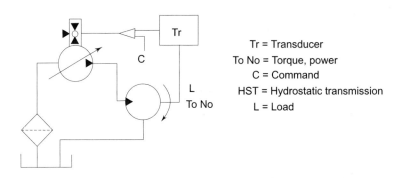

Fig. 9.17 *(b) Open circuit closed loop*

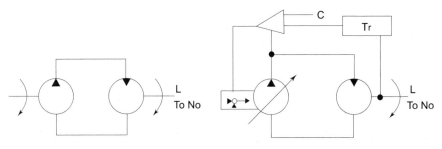

Fig. 9.17 *(c) Open loop closed circuit HST* **Fig. 9.17** *(d) Closed loop closed circuit*

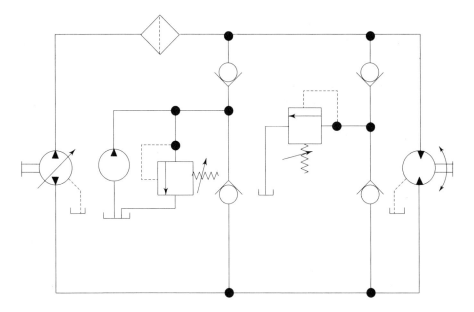

Fig. 9.18 *Closed circuit-closed loop hydrostatic system*

port without interruption and again from the motor it is directly fed to the pump.

In Fig. 9.18, a closed circuit closed loop hydrostatic transmission circuit diagram is shown. Here the variable displacement pump can vary the speed of the hydro-motor shaft as well as reverse the shaft rotation. A small replenishing pump is used to make up for every leakage which may occur in the system. Closed loop systems are compact and for reversing the system, there is no need of a direction control valve. Similarly speed control is also achieved without any flow from the control valve.

In-built braking capability is an advantage of the closed loop system over the open loop. The reason is that as soon as the pump stops pumping, the motor cannot rotate without displacing the pump. This means that the motor now tries to act as a pump and the pump as the motor which ultimately results in "braking" the system and thus free wheeling is avoided.

Though we have discussed a number of theoretical possibilities of pump-motor combinations, in common practice one uses only a variable pump and a variable motor for most hydrostatic transmission. As the same oil is being fed from the pump to the motor and then recirculated back to the pump, there is every possibility that the quantity of oil in the system may get depleted because of internal leakage. Moreover due to continuous recirculation inside the circuit the oil may get heated. So this oil needs continuous replenishment with fresh oil for which a small pump is used to compensate the leakage. This pump is called either a *charge pump* or a *booster pump*.

9.11 APPLICATION OF HYDROSTATIC TRANSMISSION

Over the years hydrostatic transmission systems are finding growing applications in mobile equipment.

Because of the advantages hydrostatic transmissions are used in varieties of applications as given below.
1. Earth moving equipment
2. Truck mounted cranes
3. Winches
4. Armored vehicles steering system
5. Railway shunting engines
6. Wheeled loaders
7. Other mobile applications, e.g. heavy automobiles, tractors, etc.
8. Ship installations, e.g. rudders, mast etc.
9. Aircraft systems, etc.

In an automobile system using rotodynamic transmissions, as the vehicle accelerates, the transmission ratio decreases smoothly, and the input shaft is either connected directly with the output shaft or the torque converter automatically begins to operate as a fluid coupling. In aircraft rotodynamic couplings are used to transmit torque from the starter engine to the rotor of the main gas turbine engine.

Figure 9.19 (a) and (b) provides the circuit diagram of a HST drive for a wheeled loader (M/s Marshall Sons and Co. Ltd., Chennai). The circuit diagram consists of a variable displacement axial piston pump. A gear pump is used as a charge or booster pump. The hydromotor, a variable displacement axial piston motor is used for the speed changing device which in turn drives the wheels through axles and propeller shafts fitted to it.

Mostly the pump is a swash plate type, the swash plate angle being controlled by four control cylinders. The angle of swash plate is decided by the load/speed of the machine. This is controlled by the charge pump which is intergrated and coupled to the main pump.

The traction device is operated by the variable axial piston motor for which the operator of the machine can pre-select the higher or lower displacement level depending on the load or speed requirement. Maximum displacement position is used for working and the minimum for road travel.

The controls to the system can be manual, hydrostatic or electronic. Selection of forward or reverse is done through an electrical switch. The machine starts moving forward or backward when the take-off speed of the engine is reached. For hoisting, tilting, steering, etc. the inching mode of the pedal control is used. The whole system is quite flexible and reduction or increase of speed is done at ease with adequate electronic support for feed back information between the various functional elements. While on the road the total disengagement of the motor may enable the vehicle to act as a normal vehicle.

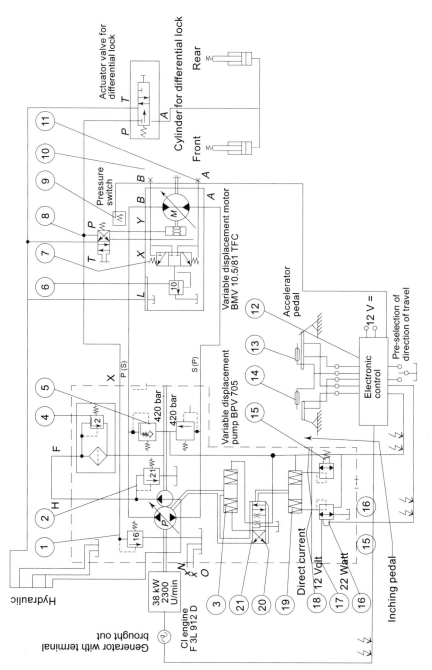

Fig. 9.19 *(a) Hydrostatic scheme AR 61 BE*

310   *Oil Hydraulic Systems: Principles and Maintenance*

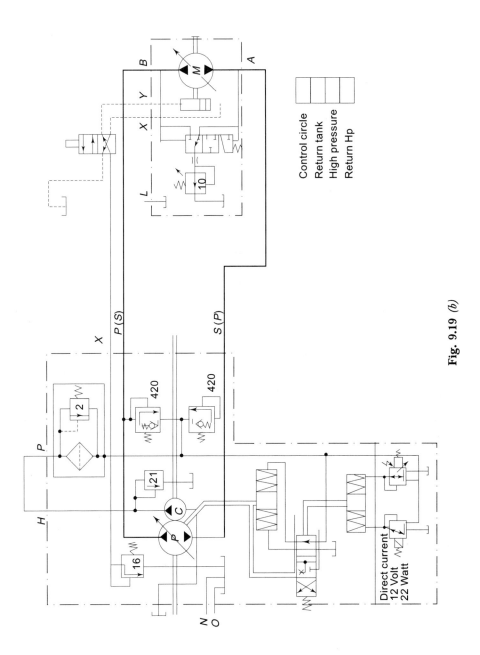

Fig. 9.19 *(b)*

The circuit diagram consists of the following elements [Refer Fig. 9.19(a)]
1. Feed valve
2. Cold starting valve
3. Positioning piston
4. Filter
5. High pressure relief valve with feeding unit
6. Scavenging valve
7. Change-over valve with minimum pressure opening limit
8. Change-over valve for on the road and cross country driving
9. Pressure switch
10. Measuring connection forward travel pressure.
11. Measuring connection reverse.
12. Electronic control
13. Travel potentiometer
14. Inching potentiometer
15. Forward control magnet
16. Reverse control magnet
17. Pressure reducing valve
18. Measuring connection for control pressure
19. Positioning cylinder
20. Pilot valve
21. Nozzles.

9.12 HYDROSTATIC STEERING

It is noticed that hydrostatic steering transmission offers certain advantages:
1. The overall efficiency is very high since there is no energy loss except for the efficiency of the individual components. Hydrostatic steering is more efficient than mechanical steering.

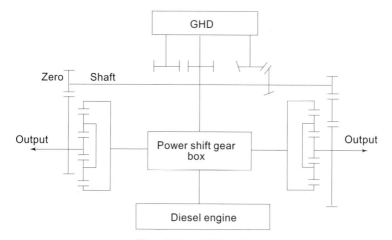

Fig. 9.20 *HST steering*

2. Hydrostatic steering transmission allows to choose an accurate and continuous turning radius with a direct relation between the steering wheel of the vehicle and the reaction of the tracts.
3. No wearing of parts and longevity of the system is high.
4. High and quick response—response time being as low as 0.3 to 0.5 s.
5. Operates at high pressure and high speed.
6. Can operate at high oil temperatures up to 160°C or above.
7. Power range achieved from 100 kW to 1000 kW.
8. Can be used for very heavy vehicle even for loads of 50–60 tonnes.

A simple block diagram of a power steering system is shown in Fig. 9.20.

9.13 TORQUE CONVERTER

During the last couple of decades torque converters are being extensively used as a means of power transmission. It is used as a connecting link between the load and the primemover. In a purely mechanical system, power transmission takes place through a gear box, chain and sprocket or belt and pulley. The function of a torque converter can be compared to the above mechanical methods but there is a specific difference. The transfer of energy by a torque converter takes place through medium of flow. In the case of gears or other mechanical methods, the power transfer takes place with a definite ratio of input to output. But in a hydrodynamic torque converter, the prime mover torque is not multiplied by a continuously varying ratio dependent on the resistance due to load. Torque converters are used when it is necessary to change the torque and develop a transmission ratio other than unity while retaining appreciable efficiency. Torque converters may be single-stage or multistage.

9.13.1 Construction

There are three main parts in a hydrodynamic torque converter:
1. An *impeller* is a bladed member connected to the engine which absorbs the engine torque and energy is transferred to the hydraulic fluid.
2. The *turbine* is also a bladed member which receives hydrodynamic force from the hydraulic fluid and converts it into output torque, i.e. mechanical energy.
3. The *guide wheel* is another main part which directs oil back to the impeller from the turbine. The oil is fed to the impeller at a specific angle so that a reaction torque which adds to the turbine, counter balancing the torque in the impeller.

The other important parts are:
4. Charging pump
5. Input cut-off clutch
6. Power take-off
7. Casing
8. Heat exchanger, etc.

Automatic multiplication of torque and adjustment of speed is made possible by the hydrodynamic action of the fluid inside the converter.

9.13.2 Principle of Operation

A general view of a single-stage torque converter is shown in Fig. 9.21. The impeller is fitted to the engine and rotates along with the fly wheel which thus imparts kinetic energy to the hydraulic oil passing over the impeller blades. The kinetic energy thus produced is absorbed by the turbine and converted to torque. The oil from the turbine enters the guide wheel where the oil undergoes a change of direction of velocity that allows generation of the reaction force. The reaction force adds to the turbine and hence multiplication of torque is effected. In a

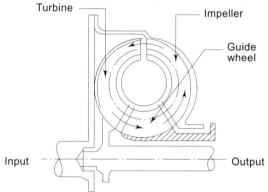

Fig. 9.21 *Single-stage torque converter*

multistage converter there are a number of turbine blades interposed between the impeller and the stator blade sections. Thus in a three-stage torque converter the blade sections of the turbine are interposed between the impeller and two stator blade sections. As the fluid passes over three stages of the turbine blades, it is called a three-stage torque converter.

While in a single-stage torque converter, the torque can be multiplied by 2 to 3.5 times, in a three-stage converter it may be 4 to 5 times. It has been noticed that around 25 to 30% of the input power gets converted to heat. A heat exchanger is therefore essentially required to dissipate the generated heat and maintain a tolerable temperature range. Similarly, a filter is needed to maintain a pre-determined cleanliness level. Figure 9.22 (a) shows a three-stage torque converter.

When the output speed is zero, the oil velocity over the blades is maximum. The fluid acts on the turbine and stator blades at the most severe angles; thereby the efficiency of transmission is reduced to zero. But the reaction forces are at a maximum value which results in a high stall torque at the output. With the load picking up speed the velocity of fluid over the blades reduces as well as the oil acts on the blades at a more favorable angle thus lowering the multiplication factor and increasing the efficiency.

A typical engine torque converter matching curves are shown in Fig. 9.22 (b).

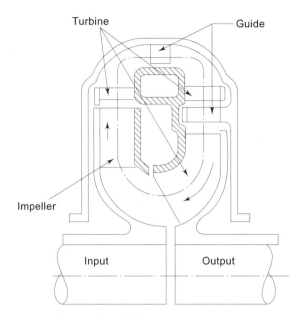

Fig. 9.22 *(a) Three-stage torque converter*

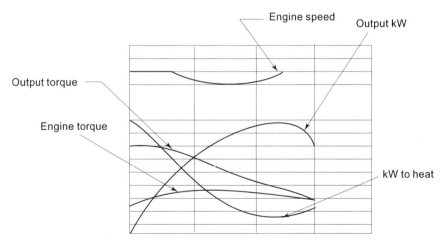

Fig. 9.22 *(b) Typical engine torque converter match curve*

9.13.3 Advantages

The advantages of a torque converter are:
1. There is no direct mechanical connection between the engine and the wheels.
2. The road and load shocks are effectively cushioned in the hydraulic medium.

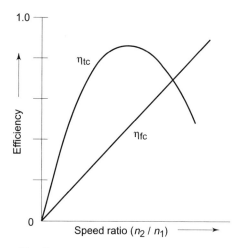

Fig. 9.22 *(c) Efficiency curves of torque converter and fluid coupling*

3. A vehicle needs the maximum tractive force at the start. The tractive efforts start reducing as the vehicle picks up. With stability in the traction force as the vehicle moves, greatly reduced traction force is required. The output torque is maximum at the start when the output speed is zero. As the output speed of the load picks up, torque multiplication reduces. The torque multiplication remains the least when the output speed increases and reaches its stable value.
4. It is known that in a vehicle with a manually operated gear box, the wheel speed is related to the crank shaft speed. This implies the resistance to movement controls the engine rpm which is turn will determine the power availability. But with the use of a torque converter, the impeller which connected to the engine is also physically separated from the turbine through oil while the turbine is connected to the wheel. Therefore there is every chance of speed of the vehicle which is influenced by the road resistance, controlling the engine rpm directly. Thus the engine continues to run at its usual high speed and ensures reasonably higher power throughout a wide range of vehicle speed.
5. In heavy mobile plants like loaders, earthmovers, etc. because of the advantage of fluid cushion, the torque converter facilitates shifting of changing gears under power. Thus the operator can have full control during sensitive operation of gear changes.
6. Torque converters are more economical than hydraulic coupling at high transmission ratio as may be seen from the curve in Fig. 9.22 (c).

9.13.4 Applications

Torque converters have been extensively used in various types of heavy machineries, e.g.
  1. Railway shunting locomotive

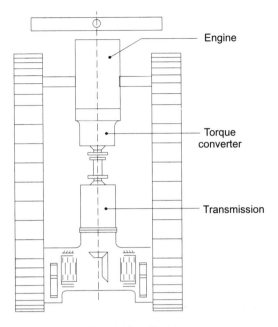

Fig. 9.23 *Crawler tractor*

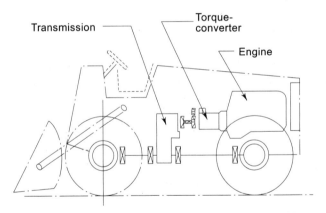

Fig. 9.24 *Front end loader*

2. Crawler Tractor (shown in Fig. 9.23)
3. Dozers Loaders (shown in Fig. 9.24)
4. Earth moving plates, etc.

Review Questions

1. How do you rate hydraulic motor? Define torque. How can one increase or decrease the torque in an axial piston motor?
2. Classify hydraulic motors on the basis of function as well as construction.

3. Explain the construction and function of a swash plate motor.
4. Explain the working principle of a vane motor.
5. How does one get a stable and continuous rotation in a hydromotor? Explain.
6. Differentiate between a hydrostatic and hydrokinetic system.
7. What is a closed loop and open loop circuit?
8. What is a torque converter? Where is it used? Explain its working function.
9. Compare a hydromotor and an electromotor.

References

9.1 Proceedings of the Regional Seminar on "Recent Advances in Fluid Power for Mobile Applications", organized by Fluid Power Society of India, Madras Chapter, Aug. 10, 1986.
9.2 Suresh, PG, "Hydrostatic Transmission", Marshall Sons & Co. Ltd., Chennai, Paper published in the seminar on "Recent Advances in Fluid Power for Mobile Applications", 1986.
9.3 Yeaple, FD (Ed.), *Hydraulic and Pneumatic Power and Control*, McGraw-Hill Book Co., New York, pp. 241–248, 1966.
9.4 Parker, *Industrial Hydraulic Technology,* Hannifin Corp., Cleveland, Ohio, USA, 1973.
9.5 Bosch, *Hydraulics Information and Data*, 1971/72, Part I, Stuttgart, Germany.
9.6 Lewis, EE and H Stern, *Design of Hydraulic Control Systems*, McGraw-Hill, 1962.
9.7 Streeter, VL, *Handbook of Fluid Dynamics,* McGraw-Hill, 1961.
9.8 Ernst, W, *Oil-Hydraulic Power*, McGraw-Hill, 1960.
9.9 Blackburn, JF, Reethof, G and Shearir, JL (Eds), *Fluid Power Control,* Tech. Press & John Wiley & Sons, 1960.
9.10 Holzbock Werner, *Why Hydraulic Drive?* Vickers Inc. Div. Sperry Rand Corp.
9.11 Nekrasov, B, *Hydraulics for Aeronautical Engineers*, Peace Publication, Moscow, Russia, pp. 254-255.
9.12 Reichel J. Dipl. Ing, "Hydraulische Antriebe in Betrieb mit Druckfluessigkeiter HFC", *Oel Hydraulik & Pneumatik,* 1991, p. 854.

10
Heat Generation in Hydraulic System

LIKE any other electromechanical system, a hydraulic system too gets affected by generation of heat which may sometimes lead to overheating of the hydraulic oil. Overheating can be a nightmarish experience for a maintenance mechanic and hence this should be at taken care of at the design stage itself. A hydraulic system designer, generally depends on the following factors to combat the effects of heat while designing the system:
1. Shape, size and type of oil reservoir.
2. Use of an appropriate heat exchanger for heat dissipation.
3. Appropriate use of oil volume in the system.
4. Optimization of system design parameters to reduce heat generation.
5. Use of appropriate size of fluid power components and their appropriate location.
6. Optimal pipe-layout of the system and reduction of unnecessary components and filters which may generate heat.
7. Proper choice of component and reservoir material, paints and protective coatings used, etc.

10.1 SOURCES OF HEAT

Though there may be a variety of reasons for heat generation in hydraulic components and system, the following sources of heat generation are important to note. A designer or a maintenance mechanic should take these into consideration in order to keep heat generation to a minimum if not eliminate altogether.

(i) Pump The pump which provides the required fluid flow to the system is the heart of the system and converts the mechanical energy into hydraulic pressure and flow. However if some energy from the electric motor to the pump is not converted to work, there will be heat generation due to the unutilized power which may be clear from the example below.

Let a pump in a system, working at 75 bar delivers 15 l/min of oil at 85% efficiency. If there is a loss of (15%) energy, this will be converted to heat, which will be equivalent to almost 0.32 kW of power as per the calculation shown below.

The output power of the pump is $= \dfrac{PQ}{612}$

$$= \dfrac{75 \times 15}{612} = 1.84 \text{ kW}$$

But input to the pump is $= \dfrac{1.85}{0.85} = 2.16 \text{ kW}.$

∴ Losses converted to heat = (2.16 − 1.84) kW = 0.32 kW

This is almost equivalent to 17.4% of the output pump power.

(ii) Friction   Losses due to friction in the pipes, joints, line fittings, etc. are also responsible for generating heat in the system.

(iii) Blockage in filter, valves and other elements   In case the movement of oil in the system pipes or motion of mechanical elements in the hydraulic components is hampered due to blockage, there will be heat generation.

(iv) Oil viscosity and other physical oil parameters   In case oil viscosity and other properties of the oil are erroneously chosen and are incompatible with the system and its working conditions, there may be heat generation in the system.

(v) Relief valve, flow control valve, etc.   Very often it is noted that system oil while passing through the pressure relief valve, flow control valves, etc. generates heat. Heat generation due to blowing over excess oil in the meter-in or meter-out circuit through a relief valve is a common experience.

Generally the relief valve is used only in a short period of the cycle. It is therefore necessary to find the maximum rate of heat generation by calculating the power in kW and then calculating the average for an entire hour.

Pressure reducing valves and pressure compensated flow control valves are other sources of heat in the system.

Pressure compensated flow controls used to reduce the flow to or from a cylinder or a motor in a circuit using a fixed delivery pump will cause the excess oil not used to move the actuator to flow over the relief valve. Sometimes it is best to use a pressure compensated pump in a system like this or use a bleed-off circuit. Some heat may be saved by designing a bleed-off hydraulic circuit as the restriction is avoided here in the main flow line and the load resistance may be less than in the relief valve setting.

Sometimes it is helpful to estimate that 15%–20% of the electrical motor power will go into heat. This will help calculate the hard-to-figure-out components or line joints in a system for heat estimation, e.g. miscellaneous valves, fluid friction in pipes, tubings or hose, mechanical friction and slippage of pumps, fluid motor and cylinder etc. In cases of marginal to high heat generation, it is a good practice to provide heat exchanger connections in the main oil return line and the relief valve return line. This will allow a heat exchanger to be added at any time, when needed.

(vi) Entrapped air in oil   Most designers prefer to use an oil volume 3 to 5 times the pump flow-rate per minute. It is a fact that air has a natural tendency to entrain into the oil. The entrained air can cause localized heating in the system and can damage system components specially pumps, due to cavitation. The oil volume in the reservoir may act to allow de-aeration of the oil to drive out the air that might have gone into the system. One must note here that effect of temperature on the air-in-solution is not much, but entrained air can be controlled more readily at higher temperatures with enhanced performance but the fluid life may get shorter. Entrained air is also responsible for damage due to cavitation. The hydraulic tank should be designed such that the working temperature could be reached quickly as soon as the machine is started. To reduce air in the hydraulic fluid, one must maintain a steady back pressure in the system. Some prefer this back pressure to be about 5 bar. Heat generated to overcome a 5 bar back pressure by 2 l/s oil flow will be 0.95 kW.

10.1.1 Effect of Heat

Heat is generated in a hydraulic system whenever oil flows from a higher to a lower pressure without doing mechanical work. This means that if a relief valve is allowing the oil to flow back to the tank and the system pressure is being maintained, the difference in pressure or pressure loss is the difference between the system pressure and the tank line pressure. Pressure losses may occur when oil flows through inadequately sized valves or pipes, kinked hoses or sharp bends in hoses and tubes. Excessive solar temperature can cause undesirable heat. An easy way to dissipate heat is to use a reservoir sized large enough to dissipate normal heat generated in a system. The normal practice followed is to size the reservoir capacity 3 times the rated pump capacity in liter per minute. The reservoir temperature should not be allowed to rise above 60°C. At higher temperatures the oil can break down and oxidation may occur creating varnish. This varnish can clog the small orifices, produce an acid which can corrode metal parts, produce sludge and cause metal parts to wear very rapidly. Worn out metal and other particles may initiate a chain reaction of chemical disintegration and may cause high oil temperature and generate heat. By prior checking, the problem can be corrected before serious damage occurs.

High oil temperature may change the viscosity of petroleum-based oil and thus may enhance leakage of oil. Therefore the oil temperature should be maintained within 55–60°C when pertroleum based oils are the media.

Hydraulic oils and component materials used in a hydraulic system have their own respective temperature limitation which makes it imperative that the system temperature should be maintained within the tolerable limit in order to protect them from heat-related failures. It is important to dispose off the accumulated heat. The life of oil at 50°C may get reduced by 30 to 50% of the oil life at 38°C.

In many hydraulic applications, energy is expended for control purposes as well as for lowering or rapid reciprocating of heavy parts. Much of the power so applied is converted to heat in the fluid. High system temperature, like high blood pressure in human body, works silently to damage the hydraulic system, i.e. life of the hydraulic fluid, components and system is significantly shortened if the system continues to operate for long periods at high temperature. If the system temperature can be maintained between 50° and 55°C, maximum fluid life may be attained. A general rule says that for every 5° to 7°C above 50°C, oil life is cut by half.

Higher temperatures reduce the viscosity of the hydraulic oil allowing greater internal leakage and damage the pump and increase the oil temperature even this may further eventually destroying the oil. Lower temperature increases oil viscosity and makes the pump cavitate. A cooler system on the other hand improves the performance of all rubber products (hoses and seals) used in hydraulic components as well as the system.

One of the advantages of the various controls used for pumps like load sensing controls, is that the pump will not generate excess heat when unloaded or partly loaded. Another solution to excess heat is to build in enough heat dissipation capacity by properly sizing the reservoir (tank capacity 3 to 5 times of pump output) and by using appropriate materials. In a cooler atmosphere one may have to depend on the process of convection to heat up the oil in the tank.

High operating temperatures are damaging specially with water-based oils because of the possibility of water evaporation and subsequent loss of fire resistance. One solution is to use a sealed, pressurized reservoir which however makes the system costlier. A slightly high temperature in the pressurized tank does not affect the performance of the fluid; on the other hand it actually avoids water evaporation.

If water evaporates out of water-glycol fluid two things may happen—1. It may increase the concentration of the polymer that imparts viscosity to the fluid, the fluid becomes more viscous resuting in less efficient operation, and 2. The fire resistance property of the fluid which is based on the water content, is reduced.

Phosphate esters can operate above 80°C though at high temperature, they lose some viscosity as the viscosity-temperature coefficient is not good. Most phosphate esters are stabilised at high temperature by the addition of antioxidants.

10.2 ESTIMATION OF HEAT RISE

In order to eliminate the effect of heat, appropriate heat dissipation measures need to be taken up while designing a hydraulic system. It is therefore necessary to estimate the heat rise in the system. For a normal oil reservoir, the loss of heat in kcal/hr can be estimated from the graph given in Fig. 10.1 at a temperature difference $\Delta T = 20 - 50°C$. The rise in temperature can be calculated from the formula given here:

$$\Delta T = \frac{Q_w}{\sqrt[3]{V^2}}$$

where ΔT = rise in temp. in °C over room temperature
 Q_w = loss of power due to heat in kcal/h
 V = volume of oil in tank in liters

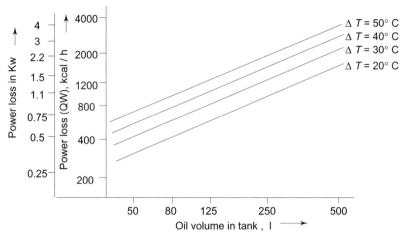

Fig. 10.1 *Heat loss estimation graphically in kcal/hr*

Heat generation across a throttle orifice or in a relief valve may be calculated if the pressure drop across the valve and the flow rate are known which can be calculated from the following formula.

$$\text{Power loss} = \frac{\Delta P \cdot Q}{612}$$

where ΔP = pressure differential across the valve, bar
Q = flow rate of oil, l/min.

In the case of British unit, the following formula can be used

$$\text{Power loss in HP} = \frac{\Delta P \cdot Q}{1714} \text{ B.Th.u/h}$$

where ΔP = pressure differential, P.S.I.
Q = Oil flow rate, gpm
1 HP = 2545 B.Th.u/hour = 42.2 B.Th.u/min = 746 watt

One can use the following formula also to calculate the heat loss (E_L) generated in a system in B.Th.u/hr.

$$E_L = 1.48 \times P \times Q \times (1 - \eta)$$

where P = pressure in P.S.I.
Q = flow rate in g.p.m.
η = system efficiency in %.

10.2.1 Heat Balance

The total heat loss (E_L) generated in a system may either be absorbed by the hydraulic oil in the reservoir or by associated system components like the oil tank, pipes, valves, etc. thereby increasing the system temperature or may be dissipated to the atmosphere or to a cooling medium used in the system. Over a period of time a heat balance may be arrived at between the heat entering and leaving the system depending on a number of complex factors like mean specific heat of oil and components, overall heat transfer coefficient and heat dissipating surface area. The rise in temperature of oil over the ambient temperature in the system is the difference between the temperature of oil minus the temperature of air.

10.2.2 Heat Dissipation through Radiation and Convection

Cooling of oil depends on the flow of air passing over the wall and on radiation of heat.

As has been stated earlier, the hydraulic tank plays an important role in the process of heat dissipation. As a thumb rule one can say that a hydraulic tank 200 to 250 l capacity located in an open place under normal ambient condition with oil temperature between 40 to 50°C may continuously dissipate 1.5 to 2 kW heat by radiation. The size of the reservoir is to be designed carefully as it is found that a larger volume of oil in the tank compared to the oil volume in circulation, may provide ample time for the return oil to cool down substantially before re-circulation and thus enhance oil life. Painting of the reservoir and proper choice of reservoir material may contribute significantly in cooling. Sometimes 60% cooling can be effected if the reservoir wall has high emmissivity. Emissivity of certain materials is given in Table 10.1.

Table 10.1 Emissivity of Some Materials

Material	Emissivity	Remarks
Oil paint	0.92 to 0.96	Good Radiation
Sheet steel (untreated)	0.65 to 0.82	Moderate
Aluminum plate (Untreated)	0.65 to 0.82	Moderate
Aluminum paint	0.27 to 0.67	To be avoided

Outside cooling by convection depends on the air flow. Sometimes adequate fins can be added to the tank wall and thus the area of heat dissipation can be increased 2 to 3 times.

10.2.3 Heat Flow

Heat flow in a system increases
(a) With higher temperature differential
(b) With increase in area across which heat flows
(c) With better heat conductivity of the wall material

The amount of heat flow (Q_W) can be calculated from the formula

$$Q_W = UA\Delta T$$

where, U = coefficient of heat transfer in cal/hr/m^2/°C
ΔT = Temperature difference, °C
Q_h = Heat in calories/hr
A = area of surface, m^2

10.2.4 Coefficient of Heat Transfer

Heat exchangers are often used in critical hydraulic systems to dispose off the heat generated in the system during circulation. Water, oil and air are generally used as the medium of heat transfer in a heat exchanger. Heat in a hydraulic system may flow through the metallic walls of valves, pipes, etc., but the quantity of such heat propagation is quite insufficient except for reservoir. Oil is a good heat insulator. A comparative view can be understood from the fact that a 1 cm thick layer of oil allows less heat flow than a 1 cm thick layer of asbestos and about 500 cm thick layer of steel. The flow of heat depends on the coefficient of heat transfer and specific heat of oil. However the coefficient of heat transfer depends least on the resistance of the metal wall. The flow of heat from the fluid mostly depends on the flow that is made to pass along the heat conducting walls. The stagnant film of oil clinging to the wall has a lot of influence on the coefficient of heat transfer compared to the metal wall which has a comparatively lower resistance. This means that using a thinner wall for better heat transfer will have no meaningful result. The specific heat of oil is 0.45 compared to 0.12 of steel, 0.22 of aluminum and 0.09 of copper. Another important factor influencing heat loss is the ambient condition prevailing outside the wall.

10.3 ROLE OF HYDRAULIC OIL TANK IN HEAT DISSIPATION

From the foregoing discussion it can be stated that the overall coefficient of heat transfer in a hydraulic reservoir may depend largely on the motion of fluid along the walls as the immovable mass of oil in the central portion of the reservoir may contribute very little to dissipate the heat. Hence to have better heat dissipation characteristics, one has to depend mostly on a favorable value of coefficient of heat transfer (u) taking into consideration the following factors:

1. Optimal design of the tank and adequate area of the heat dissipating surface
2. Viscosity of the oil
3. Degree of turbulence existing in the tank

It is found that an increase in viscosity reduces the capacity of the heat exchanger. The value of coefficient of heat transfer is influenced significantly by the presence of turbulence in the reservoir. Excessive turbulence is not good as this may not insure settling of contaminants and may lead to suction of the dirty oil during pumping.

10.3.1 Area of Heat Dissipation

The area of the heat dissipating wall can be found out from the formula provided in section 10.2.3

$$A = \frac{Q_W}{U \cdot \Delta T}$$

where Q_W = heat absorbed by oil plus heat absorbed by the wall.
The normal value of U is 150 to 300 kcal/m²h°C.

During estimation, the area of dissipation from the sides, top and bottom of the tank are to be taken into consideration. The surface area of the external plumbing may also be counted as a dissipating surface in certain cases but the bottom of the reservoir surface need not be taken into consideration unless it is exposed to free air circulation. The cooling capacity of the tank will increase in proportion to the difference between the oil temperature during system operation and the ambient air temperature which may be influenced by the location of the machine as well as the reservoir and a number of other factors.

Arrangement should be made to see that a reasonable amount of free air circulation is possible around the reservoir. In case of a critical system, a forced blast of air is directed on the side of the reservoir as it may enhance heat dissipation capacity by as much as 50% or even more.

The heated oil is generally cooled with air at least in 80% of the cases. The reason is that water is no longer as abundant and economical, air is inexpensive and readily available. Air cooled heat exchangers eliminate water usage and reduce disposal costs, use of a compact light weight heat exchanger unit is certainly advantageous to reducing the size of the total installation. The warm air from the heat exchanger helps in saving energy; the warm air is ducted to the plant and office areas for heating purposes.

10.4 USE AND APPLICATION OF HEAT EXCHANGERS IN HYDRAULIC SYSTEMS

We have already seen that due to many reasons the generated heat loss in the system increases the system oil temperature over a period of running time. If the temperature difference exceeds the specified limit and there is no scope of increasing the heat dissipating area or volume of reservoir, one may have to use a heat exchanger. Generally no heat exchanger is prescribed for hydraulic systems with a pressure less than 75 bar with adequate volume of oil in the hydraulic reservoir. However, this thumb rule is not applicable for systems using servo and other specific control systems where a change in oil viscosity may create undesirable effects.

10.4.1 Types of Heat Exchangers

Most common types are:
   (a) Air-cooled heat exchangers.
   (b) Water-cooled heat exchangers.

(a) Air-Cooled Heat Exchangers Air coolers are used where water is not readily available and the air is at least 3° to 5°C cooler than the oil. But water coolers are more compact, reliable, efficient and use simple temperature controls.

Exchangers The most common air-cooled heat exchanger is a fan-cooled oil cooler which uses a simple fan to cool down the hot oil. A simple diagram of such a system is shown in Fig. 10.2.

However, the decision to use a particular and specific heat exchanger can be taken up after taking into consideration the following factors

1. Ease of application and adaptability
2. Design configuration
3. Ease of maintenance
4. Safety and operational parameters
5. Life and operating cost
6. Noise and environmental compatibility, etc.

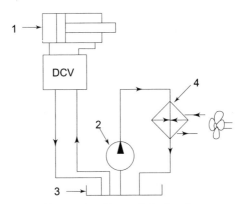

Fig. 10.2 *Fan cooled oil cooler*
[1-Hyd. cylinder/valve, 2-Pump 3-oil tank, 4-Heat exchanger]

10.5 AIR COOLING OF HYDRAULIC SYSTEM

The waste heat generated in a hydraulic system can be dissipated in various ways. Water, once a universal heat sink, is nowadays not much preferred because of global potable water crisis as well as the stringent water pollution act coupled with the cost of water. The natural substitute for water is air, which is available in abundance in nature.

The advantages of using air are:
(a) It is available in plenty.
(b) It is inexpensive.
(c) It is non-corrosive.
(d) It has better thermal capacity.
(e) The heat added to air (when exhausted to nature) does not affect the ambient air very much.

(f) Air can readily be used as a hydraulic heat sink in all hydraulic systems.
(g) Use of air as a heat exchanger is comparatively cheaper and hence economical.
(h) Oil-to-air heat exchangers can be easily used for both mobile and stationary hydraulic systems.
(i) These exchangers are available in small and modular designs also.

A simple schematic design of a oil-to-air heat exchanger is shown in Fig. 10.3.

10.5.1 Construction and Function

A simple oil-to-air cooler consists of the following main parts:
(a) Core
(b) Fan
(c) Support structure
(d) Control equipment and
(d) Driving motor

The cooling system consists of a radiator, generally called a core made of light weight finned tubes, connected at both ends by header as shown in Fig. 10.3. The rotating fan provides the flow of natural air over the finned tubes which are placed across the oil flowing system. The air flow can be either induced type or forced type. The layout of the finned tubes can be of the three types given below:
(a) Cross flow as shown in Fig. 10.3
(b) Parallel flow and
(c) Counter flow.

A design of a 2-pass counter flow type oil-to-air cooler is shown in Fig. 10.4. In comparison to a 2 or multi-pass counter flow oil cooler, a single-pass cross flow oil cooler is preferred due to low pressure drop of the fluid. But a counter flow cooler is recommended for general use as it can provide much lower discharge fluid temperature for a given thermal load. Both ferrous and non-ferrous tubes are used with round or oval cross-section. Thermal dissipation of oval tubes is higher than that of rounded tubes. Copper, brass and aluminum tubes are less subject to corrosion. For high pressure and high temperature system carbon steel or stainless steel tubes are more preferred while non-ferrous tubes can be used for

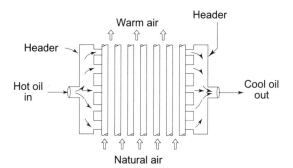

Fig. 10.3 *Oil to air cooler (cross flow type)*

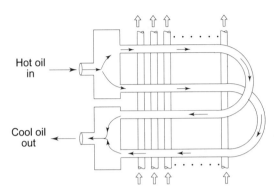

Fig. 10.4 *Oil to air cooler (counter flow type)*

low pressure systems say upto 25 bar pressure. But compared to non-ferrous tubes, the conductivity of ferrous tubes is poor (less than 50% of non-ferrous tubes) and hence an additional cooling surface may be needed to enhance cooling efficiency. Thermal conductivity of certain tube materials is shown in Table 10.2.

Table 10.2 Thermal Conductivity of Some Materials

S.No.	Material	Thermal conductivity BTh u/hr.ft. °F
1.	Pure aluminium	125
2.	Brass	63.7
3.	Copper	220
4.	Soft Steel	26.8

Aluminum has an excellent heat transfer coefficient, is light weight, has easy workability, is anti-corrosive and hence recommended for many applications.

One of the major drawbacks of oil-to-air cooler is the excessive noise produced by air separation at the fan blades and ring. A 'larger diameter-lower speed' fan may reduce the noise significantly.

Water cooled heat exchangers are also used in hydraulic systems. The heat transfer mechanism between warm oil and water is explained in Fig. 10.5. Two very important points that are to be noted before one decides to use such a heat exchanger are:

(a) Corrosion of tubes, etc. due to water.
(b) Disposal of warm water in view of environmental hazards.

However the waste heat can be used for residual activities in the industry.

10.5.2 System Approach for Cooling System Design

While designing a hydraulic system, one has to give equal importance to the design of the cooling system if one wishes to avoid undesirable thermal effects on

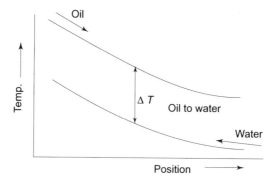

Fig. 10.5 *Heat transfer between oil and water*

the functioning of the system. However designing of a cooling system is a quite complex matter as it involves integration of large amount of technical information encompassing factors like:

- Information about the machine for which the hydraulic system is designed.
- Type and specific feature of the hydraulic circuit proposed.
- Probable sources of heat generation and heat accumulation in the system.
- Possible method of heat dissipation which may match the system.
- Proposed type of heat exchangers and their physical and mechanical characteristics as well as thermal behavior.
- Cooling fluid and coolants.
- Locational and site data where the equipment is to work, e.g. indoor or outdoor application etc.

The entire subject of cooling system design has attained more serious proportion due to the complexity involved on the above issues and designer's skill and knowledge to balance these factors for achieving optimal thermal behavior of the system. In order to design an appropriate cooling system at least the following basic information may have to be considered by the hydraulic system designer.

1. An accurate description of the machine and its functional features.
2. Description of the hydraulic circuit with its specific features.
3. Duty cycle of the system.
4. Type and process of cooling media to be used e.g. air, water, oil, etc.
5. Source and availability of the cooling media for heat transfer.
6. Purity of the cooling media, e.g. contamination, chemical properties, and flow characteristics, corrosive action of the media, etc.
7. Chemical characteristics of the hydraulic fluid used in the system.
8. Temperature of the hydraulic oil in the system—inlet temperature, outlet and running temperature.
9. Type of oil pumping arrangement and type of pumps used.
10. Hydraulic tubes and pipes used in the system and piping system design.

11. Type of hydraulic reservoir and its physical and mechanical features including overall size.
12. Quantity of oil in the reservoir.
13. Proximity of the hydraulic system to sources of heat, contamination, corrosion etc. and probability of the system getting affected due to dust, dirt and other impurities and how to protect the system from damage due to dirt related scaling, rubbing, clogging, etc.
14. How does the dust particle affect heat dissipation through the system components surface?

Apart from the physical information as above, a cooling system designer may need certain essential information about heat generation and the proposed heat exchangers, their mounting, size etc. and integrate the same with hydraulic system information in order to design a matching system.

1. Whether the heat exchanger will have water, oil or air as the cooling medium?
2. How and where the same would be mounted? Will it have close proximity to the heat generating parts of the main equipment or will it be exposed to heat source such as direct sun-rays or hot ambient atmosphere?
3. If it is to be air cooled, whether it will be forced air or natural air cooling? Will the air be moving?
4. If it is forced air, type of equipment, e.g. blower, fan, etc. to be used for forcing the air and level of vibration, shock load expected? What will be the intensity of noise? Will the pumping system area be ventilated? Will it be exposed to mechanical or structural deformation due to positioning?
5. If the system to be liquid cooled (say, water), whether adequate water is available and the quantity of water needed. Quality of water is also an important criterion.
6. One has to decide the space needed for the heat exchanger and whether adequate space will be there. Other information e.g. if the radiator receives first air or air that has been warm because it previously passed through air conditioning radiation, is also important.

Certain environment related and other general information on heat control and regulation may also be worth mentioning here:

1. Whether the waste heat from the liquid coolant could be used economically for heating parts of the plant and machinery? For the entire years? Or for the cool season e.g. winter only?
2. Method and type of temperature control to be used e.g. if there should be thermostatic regulation? Should there be automatic on-off switch or not?
3. Should there be a thermal warning device for the cooling system to signal danger status of temperature e.g. overheated or underheated?
4. Type of valving system for the heat exchanger to be introduced. Should there be a pressure relief valve for the water system to relieve the excess water pressure?

5. Nature of ambient condition and to what ambient temperature the system should be exposed may have to be indicated. It may be noted here that under a severe low temperature situation the hydraulic oil needs "Start-up" heating. Hence it is to be decided if a heater is to be introduced where the oil circulates naturally, so that the entire quantity of oil inside the oil-reservoir gets heated.
6. Whether forced circulation of the heated fluid will be enough? In the present world of acute energy crisis, the question of managing the thermal condition of the hydraulic system has attained significant importance due to the following factors:
   (a) Increasing shortage of natural oil energy as well as water.
   (b) Social awareness of common man regarding conservation of water, oil and other natural resources.
   (c) Introduction of stringent environmental protection laws by International and National Standards Body.
   (d) Social and management awareness on occupational health hazards to which workers get exposed due to pollution effects of natural resources e.g. water, air etc.
   (e) Introduction of stringent industrial and environmental laws for discharge of waste water from plant and machinery limiting the discharge temperature of industrial waste to open air in order to protect plant and water life.
   (f) Increasing cost of energy and coolant all over the world.

Under the circumstances a hydraulic system designer can not remain silent on the question of heat-related effects of the machinery and plant he is designing. Therefore lot of serious thinking has to be given as to how to protect the system as well as the surroundings from the undesirable effects of heat emanated from the designed system. However it is felt the best way to avoid heat related problems or at least to minimise them, is to design an innovative heat-efficient hydraulic circuit as given in next section. While selecting the heat removal method, one has to take note of the various points as detailed above and compare the cost benefit of each of the method e.g. how do water, air-cooled and refrigeration cooled methods including regular maintenance compare. One has also to compute the initial cost as well as the life-cycle costs of the various methods and then take the appropriate decision considering the various technological, environmental and scientific parameters.

10.5.3 Innovative Design of Circuit to Reduce Heat

Combatting heat in hydraulic circuits is a common problem in industry today. The problem of heating may become troublesome with small bore, short stroke cylinders which are cycled very rapidly. The reason is explained here. As the oil does not get enough chance to return to the reservoir, it may create a local hot spot in the hydraulic circuit. As operating speeds of machinery keep increasing, this type of hydraulic circuit is likely to cause greater heat generation problems.

Usually the cylinder and hose or piping close to the cylinder ports overheat as the volume of oil in the lines between the cylinder and valve is greater than the volume of oil used by the cylinder. Due to such overheating, seals may fail, hose life may be reduced, rise in temperatures may pose heat related danger to operating personnel who might touch the cylinder or hose or the system leakage may increase enhancing its chances to fail often. In such situations innovative circuit design may be the best solution. In this particular case the problem has been eliminated with a simple but innovative hydraulic circuit design using a double check valve as illustrated in Fig. 10.6. Hydraulic oil follows path A when the cylinder is extending and path B when retracting. The cylinder thus receives cool oil from the pump every time it cycles and the return oil gets a chance to flow back to the tank every cycle. Thus the returning hot oil can dissipate heat in the reservoir by normal methods every time the cylinder cycles.

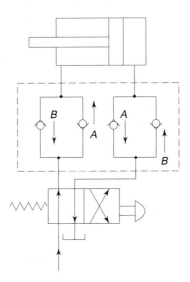

Fig. 10.6 *Innovative hydraulic circuit for heat reduction*

Review Questions

1. (a) State the basic reasons of heat generation in a hydraulic system.
   (b) What mechanisms are applied in hydraulically operated machines to dissipate the heat?
   (c) Calculate the heat loss in a pumping system delivering 40 l/min oil at 160 bar, the pump efficiency being 65%.
2. What is a heat exchanger? Describe an oil-to-air heat exchanger used in a hydraulic system.
3. How does a oil-to-oil heat exchanger differ from a oil-to-water heat exchanger?
4. What is heat flow? State the factors which influences the heat flow in a hydrosystem.

5. What is specific heat? What is the influence of specific heat in heat generation or dissipation?
6. What is the role of a reservoir in heat dissipation in a hydraulic system. Give examples.

References

10.1 "Ermittlung des Waermehaushelfs einer Hydraulicanlage", O+P Konstructions Jahrubuch 84/85, Mainz, Germany.
10.2 Snow, GA, "Controlling heat generation in small bore, short stroke hydraulic cylinder", *The Power & Control Journal,* USA, 1969.
10.3 "Thermal efficiency of hydraulic systems—Designer's choice", *Hydraulics and Pneumatics*, Penton IPC, USA.
10.4 Miller, JE, "Air cooled heat exchangers for hydraulic systems", *Hydraulics and Pneumatics*, Penton IPC, USA.

11
Hydraulic Reservoirs and Accumulators

IN any hydraulically operated machine the hydraulic reservoir plays an important role in proper functioning of the system and various components as well as maintaining the various working parameters to a reasonably desired level. Though a reservoir is relatively simple, it has to perform some important duties in the interest of the overall functioning of the system. Briefly we may summarize the functions of a reservoir as follows.

1. *Storage of oil* The reservoir acts as the main oil container for the entire system. The volume of oil to be maintained depends on the need of the hydraulic circuit and other physical characteristics of the system.
2. *Cooling of oil* The reservoir should act as a cooling device for dissipation of accumulated heat in the oil and should provide a reasonably large surface area to ease heat dissipation to the ambient environment through radiation, etc.
3. *Expansion of fluid* The hydraulic reservoir acts as a medium to provide space for expansion of the fluid due to variation in temperature. It should be in a position to accept oil drain back from the system during shut down.
4. *Separation of contaminants* The reservoir acts as a settling tank for separating the accumulated contaminants from the oil, e.g. dirt, dust, worn out particles, air, etc. It should provide air space above the oil surface to enable entrapped air to escape into the atmosphere.
5. *Structural support* In certain cases the reservoir should act as the main supporting base for the power pack, e.g. the pump-motor assembly, components like relief valves, mechanical accessories, etc.
6. *Easy access* The reservoir should be designed such that it provides easy access for removal of old and used oil and contaminants from the tank. It should facilitate easy cleaning of the interior.

11.1 COMMON TYPES OF RESERVOIRS

Non-pressurized reservoirs are the most common reservoirs used in industry. Hydraulic reservoirs commonly used in the industry today may be classified as:
1. Pressurized reservoirs
2. Non-pressurized reservoirs

As far as the use of hydraulic reservoirs in normal machine tools and other equipment is concerned, one prefers only the non-pressurized type. Pressurized reservoirs are used mostly in aviation and other critical hydraulic systems.

Non-pressurized hydraulic reservoirs can be sub-divided again into two categories:
(a) Open type reservoir
(b) Closed type reservoir

Reservoirs can also be an integral part of a machine body. But in such a reservoir, the pump assembly should be arranged on a separate fixture and also there is possibility that the machine vibration and shock will be directly transmitted to the hydraulic system. From the point of view of contamination, open type reservoirs are poor in preventing air-borne contaminants and are hence unpopular. In majority of the cases, therefore, people prefer closed type reservoirs only.

Closed type reservoirs may have variety of designs and shape, e.g. square, rectangular, spherical, cylindrical etc. But the rectangular shape is mostly used. Square type construction is used in hydraulic systems having a smaller capacity. However, there is no hard and fast rule as to the shape of the reservoir.

11.1.1 Reservoir Standards

American National Standard Non-integral industrial Fluid Power Hydraulic Reservoirs have been accepted by the American National Standards Institute (ANSI) as B.93.18M–1973. However it does not cover pressurised oil reservoirs which are pressurised up to 10 to 20 psi and are sealed for specific reason.

11.2 RESERVOIR MOUNTING AND CONSTRUCTION

The reservoir should be mounted on legs to raise the bottom of the reservoir by at least 150 mm from the ground level for better air circulation, easier servicing and draining of oil during servicing.

As the pump, drive unit, hydraulic components are mostly mounted on the top plate of the reservoir as shown in the basic design of a tank in Fig. 11.1, the top should be structurally quite rigid. It should also maintain perfect alignment and be able to absorb all shocks due to vibration and mechanical misalignment. An auxiliary plate sometimes may be mounted on the top surface. In certain designs the pump-motor combination is mounted on an L-shaped reservoir fitted to the power pack as an extended base which provides the following advantages:
1. The inlet to the pump can be fed easily due to the weight of the oil, i.e. due to gravity in addition to the suction pressure exerted by the atmospheric pressure of air.

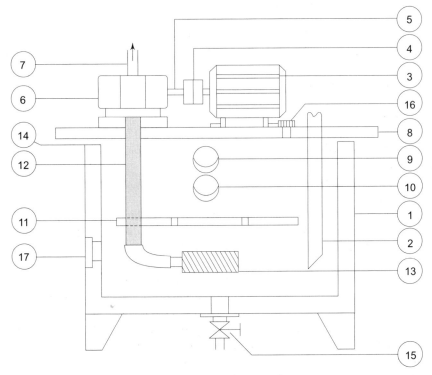

Fig. 11.1 *Basic design of a hydraulic reservoir*

2. Servicing of tank and other components can be done without disturbing the pump unit.

But L-shaped reservoirs may need more floor space which is a disadvantage considering the exorbitant cost of land specially in big cities.

An L-shaped reservoir is shown in Fig. 11.2.

Fig. 11.2 *L-shaped oil reservoir*

The common parts of a hydraulic reservoir are marked in Fig. 11.1.
1. Cold rolled steel body
2. Return line
3. Electric motor
4. Pump-motor coupling
5. Pump shaft
6. Hydraulic pump
7. Pump delivery line
8. Top lid
9. Top sight glass
10. Lower sight glass
11. Separating plate
12. Inlet line
13. Suction filter
14. Gasket
15. Drain plug
16. Filler cap
17. Side man-hole

The shape of the welded reservoir is mostly rectangular. The top lid must be made to set at the top tight with gasket against the entrance of contaminants like dust, dirt, etc. The filler cap is at the top with a strainer above 100 microns in size. Care should be taken that when the cap is removed for filling up the oil, it should not be placed on the ground so that it does not come in contact with dust/dirt, etc. Many a times, the filler cap contains the breather too. While filling the reservoir with oil, one must take note of the level gauge or sight glass so that the correct amount of oil is used to fill up the reservoir. The capacity of the reservoir should be sufficient enough to enable the system to work satisfactorily.

The pump unit is placed in many cases over the lid with proper damping arrangement. To minimize the problem of contamination, the return pipes should be separated from the inlet pipe to the pump by a separating plate. The ends of the pipes should be cut at 45°.

Manhole covers are needed in a reservoir to enable easy cleaning of the reservoir. The clean-out cover should be designed such that it can be removed and replaced by a single person easily. A design of a clean-out cover is shown in Fig. 11.3.

A suction filter is generally connected to the inlet line going to the pump. In certain cases, magnetic trappers are provided to trap solid particles of the ferrous group. They should be installed such that removal can be done easily without disturbing the system.

Generally all pipes entering the reservoir enter vertically from the top. This facilitates easy pipe disconnection. It is better not to have a common return line due to danger of producing a back pressure. For facilitating proper cleaning of the tank, the bottom of the tank should be dished out with a drain positioned with an on-off valve so that the last drop of oil can be drained out.

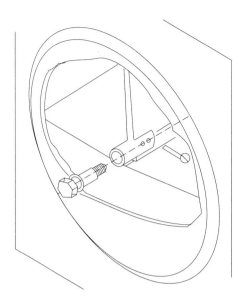

Fig. 11.3 *Clean out cover*

11.2.1 Constructional Features

As already mentioned, the reservoir is mostly made of mild steel plate having a thickness of 3 mm or above. The interior walls of the reservoir should be blasted to get rid of scale and corrosion. The welded joint should be tested for leakage. The pipe ends should not touch the base plate of the reservoir. Return line pipes and inlet pipes should be placed apart so that the pump does not suck contaminated oil from the system. The oil level indicator, air breather and filling caps should be sealed with an air-tight gasket or O-ring. The oil level gauge is mostly flush-mounted to protect it from damage. The design of the reservoir should be such that the return oil has sufficient circulation inside so that heat-dissipation can take place without any problem. This will save the system a lot of problems associated with heat effects.

11.3 RESERVOIR SHAPE AND SIZE

1. Square or rectangular shape may help better heat dissipation per unit volume.
2. A shallow cylindrical shape may not be able to use the complete surface for heat dissipation. A tall and narrow cylindrical shape may not provide enough fluid surface area for air to escape.
3. The reservoir capacity should not be less than 3 times the flow rate of the fixed displacement pump or 3 times the mean flow rate for pressure

compensated variable displacement pump. This is needed to help the fluid to rest between work cycles, to dissipate heat, to help settling of contaminants, to help decrease and reduce the risk of pump cavitation.
4. A smaller reservoir can be used if ambient conditions are favorable for cooling the oil, external oil cooling is available or if cycling is intermittent.
5. A reservoir should have additional space of at least 10% of its fluid capacity to take care of thermal expansion of the oil.

11.4 RESERVOIR ACCESSORIES

Various accessories are used along with the reservoir for different functions.
1. To indicate fluid level
2. To indicate oil temperature
3. To remove metallic and particulate contaminants
4. To route returning oil to reduce pump cavitation and enhance heat dissipation
5. To separate contaminants from entering the pump inlet by appropriate filtration
6. To facilitate periodic filling up of the tank and breathing

(i) Level indicator These indicators are to be placed at each filler and distinct marking should be provided on the sight glass to indicate high and low levels. An auxiliary float operated switch is sometimes provided to signal low fluid level. A protected column indicator is recommended as NFPA standard.

(ii) Temperature indicator and oil heater Though not recommended as a standard, thermometers or heat sensing devices are used in fluid reservoirs in the same housing as the level indicator. Thermostatic control is used to provide the indicating signal when temperature is outside design parameters.

A thermostatically controlled heater (shown in Fig. 11.4) is installed in the reservoir when the system has to work in very cold ambient conditions. It heats the oil to its normal working temperature so that the oil can function effectively.

(iii) Magnetic trappers Magnets are very often used in the oil tank to trap ferrous particles in the hydraulic fluid. A long stem magnetic trapper holds the magnets in the fluid and is mounted through the reservoir. The type shown in Fig. 11.5 mounts through the reservoir wall under minimum fluid

Fig. 11.4 *Thermostatically controlled heater*

Fig. 11.5 *Long stem magnet*

level. The magnets should be taken out periodically, cleaned and reinstalled to maintain their trapping efficiency. See Fig. 11.5.

(iv) Diffuser A flow diffuser used in the return line helps to slow down high velocity fluid from entering the hydraulic tank. Diffusers direct the return oil away from the pump inlet and thus reduce turbulence.

(v) Filler caps and breathers Filler openings should be provided in the reservoir to facilitate reasonably rapid filling at start-up, after periodic cleaning and to compensate losses of oil. The filler opening should also be provided with strainer to intercept larger particle contaminants, say from 50 to 100 microns. The filler cap or cover is permanently attached to the filler to seal the opening and in many cases also includes the breather. In the case of a pressurized reservoir, a pressure control valve which is used to control the externally entrained air over the top of the oil surface, may replace the breather permitting filtered air (up to 40 micron filtration) into the tank. However, air is released to the atmosphere when the preset pressure is reached.

(vi) Gaskets, seals and painting Top covers can be welded or screwed to the frame. Reusable gaskets are used all around the top in order to avoid entry of undesirable contaminants. Other attachments like caps, breathers, magnetic traps, etc. should also be provided with proper seals.

The exterior and interior of the reservoir should be painted. The exterior should be coated with a rust inhibitor. The paints used should be compatible with the fluid it must store otherwise the oil will degenerate chemically.

11.5 ABOUT PIPING

It has been already mentioned that the pump inlet and return lines should be on opposite sides of the baffle. The line end should not be less than 50 mm or $1\frac{1}{2}$ times the pipe ID from the reservoir bottom. The pump inlet line should be at least 50 mm above the reservoir floor and at least 75 mm below the minimum oil level. Drain lines should be led separately to the tank away from the pump inlet. These pipes may terminate above the fluid level in the reservoir to avoid siphoning and assist gravity draining. The pump inlet line should be free from unwanted joints, bends, fittings, etc.

11.6 MAINTENANCE

A primary requirement of any reservoir is that cleaning sholud be easy. For this, proper facilities should be provided so that the lid, sides or any other parts can be easily removed for cleaning. Periodic cleaning of the suction filter and other strainers should be done at regular intervals.

To facilitate easy removal of oil during cleaning, the reservoir bottom should be shaped with adequate taper so that all the fluid can drain automatically. The standard is to provide two drains at least 3/4" (19 mm) NPTF with equal sized valves at the lowest position.

Reservoir oil should be replaced at least once in a year. Prevention of turbulence and aeration is an important design criterion and to achieve deaeration, freedom from turbulence and foaming of fluid, return and suction lines are to be brought down well below the minimum oil level. To enhance the capacity of heat dissipation, the pipe ends (cut at 45°) should be directed against the reservoir wall. If the equipment is located below the reservoir, provisions must be made to break the siphon in case of disconnecting a line. In the return line, a small hole in the pipe inside the tank, immediately above the fluid level will serve the purpose. If the pipe lines enter the tank below the oil level, a shut-off valve has to be used for disconnecting lines without draining the reservoir. To facilitate better circulation of oil, a baffle is to be added between the return and suction lines. The advantage of this arrangement is prevention of continuous recirculation of the same fluid. This also helps to direct the flow inside the reservoir along the wall and may tend to equalise temperature throughout the fluid. The flow pattern across the baffle is such that foreign particles and water tend to settle at the bottom while air is permitted to escape. The baffle should not restrict free air flow above the fluid surface. Sometimes the lower end of the baffle is welded to the tank bottom and all flow must go above the baffle. This makes the cleaning of each compartment a bit difficult as each compartment will have its high flow points to clean. Magnetic plugs are used in hydraulic tanks to arrest iron-particles.

Magnetic plugs may be installed between the baffle and reservoir bottom in such a manner that oil flowing from one tank chamber into the other has to pass through a strong magnetic field to attract ferrous particles in the fluid. They should be installed from the top and they should be easily removed without disturbing other pipe lines and fittings.

11.7 INTEGRAL RESERVOIRS

While using an integral reservoir, the following points may be noted:
1. Size may have to be compromised due to limited space.
2. Heat transfer due to convection may be affected due to proximity of the machine it serves.
3. Accessibility of internal components during servicing may be hampered.
4. Difficult to isolate adjacent components or operating personnel from the reservoir heat. Special heat shielding may improve the situation.

11.8 HYDRAULIC ACCUMULATORS

The main function of a hydraulic accumulator is to store hydraulic energy and on demand make the energy available again to the system. They can be termed as the capacitance of the system. An accumulator is suitable for:
1. Hydraulic shock suppression and eliminating pressure ripple
2. Fluid make up in a closed hydraulic circuit
3. Leakage compensation
4. Source of emergency power in the case of power failure and

5. Holding high pressure for long periods without keeping the pump running. They are used in large hydraulic presses, farm machinery, diesel engine starters, hydraulically operated hospital beds, landing gear mechanism on aeroplanes, hatch cover in ships, lift, trucks, etc.

11.8.1 Types

Basically accumulators are of three types:
 (a) Dead weight type
 (b) Spring loaded type
 (c) Hydro-pneumatic type

Schematic diagrams of the types of accumulators are shown in Fig. 11.6.

Dead weight type It consists of a piston loaded with dead weight and moving within a cylinder which exerts pressure on the oil. The stroke of the piston

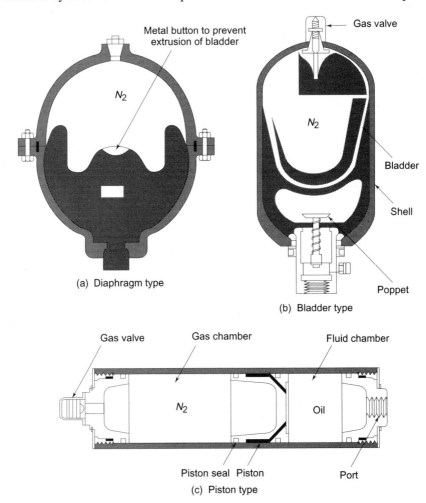

Fig. 11.6 *Types of accumulators*

pressure will remain constant because the load remains unchanged. The weights may be of some heavy material such as iron, concrete block or even water. To minimize leakage, the piston must be a precision fit in the accumulator cylinder generally honed to reduce friction and to ensure a long and trouble-free life.

However, due to their large size they are not very popular. But their advantage is that they supply oil at constant pressure.

They can serve several hydraulic systems at a time and are most often used in mill and central hydraulic systems.

Spring loaded They usually serve individual hydro-systems and operate at low pressure. A spring loaded accumulator consists of a cylinder body, a moveable piston and a spring. The spring applies a force to the piston. As fluid is pumped into it, the pressure in the accumulator is determined by the compression rate of the spring. By adjustment, the spring pressure may be varied.

$$\text{Pressure} = \frac{\text{Spring force}}{\text{Area}}$$

where, spring force = spring constant × compression distance

Advantage
They are smaller and mounting is easy.

Disadvantages
1. They supply a small volume of oil at low pressure.
2. The pressure does not remain constant, the accumulator pressure reaching its peak as the spring compresses and drops to a minimum as the springs approach free length.

Care has to be taken to see that the leakage oil is vented from the spring chamber.

Hydro-pneumatic accumulators They are the most commonly used accumulators in industry. They apply force to the liquid by using a compressed gas which acts as the spring.

Only dry nitrogen is to be used. Because of the danger of exploding an air-oil vapour, compressed air should never be used.

Types Four types of hydro-pneumatic accumulators are used:
1. Non-separator type, but not used due to possibility of foaming
2. Piston
3. Diaphragm
4. Bladder

1. Non-separator type It consists of a cylinder with hydraulic fluid and the charging gas with no separation between them.

They are generally used on die casting machines or other similar places.

They are always to be mounted vertically. Pressure-volume relationship should be such that not more than 2/3 of oil is discharged to protect the system from the gas.

2. Piston type It consists of a cylinder body and a moveable piston. The gas which occupies the volume above the piston is compressed as the cylinder body is charged with liquid. Here too pressure is a function of compression and varies with the volume of oil in the chamber.

3. Diaphragm type It consists of two metal hemispheres which are separated by a flexible, synthetic rubber diaphragm. Proper material for the diaphragm is to be selected as certain fire resistant and synthetic fluids may not be compatible with conventional diaphragm materials.

The storing action is effected by the compression of the volume of nitrogen enclosed in the diaphragm.

4. Bladder type It consists of a synthetic polymer rubber bladder like chloroprene, nitrile, etc. inside a metal (steel) shell. The bladder is filled with compressed gas. A poppet valve located at the discharge port closes the port when the accumulator is completely discharged. This keeps the bladder from getting out into the system. The advantage of a bladder type accumulator is that it responds quickly for receiving or expelling flow of oil. But a major disadvantage is that possibility of bladder failure which is to be taken into consideration.

Note: The gas charged accumulator should be pre-charged while empty of hydraulic fluid. The pre-charge pressures vary with application and depend upon the working pressure range. It should never be less than 1/4 of the working pressure.

Pressure: Let as assume P_0 = initial pressure (bar)

To protect against damage in transit, the accumulator bladder is gas-charged prior to dispatch. This pressure P_0 is of no significance for calculation purposes. Let P_1 = gas pre-charge pressure in the bladder or gas chamber when in operation. This pressure P_1 depends on the condition of use and is established by the manufacturer of the hydraulic system. It is approximately equal to $0.9 P_2$ to ensure that the bladder or diaphragm does not shut the poppet valve whenever a load change occurs, where,

P_2 = minimum operating pressure and

P_3 = maximum operating pressure

$P_1 = 0.9 P_2; \quad P_3 \leq 3 P_2$

The pressure P_3 should never exceed 3 times the pressure P_2 as otherwise with rapid compression the temperature rises too far and the bladder or diaphragm may deform.

11.9 SAFETY INSTRUCTIONS

1. Under no condition air or oxygen should be used or allowed to enter the system due to the danger of explosion.
2. The gas pre-charge pressure should lie below the minimum operating pressure.

3. Repair, connection of pressure gauge, etc. may be done only after the fluid pressure has been released.
4. Bladders should not be stored folded, but hanging by the gas valve and blown up with air up to their full length.
5. Accumulators should be fitted with robust bracket.
6. A non-return valve must always be fitted with an accumulator to prevent oil from flowing back to the pump.
7. A pressure relief valve is also to be fitted with the accumulator.
8. Due to frequent pulsation, higher pressure ratio and higher temperature, the bladder gives in and bursts.
9. The piston type accumulator is to be used for lower pressures only (100 bar) as at higher pressure prevention of gas leakage may not be effective.
10. Also this type of accumulator has to overcome frictional resistance and hence works slowly as compared to the bladder and diaphragm types.
11. The gas from the bladder will diffuse and also escape through the gas filling socket. Hence necessary examination and testing should be done once in a year to ensure gas capacity and the pre-charge pressure.
12. If the gas volume (V) and the charged pressure (P), i.e. VP is 200 kgf/cm^2, the accumulator bottle/shell should be registered according to the pressure-vessel law. If this product is more than 1000 dm^3. kgf/cm^2, the vessel should be properly tested according to the pressure-vessel rules.

Table 11.1 Comparison of Various Accumulators (Gas Type)

Type of accumulator	Bladder type	Diaphragm membrane	Cylinder piston	Flasked cylinder piston
Maximum operating pressure P_{max} (bar)	200	150	200	350
Volumetric efficiency (%)	60	60	90–96	38
Safe pressure ratio $\dfrac{P_G}{P_{max}}$	4	10	any	any
P_G = Pre-charge pressure Temperature °C	−15 + 65	+ 80	−30 + 80	− 30 + 80

11.10 PROPERTIES OF NITROGEN

N_2 forms about 90% of the air around us and is a non-poisonous gas. But nitrogen without oxygen can be suffocating so accumulators and N_2 gas bottles should be discharged in ventilated areas. A standard N_2 bottle contains 4.67 m^3 of gas and is therefore, unlikely to represent a hazard unless a number of bottles are discharged or the working space is small. N_2 is inert, i.e. it does not support combustion and does not combine easily with other elements.

11.10.1 Importance of N_2 Volume in the Bladder

The following points are to be noted:
1. If N_2 escapes from the bladder, a usual symptom in oil hydraulics is the formation of foam which will collect in the reservoir.
2. The reduced volume of N_2 will cause the pump to operate between narrower limits (reduce the working volume) and so will cut in and out more frequently. This may give rise to overheating.
3. This reduction in volume will reduce the speed of the cylinder and motors.
4. When the accumulator is discharged through the blow down valve, the gas can usually be heard bubbling into the tank.

11.11 ACCUMULATORS IN A CIRCUIT

In a simple hydraulic system consisting of a pump, an electric motor, valves and actuators, as hydraulic power is generated it is immediately used or converted to heat. With an accumulator this is not the case. Hydraulic accumulators allow the energy generated and by pump store the hydraulic energy for use at a later time.

Few important cases where accumulators are very often used by designers, are briefly described here.

1. Flow compensation   Fluid energy stored in an accumulator is made use of to compensate additional requirement of a hydraulic circuit if system demand is higher than the pump output. For example if one specific function of a machine is designed to cycle infrequently compared to its other major functions, the flow from an accumulator in the circuit, which gets charged during the pump operation, could be used for operating the specific function when the particular phase of the cycle arises. Thus the flow requirement is compensated even if the pump flow is inadequate.

2. Maintaining pressure   During the cycle of operation, a machine may experience a phase when it needs additional fluid to maintain certain specific load. For example a linear actuator used for clamping, may not need its full potential of force when the cylinder starts moving from its rest on its forward motion for clamping. But as soon as the actuator clamps down the job, its requirement of additional force is felt. In such a situation, if the pump pressure is not sufficient accumulator having sufficient stored fluid under pressure may be helpful in supplying the additional pressure to make up for any additional requirement of force.

3. Leakage compensation   Stored energy of the accumulator can be easily used to make up any possible loss of energy due to leakage or due to drop in pressure.

4. Shock suppression   Accumulators are used to suppress shock in a hydraulic circuit. A schematic circuit diagram is shown here in Fig. 11.7 to show how press circuits make use of accumulators to suppress the shock.

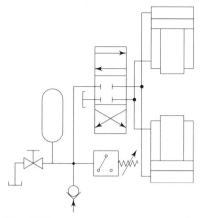

Fig. 11.7 *Accumulators to suppress shock*

Hydraulic shock in a circuit, specially in a high pressure system, may cause undesirable damage to the system. Shock may arise due to sudden stoppage of high velocity oil resulting in:
- Damage to pipings, fittings etc. as well as other internal and external parts of the system.
- Failure of seals in cylinder, valves etc.
- Oil leakage through mating surfaces
- Drop in pressure

It may be noted here that in case of sudden surge in hydraulic system, it is better to use a dead-weight type accumulator if size permits and the accumulator is matching with the machine size. This is because the sudden surge causes the dead weight to ride up and down only. But in a bladder type accumulator, the bladder may fail even though it is more compact and aesthetically much better compared to a dead weight type accumulator.

5. Application in press circuit As already stated, accumulators are very often used in hydraulic circuits for forming tools and other presses. An example is shown in Fig. 11.8 where the table of a press machine is to be operated by a hydraulic cylinder and to be maintained in position with sufficient pressure for a long time during the forming operation.

From the diagram we can see that there are two pumps—one for low pressure (2) and the other for high pressure (4) each having their individual prime movers 3 and 5 respectively. The press table is held by a single acting cylinder 1.0 which is operated by oil in its forward motion but its movement is controlled by the weight of the table. Its forward (i.e. lifting) motion is actuated by the direction control valves 14 and 16. During the valve position zero (0) of the dc valves 14, 15 and 16, both the pumps unload their flow into the tank (1) at no load pressure.

When DC valve 14 is actuated to its position I, the flow from pump 2 moves to lift the table through the pressure operated non-return valve (20). When the table is in its operating position, flow from pump 4 flows to cylinder 1.0 through the

348 *Oil Hydraulic Systems: Principles and Maintenance*

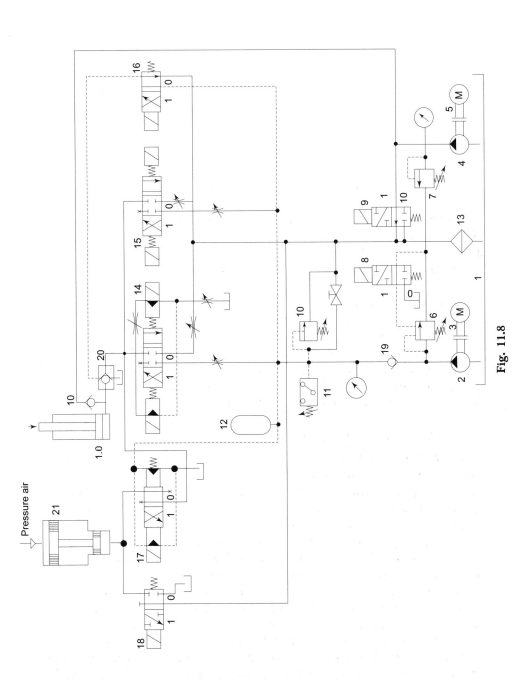

Fig. 11.8

check valve (19). The accumulator (12) gets charged when the pump (2) starts. Part of the oil is also flown to an air-oil pressure converter which boosts the pressure and during the forming operation when the press table may have a tendency to sink down from its forming position due to the excessive forming force, the accumulator flow gets discharged along with the flow under pressure from the air-oil converter with the actuation of dc valve 17 and 18 and thus help maintain the table position with sufficient force.

11.12 IMPORTANT POINTS TO NOTE REGARDING ACCUMULATOR SELECTION

1. Piston type accumulators do not react as quickly as a diaphragm or bladder type accumulator, but they can be mounted in any position, adapted for high temperature applications and catastrophic failure is mostly uncommon.
2. Diaphragm type accumulators are commonly found in the aircraft industry.
3. Bladder type accumulators have to be mounted vertically.
4. An inert gas like dry N_2 is used. Compressed air is never used because of danger of exploding.

11.12.1 Circuit Efficiency

Sometimes it is necessary to know what should be the efficiency of an accumulator circuit. Circuit efficiency with an accumulator is certainly not 100%. A pump charges an accumulator at a flow rate (Q) from a minimum system pressure P_1 to a maximum pressure P_2

$\therefore$ Input power $= P_{in}$ = flow rate × Avg. pressure

$$= Q \cdot \frac{P_1 + P_2}{2}$$

The output power P_{out} of the accumulator is a function of flow control and load value of the pump flow rate. Assuming that the flow control is set at the same value as the pump flow rate, the output power is

$$P_{out} = Q \cdot P$$

$\therefore$ Efficiency $= \eta = \dfrac{P_{out}}{P_{in}} = \dfrac{Q \cdot P_1}{Q\left(\dfrac{P_1 + P_2}{2}\right)} = \dfrac{2P_1}{P_1 + P_2}$

A normal practice would be to keep the circuit efficiency to approximately 70%.

11.13 TESTING OF ACCUMULATORS

Testing is to be done as follows
1. Accumulators should be tested at least once in 4 years.
2. Openings should be of sufficient size to enable the interior to be examined.

3. The accumulator should be tested hydraulically to the required pressure by:
   (a) A water jacket capable of indicating the temporary and permanent stretch whilst under pressure. The permanent stretch should not exceed 10% of the total stretch.
   (b) A plain hydraulic test when the permanent deformation can be ascertained by measuring the accumulator before and after the test.
   (c) The maximum ovality permitted is 2% of the diameter.
4. The accumulator should be thoroughly dried after the test if the medium used is water for testing.
5. Re-test date should be marked on the body.

Review Questions

1. How would you classify hydraulic reservoirs? Differentiate between open and closed type reservoirs.
2. In certain hydraulic systems, pressurised reservoirs are used. Why? Where is such a reservoir used?
3. State the function of the following
   (i) Magnetic trapper
   (ii) Diffuser
   (iii) Level indicator.
4. How does an integral reservoir differ from a normal hydraulic reservoir?
5. What is an accumulator? How does it differ from a reservoir?
6. State various types of commonly used accumulators.
7. Compare a spring loaded accumulator with a gas charged accumulator explaining their merits and demerits.
8. What type of gas is used in a gas charged accumulator? Why is oxygen not used?
9. What is a bladder type accumulator? What precautions are generally taken when one uses a bladder type accumulator?

References

11.1 Bosch, Robert, *Bosch Hydraulics Information and Data* 1971/72, G mbH, Stuttgart, Germany.
11.2 Schneider, RT, "What you should know about hydraulic reservoirs", *Hydraulics & Pneumatics*, Penton IPC, USA 1978.
11.3 Publication from m/s Vickers Sperry, USA.
11.4 Taschen Buch m/s Herion-Werke, kg, 7 Stuttgart 1, Germany, 1969, p. 294.
11.5 Stewart Harry L, *Hydraulic & Pneumatic Power for Production*, Industrial Press Inc. NY, USA.

12
Design of Hydraulic Circuit

HAVING already understood the basic facts relating to system elements and components that are used in a hydraulic system, let us now have a close look at how a hydraulic system is designed and constructed. The main objective of a hydraulic system designer is to find a specific solution to a position, power, sequence or speed problem of any mechanical system such as a machine tool, a hydraulic crane or similar other mechanical or electro-mechanical implements that are commonly used. Common problems that confront a designer may be the movement of a machine tool table, a machine slide, a press ram or mechanical transfer of load, etc. For example, in the case of a machine tool, a hydraulic designer may have to design a composite system such as a clamping device for the workpiece, loading the job, feeding the material and tools, cutting, pressing or other metal cutting or metal forming operations, ejecting or withdrawing the jobs performed or tools used, etc. each of which may constitute a work cycle. The total machine cycle is the summation of all these cycles put together and a hydraulic system designer needs to synthesise all these functions taking into consideration the following factors:
1. Conformity to the desired functions
2. Level of performance efficiency
3. Safety and reliability of the system
4. Ultimate cost benefit in terms of maintenance and higher life expectancy.

A hydraulic control system is used to control the speed and position of the load with a driving force which comes from an actuator—either linear or rotary. The main parts which are essentially needed to design and construct a system may be listed as:
1. *Hydraulic power pack* comprising of the pump, drive motor, mechanical couplings, oil reservoir, strainers, filters, coolers, etc.
2. *Hydraulic control elements* like various types of direction control, pressure control and flow control/flow regulation valves, one-way valves,

servo valves, valve controllers, electrical/electronic control systems, actuators, e.g. cylinders, hydro motors, etc.
3. *Power drive units* comprising of cylinders, motors, etc.
4. *System accessories* such as pipes, hoses and other fluid conductors, filters, accumulators, boosters, related mechanical elements, etc.

A hydraulic designer has to decide not only the type of components to be chosen for the desired function, but he also has to analyze their mutual compatibility so that the ultimate design is comprehensive enough in terms of functional efficiency as well as reliability and maintainability of the constructed system.

12.1 HYDRAULIC CIRCUITS

Figure 12.1 illustrates various hydraulic components mutually interconnected by pipes depicting the hydraulic system of a surface grinding machine. It shows a table hydraulic with the cylinder, control valves, pipes, tubes and other elements. Some portion of the figure is made in sectional drawing for easy understanding. In order to avoid complicated drawings it is the practice in hydraulics and pneumatics to use symbols for each element and a circuit diagram is drawn with the symbols only.

A circuit diagram may be defined as the graphic representation of the hydraulic components in a hydraulically operated machine. It gives us an idea how control valves, actuators, pumps, etc. are interconnected. The symbols used in a hydraulic system as per 1S:7513–1974 are shown in Fig. 12.2. The hydraulic circuit diagram of a grinding machine drawn symbolically is shown in Fig. 12.3.

12.1.1 Design of Hydraulic Systems

Design of a hydraulic system starts with the mode of energy transfer in the system. The energy transfer in the system is done by the pump which draws the oil from the reservoir and transfers the same to the system. We know that energy transfer occurs due to changes in potential energy inside the reservoir and as the oil starts moving onward, there are further losses. This has been graphically illustrated in Fig. 12.4. This will help the designer to assess the exact energy requirement of the system.

12.2 MANUAL OR AUTOMATIC HYDRAULIC SYSTEMS

A simple machine tool bed can be operated either manually or automatically by a hydraulic system. The system oil is kept in a reservoir from which it is pumped to the hydraulic cylinder. The pressure oil from the pump to the cylinder is controlled for the respective operation by means of control valves.

The circuit diagram in Fig. 12.5(a) shows how to control a machine element with a double acting cylinder and a 4/2 direction control valve. Here the cylinder is expected to generate to and fro motion dependent on the valve condition. The system pressure is maintained by the pressure relief valve. All return lines from

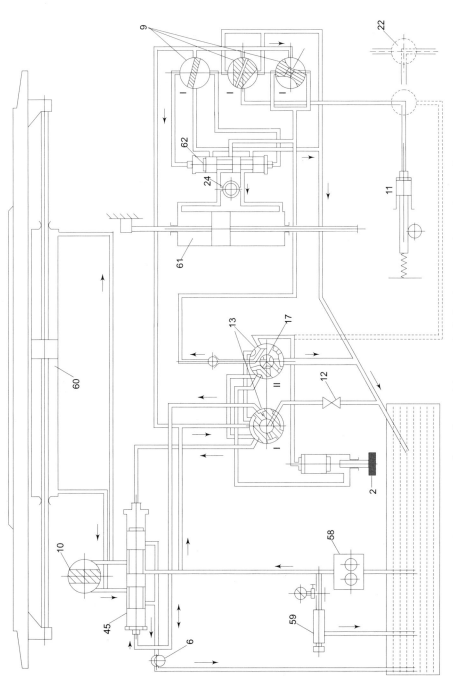

Fig. 12.1 Mechanical way of drawing the hydraulic circuit of a grinding machine

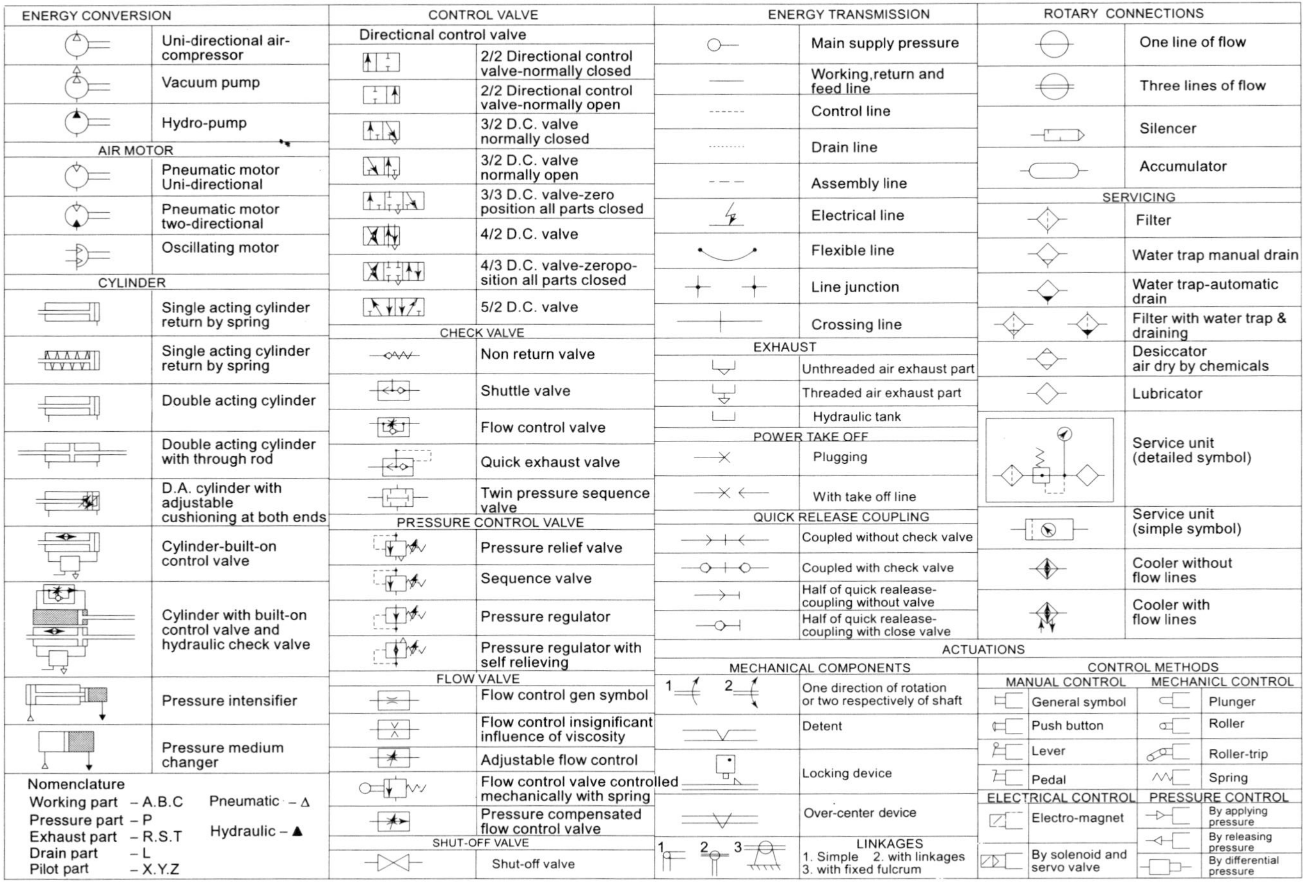

354 *Oil Hydraulic Systems: Principles and Maintenance*

Fig. 12.2 *ISO-symbols for fluid circuits*

Design of Hydraulic Circuit 355

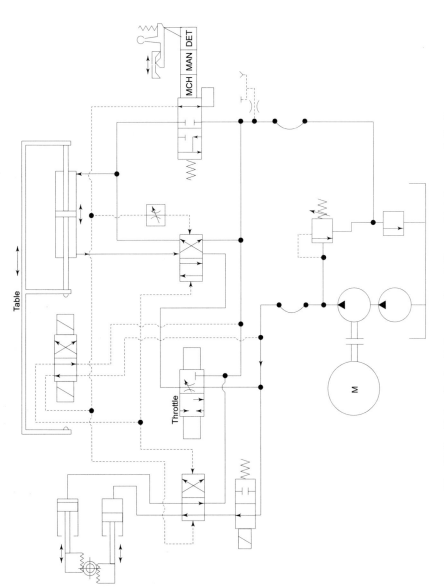

Fig. 12.3 *Hydraulic circuit diagram of a grinding machine drawn symbolically*

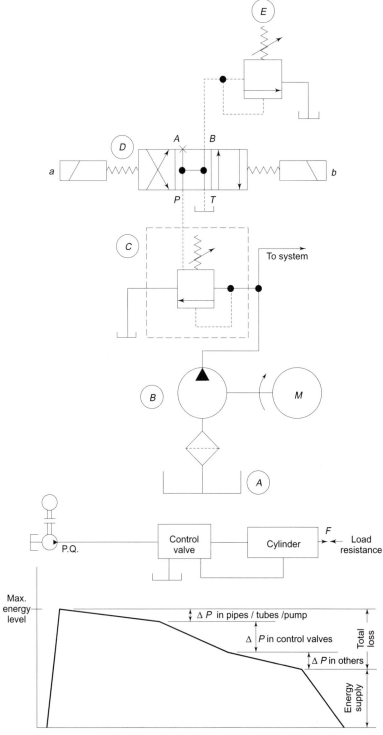

Fig. 12.4 *Energy losses in a hydraulic circuit*

Design of Hydraulic Circuit 357

the system are brought back to the reservoir. The control valve could be actuated by various manual, hydraulic pneumatic or electrical elements as shown in Fig. 12.5 (b). One of the problems of such a circuit is that the cylinder cannot have an intermediate position if we use a 4/2 DC valve. However, with a 4/3 or 5/3 DC valve one can have an intermediate position also as shown in Fig. 12.5 (c).

In the circuit diagram in Fig. 12.5 (c), the cylinder movement is controlled by a 5/3 direction control valve (DCV) with solenoid control. The electrical system

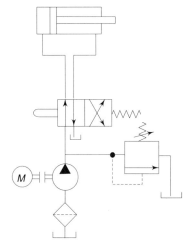

Fig. 12.5 *(a) Controlling a DA cylinder*

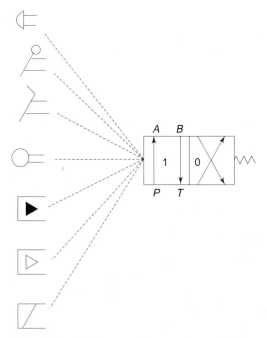

Fig. 12.5 *(b) Various valve actuation possibilities*

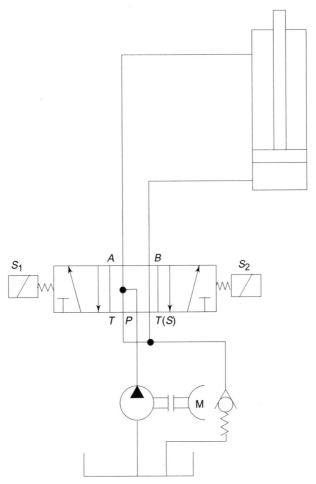

Fig. 12.5 *(c) Solenoid control of a hydraulic cylinder*

used to control the solenoids could again be a simple manual push button type or could also be made automatic by using limit switches. It may be noted here that the addition of a non-return valve in the return line to provide some resistance to the flow of oil through the return line will protect the cylinder from abrupt fall when both the return lines of the DC valve are connected to the tank in the null position of the valve (zero position of valve).

12.2.1 Velocity Control

To effect control of velocity of an actuator, one may use various types of flow control or throttle valves. Both ordinary flow control valves or compensated flow valves are used depending on the degree of control accuracy needed for the machine element in question.

Though the speed of an actuator in a hydraulic system is controlled by employing a flow control valve in the system, while discussing this particular phenomenon, we must remember that application of a variable displacement pump is most suitable in the case of a single load to be controlled. According to experts if the number of loads to be controlled is more, then individual speed control is resorted to by most of hydraulic system designers. The simplest method of speed control is however by throttling the flow as shown in Fig. 12.6 (a). In most machine tools and other mechanical equipments, speed control is adjustable in order to increase the working versatility of the machine. The flow control valve may be used either as a meter-in or a meter-out circuit. In both cases excess flow is discharged over the relief valve [See also Fig. 12.6(b)].

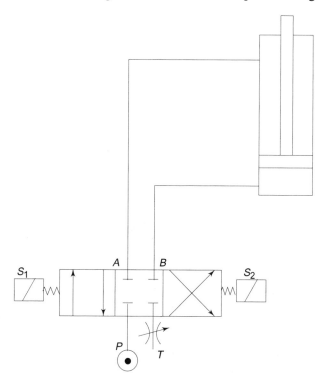

Fig. 12.6 *(a) Throttle valve in a hydraulic circuit*

12.2.2 Meter-in, Meter-out and Bypass Control

In a meter-in circuit or primary control the valve is placed between the pump and the load actuator and controls the flow rate of the oil into the cylinder whereas in a meter-out circuit or secondary control, the flow control valve is placed in the return line limiting the flow rate of oil from the cylinder.

Apart from the above two cases, the valve can be connected in parallel to the actuator controlling the rate of flow by bleeding-off a portion of the oil back to the tank. This is called bleed-off or by-pass control. At this point it may be necessary to know the various characteristics of each of the flow control techniques.

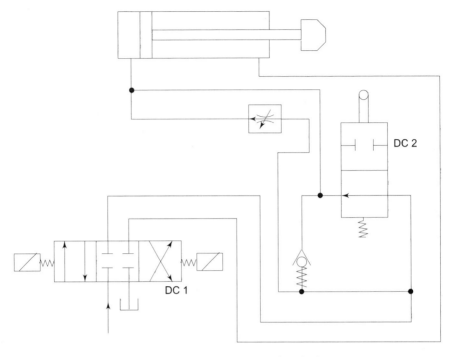

Fig. 12.6 *(b) Speed control of a cylinder*

(i) Meter-in Circuit Here [Fig. 12.7(a)] only the rate of flow to the actuator is controlled. The following specific characteristics are noteworthy:
  (a) There will be a drop in oil pressure fed to the cylinder due to throttling effect of the flow control valve. Hence for a low pressure system one should be careful to select a meter-in control.
  (b) Finer speed control is possible.
  (c) Heat generated due to throttling is fed to the actuator.
  (d) Suits well for resisting load actuated by the cylinder, but should not be used with hydromotor for its speed control.

(ii) Meter-out Circuit Here [Fig. 12.7(b)] the oil returning from the cylinder is only controlled. The following characteristics may be noted when a meter-out circuit is used:
  (a) No loss of pressure to the actuator as the full oil pressure is fed to the cylinder. Even at no-load the actuator is subjected to maximum pressure.
  (b) The actuator movement is more stable.
  (c) Heat generated due to throttling is fed to the oil reservoir.
  (d) Suits both over running loads as well as speed control of hydromotors. As the cylinder is fed with the entire pump pressure, there is possibility of higher friction loss.
  (e) Pump works against the maximum pressure.
  (f) Provides positive speed control of the cylinder.

(iii) Bypass Control   Here the pump delivery is bypassed to the tank at system pressure. The specific characteristics are:
(a) Does not allow positive speed control.
(b) Adjustment of speed is up to average value only.
(c) Does not suit accumulator circuits.
(d) Speed fluctuation if flow rate of pump fluctuates.
(e) Heat generated due to throttling is fed to tank.
(f) Pump works against the load resistance only.

Meter-out circuit is most popular due to its positive speed control capacity and stability and is hence universally accepted. Schematic circuit diagrams indicating application of meter-in, meter-out and bleed-off circuits for speed control have been shown in Figs 12.7(a), (b) and (c).

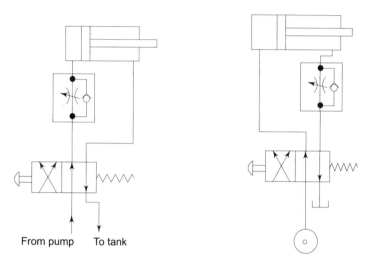

Fig. 12.7   *(a) Meter-in circuit*   **Fig. 12.7**   *(b) Meter-out circuit*

12.3 REGENERATIVE CIRCUIT

A regenerative hydraulic circuit is shown in Fig. 12.8. Regenerative circuits are sometimes used to obtain equal speeds in both directions in a differential cylinder.

Here the direction control valve (a 3/2 DCV in this case) is used in such a way that the return line from the differential cylinder instead of being connected to the tank line, is connected back to the supply line of the direction control valve as shown in the figure. Such a system may appear to lock up the piston hydraulically. But in practice the return oil from the rod end of the cylinder adds up to the pump delivery to the piston end increasing the speed of the cylinder in its forward motion.

If a 2:1 cylinder is used in the system the cylinder's speed will be the same in both directions. A 2:1 cylinder has a piston rod with a cross-sectional area equal to one-half of the piston area. Since in a 2:1 cylinder the discharge fluid from the

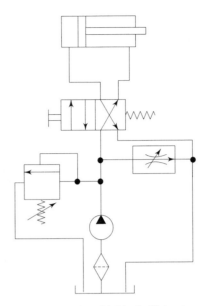

Fig. 12.7 *(c) Bleed-off circuit*

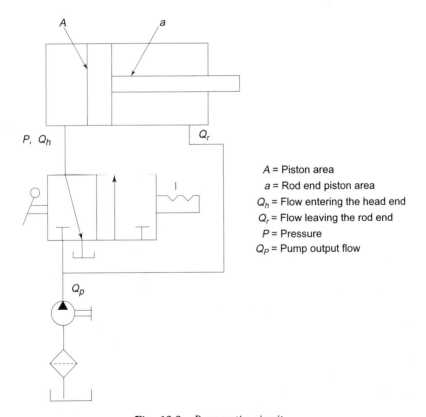

A = Piston area
a = Rod end piston area
Q_h = Flow entering the head end
Q_r = Flow leaving the rod end
P = Pressure
Q_p = Pump output flow

Fig. 12.8 *Regenerative circuit*

piston rod end is always half the volume entering the cap end, only the volume which is filled by the pump flow is the other half of the cap end volume. To retract the piston rod, the direction control valve is actuated, and the cap end of the cylinder is drained to the reservoir. The pump discharge and pressure is directed to the rod end side. Since the pump is filling the same volume as at the cap end side, the rod retracts at the same speed. Use of a 2:1 cylinder to regenerate flow does increase the rod speed but it also reduces the output force of the cylinder.

12.3.1 Model Calculation of Regenerative Circuits

From Fig. 12.8 we can see that force available to push rod against a load = $F = P \cdot (A - a)$, where A = area of cylinder piston, a = area of rod cross section.

As the piston moves, the fluid from the rod end (Q_r) is pushed out and joins the pump flow.

The speed of the piston is proportional to Q_h/A where Q_h is the quantity of flow to the head-end side which includes the flow leaving the rod end.

∴ The flow entering the head end (piston end) of the cylinder = $Q_h = Q_p + Q_r$

But $\quad \dfrac{Q_h}{A} = \dfrac{Q_r}{a}$ or $(Q_r) = Q_h \cdot \dfrac{a}{A}$

∴ We can write $Q_h = Q_p + Q_r = Q_p + Q_h \cdot \dfrac{a}{A}$

∴ $\quad Q_h - Q_h \cdot \dfrac{a}{A} = Q_p = Q_h \left(1 - \dfrac{a}{A}\right) = Q_h \cdot \dfrac{A-a}{A}$

∴ $\quad Q_h = Q_p \dfrac{A}{A-a}$ or $\dfrac{Q_h}{A}$ or $\dfrac{Q_p}{A-a} = \dfrac{\text{Pump supply}}{\text{annular area}}$ = regenerated speed

Example 12.1 *Let us take a cylinder having a piston diameter of 70 mm and piston rod diameter of 50 mm connected to a regenerative circuit where the pump flow rate is 25 l per min and the working pressure is 100 bar. Show that both the forward and return speed are almost equal in such a regenerative circuit.*

Solution

The area of the cylinder bore = $\dfrac{\pi}{4} \cdot 7^2 = 38.5 \text{ cm}^2$

The area of the piston rod = $\dfrac{\pi}{4} \cdot 5^2 = 19.6 \text{ cm}^2$

∴ The annular area = $38.5 - 19.6 = 18.9 \text{ cm}^2$
The force developed for forward movement = $F_1 = 100 \times 38.5 = 3850$ kgf
The force of retraction = $F_2 = 100 \times 18.9 = 1890$ kgf.

If no regenerative circuit is designed, the speed of the rod during extension = $\dfrac{25 \text{ l/min}}{38.5 \text{ cm}^2} = 10.08 \text{ cm/s} \simeq 10.1 \text{ cm/s}$

Speed of cylinder retraction = $\dfrac{25\ \text{l/min}}{18.9\ \text{cm}^2}$ = 22 cm/s

But if the system is designed with a regenerative circuit, the forward speed will be

$$= \dfrac{\text{pump supply}}{\text{rod area}} = \dfrac{25\ \text{l/min}}{19.6\ \text{cm}^2}$$

$$= 21.2\ \text{cm/s}$$

The return speed = $\dfrac{25\ \text{l/min}}{18.9\ \text{cm}^2}$

$$= 22\ \text{cm/s}$$

This shows that the regenerative circuit may equalize both the forward and return speeds of the cylinder.

12.4 USE OF CHECK VALVES IN HYDRAULIC CIRCUITS

Check valves are used in many places in a hydraulic circuit.

12.4.1 Pump Inlet Line

A check valve in the inlet line to a pump, shown in the circuit diagram in Fig. 12.9 (a), serves an important function in a hydraulic system, especially in simple hydraulic equipments like lifters, jacks, simple machines etc. An inlet line check valve keeps the oil from draining out of the pump when it is stopped, thus keeping the pump primed. This is especially important where the pump is high above the reservoir, that is, when there is a high suction height to the pump. For pump inlet line check valves, the valve spring should be selected carefully to have a low spring rate. In general and to ensure smooth inlet flow, spring generated suction/inlet pressure should not exceed 3.5 to 5 bar.

12.4.2 Pump Outlet Line

A check valve in the output line of a pump prevents system oil from flowing back into the pump as shown in the circuit diagram in Fig. 12.9 (b), especially when the output line is vertical. Because the check valve is mounted in the pressure line, the

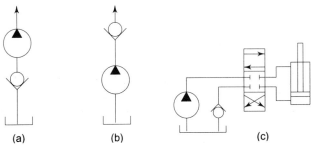

Fig. 12.9 *(a) and (b) Application of check valve in pump inlet and outlet, (c) Check valve with a counter balancing effect*

spring force selected for this application is not as critical as in the case of an inlet line check valve.

12.4.3 System Return Line

A check valve in the system return line to the tank, provides a counterbalancing effect on the actuator load keeping the cylinder and the load from extending and returning freely. The circuit diagram in Fig. 12.9 (c) represents this. The check valve spring should be selected carefully to conform to the return line pressure of other system return lines.

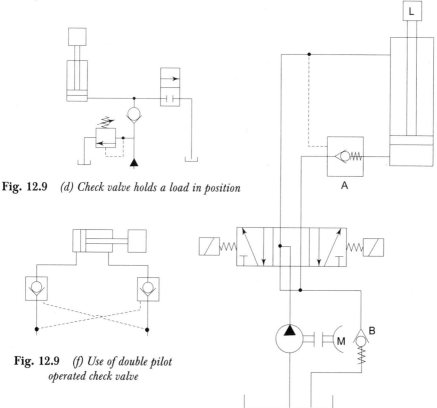

Fig. 12.9 *(d) Check valve holds a load in position*

Fig. 12.9 *(f) Use of double pilot operated check valve*

Fig. 12.9 *(e) Use of a pilot operated check valve*

12.4.4 Cylinder Position Lock

In a hydraulic system designed to position a heavy load precisely at any desired point, oil leakage through the direction control valve will allow the cylinder drift out of position. To ensure that the actuator stays at the predetermined position, an in-line check valve or pilot operated check valve may be used. In Fig. 12.9 (d) a check valve holds the cylinder in position until the 2-port, 2-position direction valve is shifted. In circuit 12.9 (e), the pilot operated check valve (A) holds the load (L) and prevents the cylinder from retracting until the head end of the cylinder is pressurized. Note however, that the cylinder is not locked and that it can be extended. Return line check valve (B) provides back pressure to the load, counterbalancing it.

A double pilot operated check valve, as shown in the circuit in Fig. 12.9 (f), locks the cylinder for both extension and retraction by blocking flow into and out of the cylinder. The load is locked as long as the cylinder seals remain effective and neither lines to the cylinder are pressurized.

12.4.5 Pressure Compensated Flow Control

In the circuit diagram shown in Fig. 12.10, four check valves have been arranged to form a bridge with a pressure compensated flow control valve at the center. In this arrangement extend and retract motions of the cylinder are controlled, that is, flows to the cylinder are metered in and out, as fluid to and from the cylinder always flows through the flow control valve in the same direction.

Two forward and return velocities are possible if a direction control valve is used in the circuit as a bypass. By applying a pressure control valve, the piston can be held under back pressure.

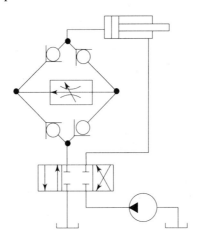

Fig. 12.10 *Check valves in a bridge*

12.4.6 Flow Equalizer

In Fig. 12.11, two cylinders have been engaged to lift a load at the same speed. By arranging the two check valves and throttle valves in the figure above, both the cylinders will extend at the same speed regardless of load resistance differentials they may have to overcome.

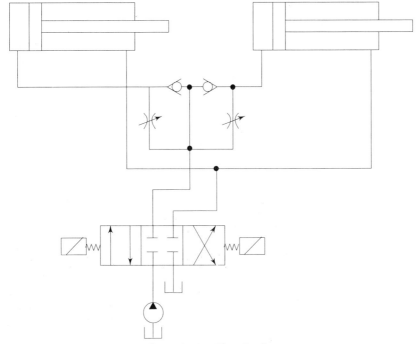

Fig. 12.11 *Flow divider*

12.5 SELECTION OF PUMP

Pump selection in a hydraulic circuit will basically depend on two factors like the maximum pressure and maximum flow requirement within a specific time frame which one can decide from the pressure-time and flow-time diagram. As far as the selection of a pump is concerned, there may be three alternatives:
1. A single fixed-displacement pump or
2. Two fixed-displacement pumps in tandem or
3. A single fixed-displacement pump with a suitable accumulator.

In many hydraulic circuits one uses two pumps to reduce or minimize the overall power. Let us take a pump with a flow rate of 200 liters per minute working at 100 bar pressure. The pump here will supply 32.3 kW $\left(\dfrac{200 \times 100}{612} \text{ kW}\right)$. Instead if one uses a hi-lo system, the system power will be less. Let us assume one pump has a flowrate of 150 lit/min and the other a flow rate of 50 lit/min in a hydraulic circuit which has to advance at a rapid rate and then from a specific point of forward motion it provides the feed movement which needs 45 lit per min oil. When such a hi-lo system is used, both the pumps may provide oil during rapid advance. But during the feed-rate the 150 lpm pump is dumped to the tank through an unloading valve and the 50 lpm pump develops the required flow for the feed rate delivering a power equivalent of $\dfrac{50 \times 100}{612} = 8.17$ kW which means a substantial saving of power.

Figure 12.12 (a) illustrates a schematic arrangement of the pumps and unloading valves in a hi-lo system.

Flow isolation   In a hi-lo pump circuit, a check valve between the pumps isolates high pressure flow from low pressure flow, thus protecting the low pressure components, from damage as shown in Fig. 12.12 (b).

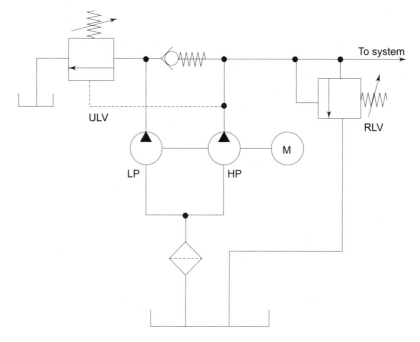

Fig. 12.12   *(a) Schematic diagram of a hi-lo system*

12.6   STANDARDS IN CIRCUIT DIAGRAM REPRESENTATION

The following points need to be observed:
1. A circuit diagram is the technical representation of a hydraulically operated system which is prepared with symbols according to the ISO/R 1219 standard. A circuit diagram is drawn without consideration for the actual physical arrangement of the hydraulic components in the system. The components are normally drawn from bottom to top in the direction of energy flow, e.g. in the direction of tank, pump, control elements, actuator.
2. With an electrohydraulic control system, separate circuit diagrams are drawn for electrical and hydraulic control with signal and control elements like limit switch, solenoids, etc. included in both circuit diagrams.
3. The circuit diagram should be drawn keeping the valves in their initial or normal position.

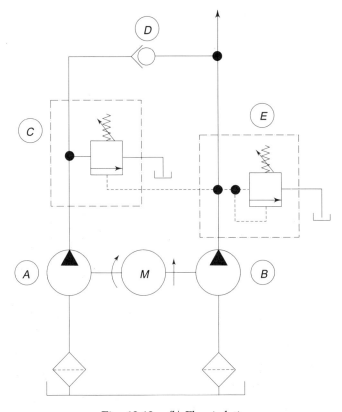

Fig. 12.12 *(b) Flow isolation*

4. Cylinders, direction control valves, etc. are represented in the circuit diagram in the horizontal position.
5. All symbols to be used in the circuit diagram should conform to international standards (i.e. ISO symbols which are equivalent to Indian Standard IS 7513–1974).
6. All hydraulic and electrical/electronic components used in the circuit diagram should be provided with systematic identification number. For example an actuator in a circuit (cylinder) can be given a number 1.0. If the number of cylinders is more, the cylinders may be given number as 1.0, 2.0, 3.0, ... etc.
7. All direction control valves can have distinct positions, for example, a 3/2 direction control valve (DCV) has 3 ports and two positions. The positions are designated as 0 and 1. This can be explained as follows

"0"	"1"
OFF	ON
Closed	Open
Pressureless	Pressurized
False	True

Hence for a 3/2 DCV one can state that the valve position will be 0 when the valve is closed or unoperated (i.e. P is closed to A and A is open to T) and the position will be 1 when the valve is operated and opens i.e. P opens to A and T closes). A 4/3 DCV using the same analogy has 4 ports and 3 positions. The positions are designated as 0 (the middle or neutral position), 1 and 2 (the two extreme operated positions). While drawing the symbol in a circuit, the valve positions should be written in the position box as shown in Fig. 12.5 (b).

8. Direction control valves which operate the cylinders should be given appropriate identification numbers. For example, the main direction control valve operating a cylinder 1.0, may be given a number 1.1 and for cylinder 2.0, the DCV can be numbered 2.1 and so on.
9. Other valves which pertain to cylinder 1.0 may be numbered as 1.2, 1.3, 1.4, etc.
10. All solenoids may be numbered as S_1, S_2, S_3, etc. and all limit switches may be given designations such as L_1, L_2, L_3, etc.
11. Pumps and the components in the power pack can be numbered as 0.1, 0.2, 0.3, etc.
11. All numbering given to represent various elements in the circuit diagrams should be referred to in the component list and on the identification tags of the components in the actual machine. This will help easy identification of the components during installation, commisioning, repairing, servicing and trouble-shooting.
12. All line connections are numbered with letters as per the ISO Standard (P, A, B, T, R ... etc.).
13. For easy identification the following color codes may also be used to identify the control lines.
    (a) Pressure line—continuous red line.
    (b) Control pressure lines for valves, etc.—broken red line.
    (c) Return line or pressureless line—continuous blue line.
    (d) Return control oil line—broken blue line.
    (e) Feed and suction line—continuous green line.
14. The circuit diagram should give the following technical details as far as possible:
    (a) *Oil reservoir*—Type of reservoir, type of oil, kinematic viscosity of oil at standard temperature, maximum capacity in liters, make, etc. If possible, the VI number should also be indicated.
    (b) *Electric drive motor*—Type, make, power in kW, power details and related parameters, speed in rpm, etc. may be provided.
    (c) *Pump*—Type, flow rate in l/min, maximum pressure in bar or in ISO unit, etc.may be indicated.
    (d) *Pressure relief valve*—The set pressure and pressure range if indicated will be helpful for service personnel.
    (e) *Direction control valve*—Nominal size, solenoid voltage, control pressure, pilot pressure, drain line, etc. may be provided.

(f) *Solenoids*—Supply voltage and other electrical parameters may be given.
(g) *Cylinder*—Piston rod diameter and stroke length (e.g. 200/140 mm × 430 mm) may be helpful.
(h) *Hydraulic motor*—Speed in rpm, capacity, torque in Nm, uni-or bi-directional, etc. are essential information.
(i) *Accumulator*—Type, type of gas used, gas filling pressure in bar, size and capacity in liters, bladder material, etc. may be provided.
(j) *Pipes*—Outside diameter and wall thickness, e.g. 12 × 1.5 mm may be given.
(k) *Hose*—Nominal size, hose material, type of braidings, etc. may be given.
15. Functional diagram may be provided along with the circuit diagram specially when the circuit has been designed for sequential operation involving a number of actuators and valves.
16. The hydraulic circuit diagram should be provided with the corresponding electrical/electronic circuits.

12.7 SPEED VARIATION IN CYLINDER MOTION

Figure 12.13 shows the circuit diagram of a hydraulic system having a double acting cylinder. The system requirement is that the cylinder (1.0) should have a rapid advance motion till it comes in contact with the 2/2 direction control valve (DCV) 1.7, and moves further with the feeding motion at a slower rate and at the end of the stroke it returns with a rapid rate. The main direction control valve (1.1) is a 5/3 solenoid operated spring centred valve, with its three positions marked as 1, 0, 2.

A Hi-lo pump (0.1 and 0.2) as explained earlier is used with an unloading valve (0.7) and the pressure relief valve (0.6) controlling the pump pressure. When the direction control valve 1.1 is in its position 1, the cylinder can have a rapid motion as oil from the pump can directly move to the cylinder through the 2/2 DCV which is in its open position (1.7). But as soon as the piston rod actuates the ball roller of the 2/2 DCV (1.7), it closes stopping flow of oil through it and allowing oil to move to the cylinder only through the pressure compensated flow control valve (1.5), thus reducing its speed. At the end of the stroke, as the piston rod comes in contact with the limit switch L_2, which energizes the solenoid S_2 reversing the main direction control valve (1.1) position and the cylinder returns rapidly.

It may be noted here that initially both the pumps deliver oil to the cylinder thus ensuring rapid advance motion. But when the oil flow is stopped through the 2/2 DCV, the high volume pump is unloaded and the high pressure low volume pump delivers oil alone to the system and feed motion of the cylinder at a slower rate takes place. While returning, both pumps deliver oil. Readers may note the presence of the filter (1.4) before the pressure compensated flow control valve

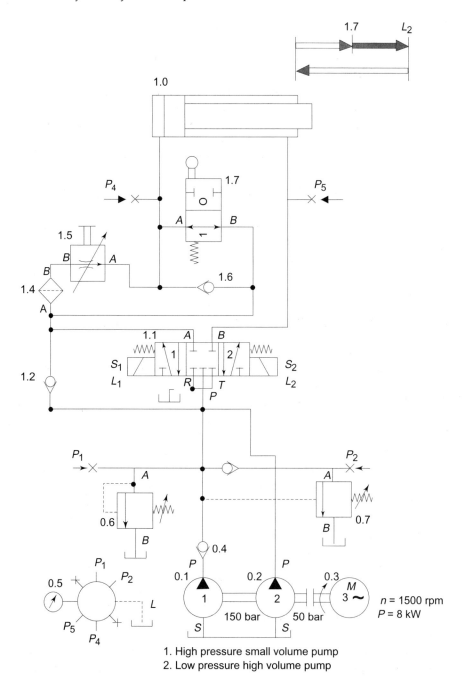

Fig. 12.13 *Circuit diagram for rapid advance, slow feed and rapid return system*

and the check valve 1.6 parallel to valve 1.7 as well as the cylinder movement diagram provided.

12.8 SOME BASIC CIRCUITS

1. Tandem center valve A schematic circuit diagram showing the application of a tandem center valve is shown in Fig. 12.14. As discussed earlier, there will be a pressure differential from P to T of about 3 to 4 bar because the P-port and T-port of tandem center valves are seldom located near each other. With the P-port at the center and T-port at the extreme, the passages are connected by means of a cored passage within the spool at the center position of the valve. This creates the pressure differential as stated above.

Here the valves are connected in series so that the actuator can act independently and the pump can also be unloaded at the same time when the valve is in its neutral position.

2. Sequential operation Figure 12.15 illustrates a schematic circuit diagram of a hydraulic system with sequential operation. Here cylinder 1.0 on its advance stroke actuates valve 2.1 to start the forward movement of another cylinder or bank of cylinders (2.0).

Figure 12.16 shows the application of a closed center valve in a hydraulic system.

3. Indirect control Figure 12.17 shows a schematic circuit diagram of how a hydraulic cylinder can be operated by a pressure (oil) operated valve which is actuated by a solenoid operated valve (indirect control).

4. Clamping force The circuit diagram in Fig. 12.18 shows a hydraulic clamping fixture for large mechanical or machine parts which need a high clamping force. Here also, for initial advance movement of the clamping cylinder one may not need a high clamping force but during the clamping operation, the force required may be very high. Hence here also a hi-lo pump can be used. To construct the circuit diagram, two double acting cylinders have been used in parallel as shown, with the control coming from a manual lever operated detented 4/2 DC valve.

12.8.1 Hydro-copying Circuit

1. Copy turning As already stated, hydraulic system is extensively used for copy turning and copy machining process. The principle of hydrocopying is shown in Figs. 12.19 (a), (b) and (c). For copy turning one can use a 2/2 or 3/2 or 4/2 direction control valve. Three such schematic circuit diagrams and the control probes are shown here in Fig. 12.19(a), (b) and (c) to illustrate the copy turning process.

Apart from copy turning, the hydraulic system is extensively used in copy milling also. Figure 12.20 illustrates a hydraulic circuit diagram of a punch shaping machine. In this circuit the movement of the machine table is controlled by the cylinder 2.0 whereas the probe stylus is operated by cylinder 1.0 whose piston rod is fitted to the valve body 1.1, the stylus being connected to the valve spool which while moving along the job profile guides the cutter to shape the

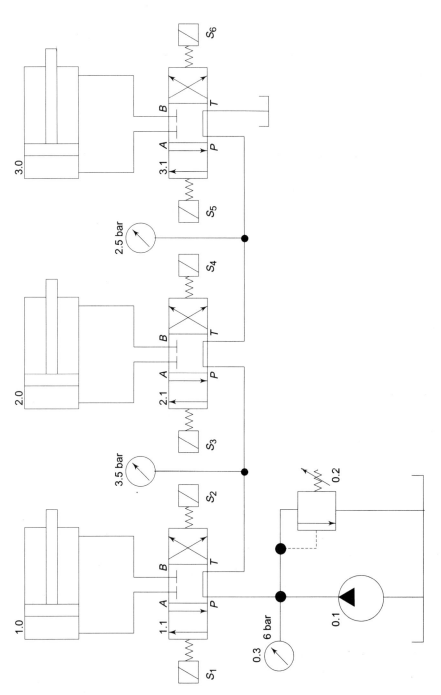

Fig. 12.14 Application of a tandem center valve

Design of Hydraulic Circuit 375

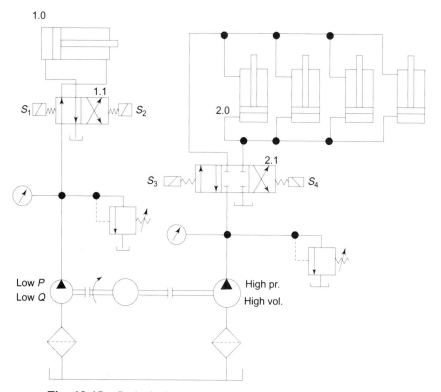

Fig. 12.15 *Bank of cylinders operated by a single 4/3 tandem center valve*

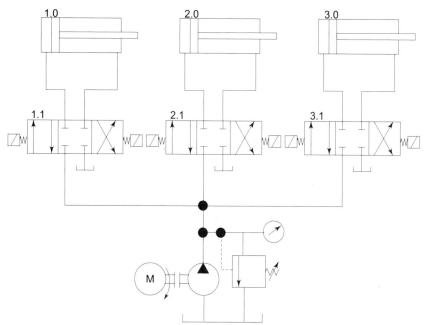

Fig. 12.16 *Use of closed center valve*

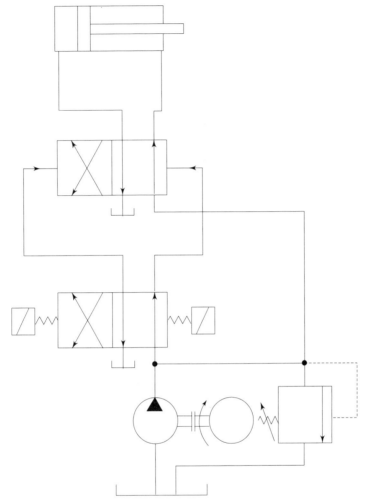

Fig. 12.17 *Indirect control of hydraulic circuit*

desired contour of the job. The system working pressure is controlled by the two pressure control valves used in the circuit.

2. Three coordinate copy milling In Fig. 12.21 (a) we find three axes X, Y and Z of a copy milling (die sinking) machine. The hydraulic circuit diagram of a three coordinate hydraulic copy milling machine is shown in Fig. 12.21(b) (courtesy— m/s Geb. Heller Maschinen Fabric, Nuertingen, Germany). From the circuit diagram it is seen that the movement of the axes X, Y and Z is controlled by three hydraulic motors. The motor shafts could be connected to finely machined precision screw spindles for smooth travel of the machine tables and tool. The X-axis motor controls the longitudinal movement of the table while the transverse motion is controlled by the Y-axis. The vertical movement of the machine is

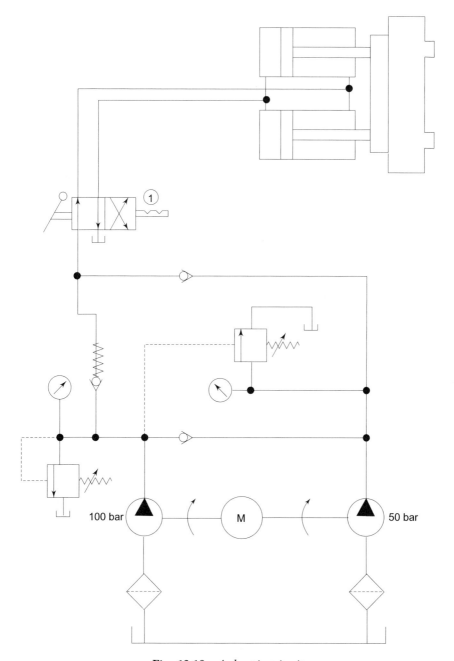

Fig. 12.18 *A clamping circuit*

controlled by the Z-axis. The three hydromotors are controlled by three electrohydraulic servo valves 1.1, 2.1, and 3.1 respectively. Any deviation in the screw spindle movement can be electronically sensed (not shown in the diagram) and compared with the valve signal. The variation, if any, is amplified and fed to

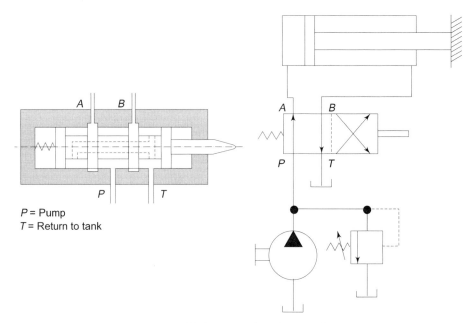

Fig. 12.19 *(a) Copy control with 4 control edges*

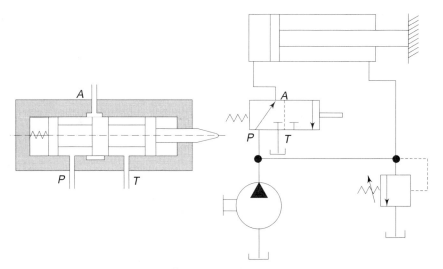

Fig. 12.19 *(b) Copy control with 2 control edges*

the valve control for correcting the error. Such hydraulic systems are very sensitive and positional accuracy is to the tune of ± 0.005 mm. Very fine spindles are used for ensuring positional accuracy.

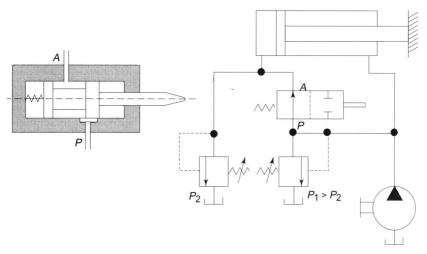

Fig. 12.19 *(c) Copy control with 1 control edge*

12.8.2 Circuit Diagram with Working Speeds Besides Rapid Approach and Return

To make a hydraulic actuator move in two or more different working speeds during feed movement besides rapid approach and rapid return stroke one may arrange two flow control valves in series as shown in the circuit diagram in Fig. 12.22. However readers may note the possibility of leakage and other losses due to use of additional fittings with simple flow control valves including probability of undesirable heating of the system.

As soon as the DC valve 1.1 is in position 1 and limit switch L_4 actuates DC valve 1.5, cylinder 1.0 starts its forward motion when the return oil from cylinder returns through dc valve 1.5 thereby providing a speed depending on the full pump supply. During its travel path, the piston rod actuates the limit switch L_3 which actuates in turn dc valve 1.6 to its position 1 thus ensuring the return oil to pass through flow control valve 1.3 and dc valve 1.6. This slows down the cylinder speed as designed by setting the flow control valve 1.3.

In case one wishes to have further slow speed of piston rod, dc valve 1.6 when deactuated to its position 'O' the return oil may be allowed to pass through flow control valve 1.4 arranged in series with flow control valve 1.3 but set at variant aperture opening to vary the oil flow. At the end of the stroke limit switch L_2 actuates direction control valve 1.1 to return the cylinder back to its original position.

12.8.3 Servo Control for an Extrusion Press

Figure 12.23 shows the schematic hydraulic circuit diagram of an extrusion press. The extrusion of rods out of billets, etc. is done by a water controlled ram (1). The throttle valve (8) which regulates the ram extrusion speed is controlled by an oil cylinder (9) actuated by a servo valve (11) and its control unit (13). The single

380 *Oil Hydraulic Systems: Principles and Maintenance*

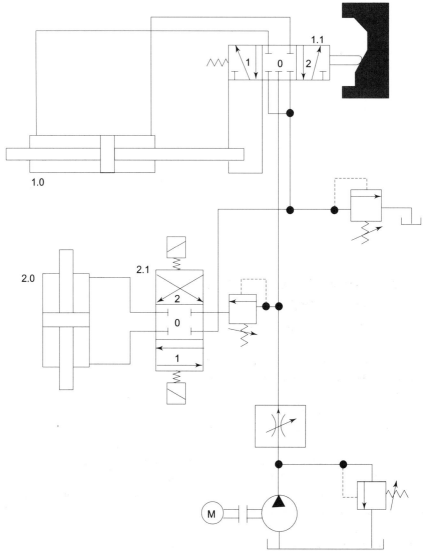

Fig. 12.20 *Hydro-copying circuit diagram for a punch shaping machine [Courtesy—m/s Gack Punch Shaper, Germany]*

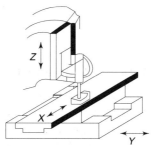

Fig. 12.21 *(a) Axes arrangement of three coordinate milling machine*

Design of Hydraulic Circuit **381**

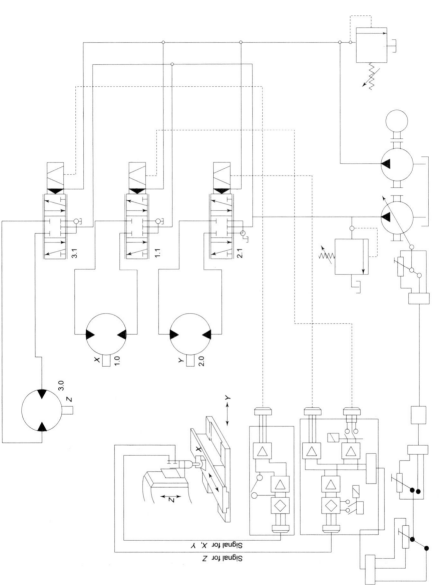

Fig. 12.21 (b) Servo control system of a three coordinate copy milling machine

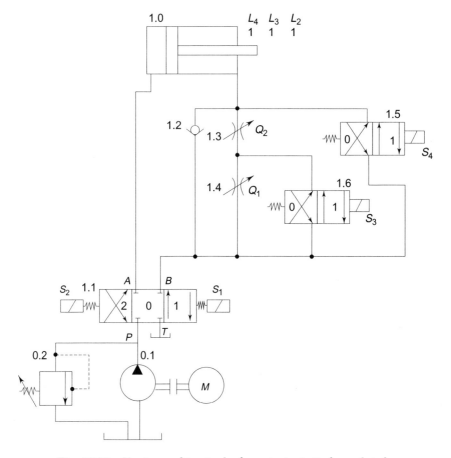

Fig. 12.22 *Varying working speeds of an actuator in its forward stroke*

acting extrusion ram is returned to its original position by using external cylinders (not shown in the diagram). The ram velocity feedback is sensed by a ram velocity transducer (6), i.e. a tachometer and cylinder position transducer.

12.8.4 Accumulator circuit

A simple schematic circuit with an accumulator is shown in Fig. 12.24. For loading and unloading the accumulator one may use a pressure switch.

12.8.5 Microprocessor Control in a Hydraulic System

Figure 12.25 illustrates the hydraulic circuit diagram of a sequencing system of actuators. The system here consists of an entire electrohydraulic servo system making it possible to couple the hydraulically powered elements with the advantage of electronic control loops with highly dynamic performance capability with a programmable digital control system employing a microprocessor.

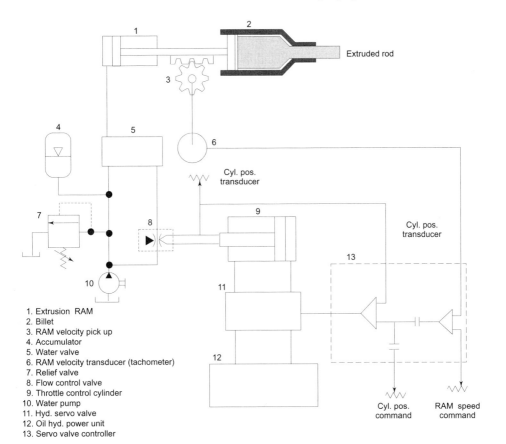

Fig. 12.23 *Servo controlled extrusion press circuit*

12.9 FUNCTIONAL DIAGRAM

A hydraulic circuit diagram comprising a good number of elements can at best represent the systematic physical arrangement of the various elements that make the system which may be quite complex. But in order to have a distinct and clear cut idea about the functioning of the system and its sequence of operation, a functional diagram of the system is very useful. A functional diagram not only provides a clear picture of the functioning of the system as a whole but also the relational position of its individual components and their mutual interaction. For trouble-shooting, machine maintenance, servicing and repair work, a functional diagram has been found quite handy for quick and effective trouble-shooting.

A fluid power circuit may consist of a number of control chains and each control chain may have the following members:
1. Signalling elements such as a push button, limit switch, pressure switch, etc. which are operated manually or automatically in isolation or in a repetitive sequence.

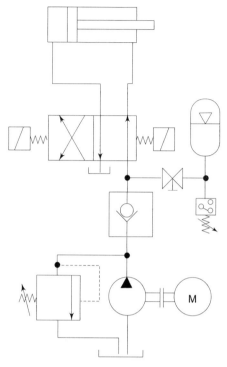

Fig. 12.24 *Accumulator circuit*

2. Controlling members such as contactors, direction control valves to control the fluid energy as per the signal received from the signalling members according to the required logic to direct the positional members to change their respective positions.
3. Positional members such as direction control valves (the main actuator element) to control the flow of energy and direction of flow in order to change the position of the actuator.
4. Driving members such as cylinders, hydromotors, etc. used to change the position of the working unit.
5. Working units such as the various subsystems of a machine like the clamping unit or feeding unit which are mechanical systems to be driven by hydraulic actuators.

At this point it may be necessary to have a look at the valve position in relation to the cylinder position. We have seen in Fig. 12.5 (b), the symbol of a 4/2 Direction Control Valve (DCV). All control valves, as we have explained earlier, may have distinct positions depending on various parameters such as the type of valves, their construction, actuation art, etc. The 4/2 DCV in question has two positions—position '0' and position '1'.

In position 0, P port is connected to port B and port A to port T.
In position 1, port P is connected to A and port B to port T.

Design of Hydraulic Circuit 385

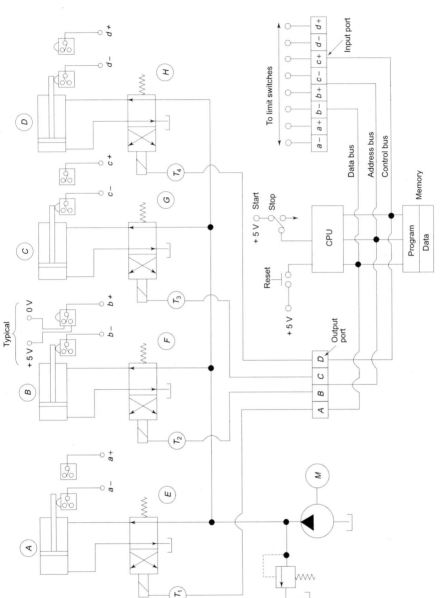

Fig. 12.25 Four-cylinder sequencing circuit and microcomputer system

While position 1 is attained by operating the valve actuating element, position 0 is attained in this case by means of the spring as soon as the force of actuation is removed from the valve actuation element. Hence this solenoid operated spring return 4/2 DC valve is a binary logic element one position òf which is spring based. Similarly a 3/2 DC valve is also a binary logic element. There are a number of such hydraulic logic elements and these can be used to design hydraulic circuits in a logical sequence. In designing such logic circuits designers need various mathematical tools and mapping techniques such as Karnaugh-Veitch map, time-travel diagram, etc.

A hydraulic cylinder may take generally two positions—position 1 when the piston rod is at rest, i.e. the rod does not move out and position 2 when the piston rod makes the forward or advance stroke (i.e. the cylinder working position).

12.9.1 Types of Diagram

As mentioned earlier, a circuit diagram may not adequately represent the functional sequence of various linear or rotational actuators and their controlling elements including their interrelation and interdependence to each other. The main purpose of a functional diagram is to project a clear and distinct sequence of operation of each element that makes the hydraulic system. A functional diagram has been found to be a very unique tool and very helpful to conceive and design the hydraulic system. Moreover during trouble-shooting of a hydraulically operated machine, functional diagram has been found to be quite helpful to pinpoint the faulty element just by tracing the sequence of signal and function line in a specific operation cycle.

Varieties of movement diagrams may be used in a control system. It can be drawn in one coordinate as shown in Fig. 12.26 (a).

On operating the push button 1, cylinder c_1 moves rapidly until it actuates limit switch b_1 which actuates a control valve (not shown in the diagram) and the cylinder speed gets reduced until it comes in contact with limit switch b_3 and thus cylinder c_1 returns to its original position. The next cycle starts again with actuation of b_2.

A single coordinate movement diagram depicts a clear picture of the working condition or status of the driving unit but it becomes difficult to provide the picture of the positioning or controlling members. Neither does it give a clear idea about the speed or other physical parameters. In order to avoid such problems one can use a two-coordinate functional diagram such as:
(a) Travel-time diagram and
(b) Position-step diagram.

In a travel-time diagram, the X-coordinate shows the time and the Y-coordinate shows the amount of travel of the cylinder movement, valve-spool movement, etc. (shown in Fig. 12.26 (b)). But the stroke length of any hydraulic element specially the actuator, may have wide variation creating problems in drawing it uniformly and hence to get over this problem one may go for a position-step diagram where the cylinder or valve stroke positions replace the travel in the

Design of Hydraulic Circuit **387**

Y-axis and the time scale in the X-axis is replaced by the step (as shown in Fig. 12.26 c).

In the position-step diagram the sequence of operations is divided in a number of steps which are expressed in numbers 0, 1, 2, etc. Cylinder/motor positions are indicated as 1, 2, etc. A time-scale however, may be indicated parallel to the step line. In the position scale for a cylinder position 1 means cylinder has retracted while position 2 means cylinder has advanced to its forward position.

The functional lines are drawn in thick lines and they determine the status or position of the working or driving unit, control or signal members during the sequence of each operation of the whole unit. Any change in position of a member has to start and stop at a corner of the squares as shown in Fig. 12.26 (c). The functional diagram is also termed a step-sequence diagram or a sequence diagram.

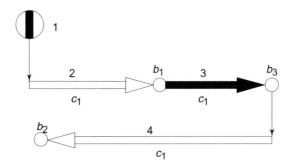

Fig. 12.26 *(a) Single coordinate movement diagram*

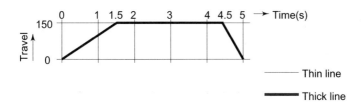

Fig. 12.26 *(b) Travel-time diagram*

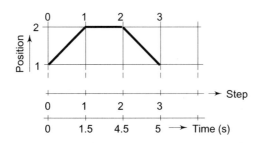

Fig. 12.26 *(c) Position-step diagram*

12.10 APPLICATION OF FUNCTIONAL DIAGRAM

A functional diagram of a hydraulic system is shown in Fig. 12.27 (a). At step 0 the solenoid S_1 of the main directional control valve (DCV 1) and the solenoid S_3 of DCV 3 are both energised. Both the valves will change their position from 0 to 1 and oil will move to the cylinder via the check valve of the non-return flow control valve 2 thus providing the cylinder an initial rapid motion. As soon as the direction control valve DCV 3 is de-energized at step 1, DCV 3 changes its position back to position 0. At this point depending on the status of the flow control valves 2 and 5, the cylinder may have a slower speed until it reaches its end of stroke at position 2 and step 2. From the diagram we can see that at step 2, the DCV1 solenoid is de-energized, the valve changes its position to 0. From step 2 to step 3 the cylinder does not change its position and stays at position 2 only while the DCV 1, 3 and 4 are all in the position 0. At step 3, the DCV 1 is actuated by energizing the solenoid S_2, and cylinder starts retraction at a slow speed till the step 4, where the DCV 4 is also energized by solenoid S_4 and the cylinder starts rapid return until it reaches its initial position 1 when all the valves reach their respective normal position, i.e. position 0. The hydraulic circuit diagram is shown in Fig. 12.27 (b).

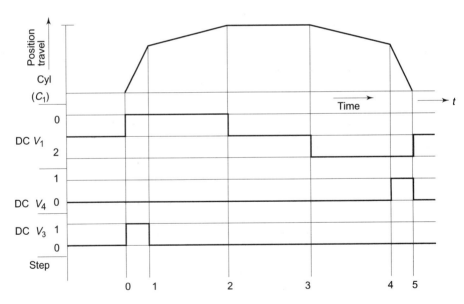

Fig. 12.27 *(a) Functional diagram of a hydraulic circuit*

12.11 ELECTRICAL CONTROL OF HYDRAULIC SYSTEMS

We may refer to Fig. 12.28 where a hydraulic circuit diagram is shown for a DA cylinder. In 0 position the cylinder moves back and stops at dog (not shown in

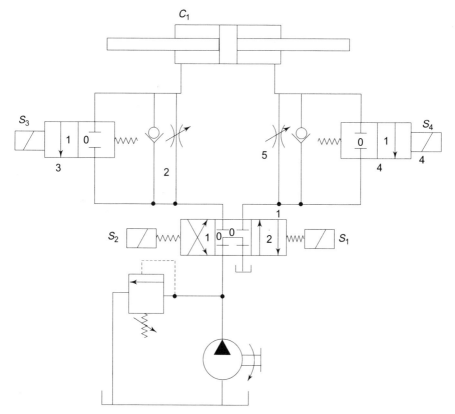

Fig. 12.27 *(b) Hydraulic circuit diagram as per the functional diagram*

the circuit diagram). Solenoid S_1 brings direction control valve 1 in position 2 and cylinder C_1 moves forward and stops on dog. As long as solenoid S1 is energized, the cylinder remains in forward position. If pressure is too high above the set pressure of the pressure relief valve the relief valve will crack and the oil will flow to the tank through the pressure relief valve PRV 1. The electrical circuit diagram of the electrical control of the solenoid is shown in Fig. 12.28(b).

Figure 12.29 depicts a hydraulic circuit which shows how relay technique is used to control direction control valve. Here a hydraulic motor has been used to reciprocate a machine tool table. When solenoid S_2 is operated, the table starts its forward movement. At the end of the table movement, it encounters the relay which in turn actuates the solenoid S_1 to reverse the table movement.

Review Questions

1. Differentiate between direct and indirect control. Draw simple hydraulic circuit diagrams of both and explain the differences.
2. Draw a hydraulic circuit diagram of a hydraulic system having a double-acting cylinder which has a rapid approach speed, then a slow feed motion and at the end of the stroke the cylinder returns rapidly.

390 *Oil Hydraulic Systems: Principles and Maintenance*

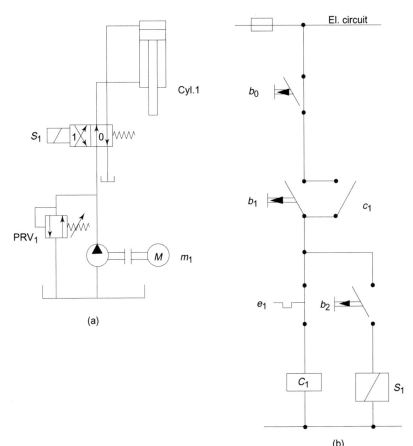

Fig. 12.28 *Electrical control of hydraulic cylinder*
(a) Hydraulic circuit (b) Electrical control diagram

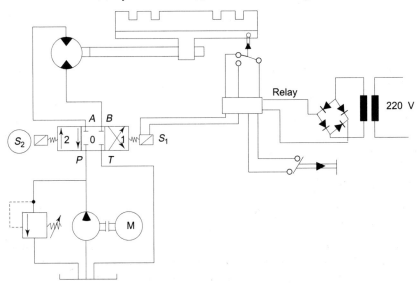

Fig. 12.29 *Hydraulic circuit with relay*

Design of Hydraulic Circuit   391

3. What is a functional diagram? How does it differ from a circuit diagram? What are the advantages of such a diagram while trouble-shooting?
4. Show the use of a check valve in a circuit as a flow divider.
5. Show the application of tandem centre valves in a hydraulic system.
6. What is a hi-lo pump? Show the application of such a system with a circuit diagram.

References

12.1 Proceedings of Short Term Course on Industrial Hydraulics, IIT, Chennai, India.
12.2 Wolfgang Leiber, *Hydraulische Anlagen*, Carl Hanser Verlag, Muenchen, Germany,1966.
12.3 Majumdar, SR, "Check Valves—Simple Designs Provide Sophisticated Performance", *Hydraulics and Pneumatics*, USA, March/1984.
12.4 "Understanding Hydraulic Circuits", *Fluid Power*, Fluid Power Society of India, Bangalore, India.
12.5 Robert Bosch, *Bosch Hydraulics—Information and Data*, GmbH, Stuttgart, Germany, 1971-72.
12.6 Singaperumal M, *Design of hydraulic circuits*, IIT, Chennai, India.

13
Seals and Packings

LEAKAGE in a hydraulic system is generally a common occurrence which needs to be arrested in order to keep the system running at optimum efficiency. A hydraulic component designer has to use lot of precaution at the design stage itself if one desires to have an optimally leakage free system. However, it is very difficult to design an absolutely leakproof system due to the sheer physical nature of construction of the hydraulic and fluid power elements and their susceptibility to wear and tear resulting in ultimate leakage. A well-oiled preventive maintenance system may ensure reliable operational efficiency of the system but without stringent quality control with regard to seals and sealing at the design stage itself, no fluid power system will work flawlessly at a latter date during its life cycle. It is interesting to note that a tiny part like the tiny seal plays a very important role in ensuring high functional reliability of the whole system. Leakage in a hydraulic system could be both internal and external and proper seals engineered to the correct specification and application will prevent leakage effectively. In Fig. 13.1 use of various seals in a hydraulic cylinder is shown.

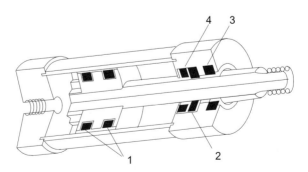

Fig. 13.1 *Use of seals and wipers*
1-Deep poly pack   2-Modular bearing   3-Wiper ring   4-Deep poly pack
[Courtesy–Parker Packing Co., Utah, USA.]

Seals are used in both static and dynamic portions of hydraulic components. Truly speaking, a major part of the application of seals involves the static part. Seals are used to seal the static mating parts between valve bodies, pumps, reservoirs, pipes, couplings, etc. Similarly dynamic parts of hydraulic components such as pump and hydromotor shafts, bearings, cylinder piston and wall, valve spools and valve bores having relative movement under varying speeds and pressure are also protected by the use of dynamic seals. Dynamic seals encounter varying extraneous requirements whose severity increases with pressure. However it is thanks to the development of seal technology and sealing methods that today a hydraulic system can function at a much higher pressure as compared to earlier years. Seals are basically used to

(a) arrest both internal and external leakage of oil/gas/steam and
(b) prevent dirt, dust and other foreign particles from entering the system.

In order to increase sealing efficiency, a component designer has to consider these broader aspects in this vital sphere, i.e.

(a) Characteristics of the mating surface
(b) Characteristics of the medium
(c) Design and nature of the seal and seal material.

Seals play an important role in the fluid power system and with the tremendous development that has taken place in this field, fluid power systems could be used at much higher pressures than earlier and under any circumstances and environmental conditions. In any pressure actuated system, seal is a must—be it rotary or reciprocating, static or dynamic system, steam, hydraulic or pneumatic medium. The seal is a tiny part in a giant machine but is such a critical element that for want of it the complete machine may remain inoperative increasing its idle time.

13.1 CLASSIFICATION OF HYDRAULIC SEALS

The seal is an agent which prevents leakage of oil from the hydraulic elements and protects the system from dust/dirt. The major function of the seal is to maintain pressure, prevent loss of fluid from the system and to keep out contamination in the system to enhance its working life and functional reliability over a longer period.

13.1.1 Categorisation of Seals

We may categorise seals as per methods of sealing, area of application, seal material, geometrical shape of the seal used in the system, etc.

According to the methods of sealing, seals can be categorised as positive sealing or non-positive sealing.

In certain dynamic seals, wipers or covering boots are used to keep away dirt and other foreign materials, e.g. use of rod wiper. Figure 13.2 illustrates the use of static and dynamic seal in a hydraulic cylinder.

394 *Oil Hydraulic Systems: Principles and Maintenance*

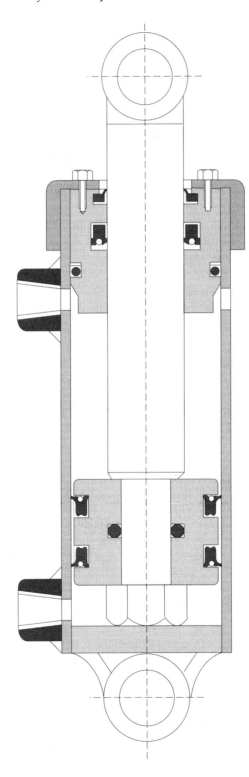

Fig. 13.2 *Static and dynamic seals in a hydraulic cylinder*

More details about static and dynamic seals are discussed subsequently. Sealing may be (a) *Positive* sealing which prevents even a minute amount of fluid from getting past and (b) *Non-positive* sealing which may allow a small amount of internal leakage such as the clearance of a spool in its bore to provide a lubricating film. As has been mentioned earlier seals are also classified by the material used for their fabrication, application or use and also by configuration, geometric shape etc., however commonly they are classified as: (a) Static seals and (b) Dynamic seals.

13.1.2 Static Seals

They are used when no relative movement occurs between the mating parts. The seal is usually compressed between two adjacent parts securing the two stationary parts together by fasteners, e.g. a seal may be used between the pump housing and the end plate.

13.1.3 Dynamic Seals

They are used between the surfaces of hydraulic parts where movement occurs and controls both leakage and lubrication. In certain dynamic seals wipers or covering boots are used to keep away dirt and other foreign materials, e.g. use of rod wiper. The motion encountered by dynamic seals may be either reciprocating or rotating or a combination of both.

13.1.4 Classification of Seals According to Shape Configuration

Very often seals are classified by the geometric shape or configuration they take during fabrication. According to shape seals may be categorized as:
1. O-ring seal
2. Quad-ring seal
3. T-ring seal
4. V-ring seal
5. U-cup ring
6. Hat ring

A diagram of these seals is shown in Figs 13.3 (a) and (b).

Irrespective of the form they can be used both as static and dynamic seals. They can also be used either as a single seal or in multiples.

13.1.5 External or Internal Sealing

In certain applications seals are used in double-acting cylinders or in valve spools to prevent leakage between areas of differential pressure. If leakage occurs between the contact surface of the seal and the mating part, this may result in:
1. Loss in volumetric efficiency
2. Loss in system power
3. Lack of control and loss of sensitivity and accuracy in response

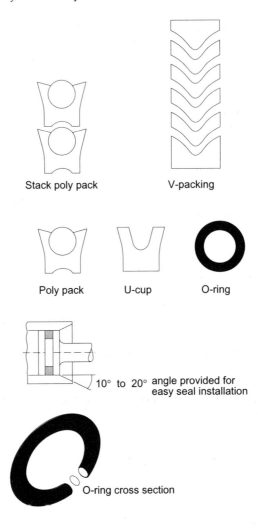

Fig. 13.3 *(a) Cross-sectional view of types of seals*

4. Generation of undesirable heat
5. Resultant system malfunctioning

Lack of lubrication may not be a problem since oil is present on both sides of the sealing ring. The seals mentioned above are mostly termed as internal seals. External seals provide a seal both against differential pressures, that is between the internal system pressure and external atmospheric pressure, as well as control lubrication of the seal itself and adjacent bearing support. This controlled leakage of fluid from the system thus lubricates and extends the life of the adjacent moving parts.

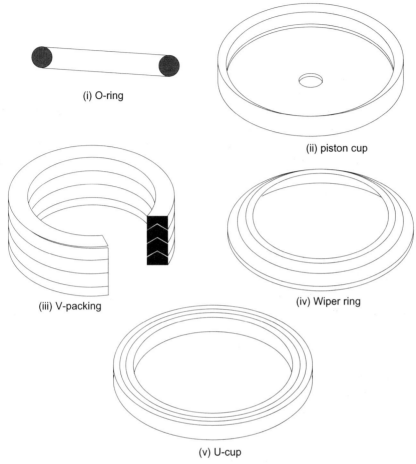

Fig. 13.3 *(b) Seals are classified as per their geometric shape*

13.1.6 Nomenclature of Seals as per Seal Material

Very often seals are classified according to the material they are made of e.g. a seal made of leather may simply be termed as a leather seal. The most commonly used seal materials are as follows:
  (a) Leather
  (b) Metal
  (c) Polymers, elastomers and plastics
  (d) Asbestos, nylon, etc.

13.2 FACTORS FOR SEAL SELECTION

There are lot of physical conditions which need consideration before an engineer opts to use a particular seal or seal material for the specific application. The governing factors are:

1. Working pressure and pressure range
2. Environmental condition
3. Fluid medium
4. Dynamic or static application
5. Temperature of the system
6. Functional reliability and expected life.

But in order to generate a failsafe sealing system, the above factors may not be enough. One has also to look for the coveted physical and chemical properties of the seal materials used. A thorough look at the following physical properties especially in case of elastomeric seals is therefore very essential for the designer as well as for the maintenance engineer in order to make an optimum choice. Some of these properties are:

(i) Hardness
(ii) Friction
(iii) Volume change
(iv) Compression set
(v) Tensile strength
(vi) Elongation probability of change
(vii) Tear strength and abrasion resistance
(viii) Thermal effects and heat resistance
(ix) Squeeze and anti extrusion property
(x) Stretch/tensile strength
(xi) Coefficient of thermal expansion
(xii) Permeability
(xiii) Oil-compatibility
(xiv) Ageing
(xv) Corrosion resistance
(xvi) Ozone and weather resistance
(xvii) Electrical properties

Some of these physical properties which affect the hydraulic system are discussed here.

(i) Hardness Hardness of elastomeric seals is a very important property. Resistance of an elastomer and other materials to indentation is defined as hardness. Elastomeric seal hardness is measured with an instrument called a *durometer*. This is done using the "shore A" scale. This scale measures hardness on a scale ranging between 0° and 100°.

Higher durometer readings indicate a greater resistance to denting and denotes a harder material; lower durometer readings indicate less resistance to denting indicating a softer material. It is better to use softer materials for lower pressure and harder material for higher pressures. For common hydraulic applications, a durometer hardness of 70° on scale A is mostly used for elastomeric seal materials. A shore A hardness of 80° is used in rotary applications to eliminate tendency towards side motion and bunching in the groove. Softer 50°–60°

hardness, O-ring and square cut gasket may be used with static seals in rough surfaces. Generally, harder 80° and 90° shore A materials have less wiping action at the surface and these are more prone to breakage during installation. An advantage with the use of harder sealing materials is the reduction of breakaway friction since softer materials have greater tendency to deform and flow into surface irregularities at the place of contact between the seal and the moving part. This helps in better sealing and is quite useful for use as lip seals, etc. where wiping action is also as important.

(ii) Volume change–swelling and shrinkage Volume change is defined as the increase or decrease in size of an elastomer as a result of continuous contact of the elastomer with the hydraulic medium. Generally most seal materials have a tendency to increase in size due to swelling when the seal is in constant touch with the oil. Swelling is therefore defined as increase in volume due to immersion of the seal in oil.

While a certain amount of swell may be desirable to compensate for wear and improve seal effectiveness, excess increase in size or swell is undesirable specially in dynamic seals where friction and the tendency to abrade are increased by addition of volume and accompanying softness. The maximum allowable swell for dynamic O-ring seals, is within 20%. 40–50% swell may be allowed for fixed or confined seals providing the fluid does not extract plasticisers which would cause the seal to dry out and leak.

(iii) Compression set It is the tendency for an elastomer to lose its resilience and is expressed as the ratio of the loss in thickness of an elastomer after it has been deformed to the original thickness. In mathematical terms, compression set is expressed as a percentage of the original thickness.

$$\therefore \text{Compression set} = \frac{\text{Loss in thickness}}{\text{Original thickness}} \times 100\%$$

Compression set is also defined as the permanent distortion of a rubber or elastomeric material after compression at a specified temperature for a period of time (i.e. 25% compression at 100°C for 72 hours).

To determine compression set, an O-ring is compressed between two heated plates for a given duration of time and then the thickness is measured to determine recovery. Temperature and the size of the O-ring are important factors which affect compression set. Increase in temperature tends to increase compression set whereas increase in the O-ring size tends to decrease compression set. As very often the seals are subjected to the effects of the fluid which cause swell, the final squeeze will be:

Final squeeze = initial squeeze + swell – compression set.

(iv) Tensile strength Tensile strength, elongation and tear strength are properties which may also affect the operation of the seal due to physical contact and relative movement which tends to stretch, abrade, tear and wear the seal. *Modern polymer compounds* are available nowadays in tensile strength value

exceeding 70 kg/cm² or more. The ultimate elongation of a seal is defined as the maximum length to which an elastomeric seal may stretch before failure and separation. It is expressed as a % of free length of the seal.

$$\text{Ultimate elongation} = \frac{\text{Separation length}}{\text{Free length}} \times 100$$

It is also one of the measures used in testing for quality control purposes. Elongation values necessary to install an O-ring in gland over small pistons without breakage must often exceed 100%.

(v) Permanent set It is defined as the permanent distortion of rubber after elongation (i.e. 150% for 10 minutes at room temperature).

(vi) Adhesion It is defined as the susceptibility of rubber to stick to a contact surface.

(vii) Aniline point It is defined as the temperature at which fresh aniline may react with the oil. A high aniline point oil causes less swell of the rubber.

(viii) Squeeze Squeeze is the diametral compression of O-ring between two mating surfaces of the gland.

Example 13.1 *A static O-ring has a diameter of 7 mm and is given an initial squeeze of 15%. The swell of the seal equals to 25% increase in squeeze and the compression set is 10%. What will be the final squeeze % of the seal?*

Solution We know that

Final squeeze = initial squeeze + swell − compression set.

$= 7 \times 0.15 + 7 \times 0.25 - 7 \times 10$

$= (7 \times .40 - 7 \times 0.10)$

$= (7 \times 0.30) = 2.1$

∴ % of final squeeze $= \dfrac{2.1}{7} \times 100 = 30\%$

13.2.1 Common Seal Materials

Various metallic and non-metallic materials are used for fabrication of seals used in hydraulic systems. Leather, metals and elastomers are very common seal materials. Very often seals are also made out of compounds of the base material or even out of reinforced fabrics of compounds for specific hydraulic applications. Material hardness is an important characteristic for determining the specific application and use. Dynamic seals need to be softer as this property helps them to seal better through compression. For dynamic applications various types of materials can be used but basically fibers are used more along with wire reinforcements, metal foils, mica, graphite and Tetrafluoroethylene (TFE). They may be braided, woven and laminated with the addition of lubricant bearing materials. For better and effective sealing, homogeneous materials are better suited rather than reinforced fabrics. Materials selected for seals must be compatible with the operating conditions and temperatures of the fluid system during both its period of operation as well as when the system is idle. Let us discuss in brief the most commonly used seal materials with their specific properties and application. But before that it may be also necessary to know the physical properties that govern their use.

(a) Metal Though not a very common sealing agent for normal low pressure hydraulic system, there are certain applications in which metals are used as seal material. Though they are not resilient in nature they have extensive use as very good sealing materials for high pressure and high temperature systems. They may not be as suitable for low pressure systems. There are cases where metals are found to be suitable as a sealing medium for extremely low temperature applications also. However metal seals are rarely used for dynamic applications as a fluid tight chamber is difficult to create with metal seals. Incapability of metals to deform readily is the biggest drawback in using metal seals. Because of this, metal seals may need a heavy structure to transmit the load to the seal to form a required sealing chamber. Accurate matching and tolerances are also a pre-requisite for the use of metal as seals.

The rigidity of metal seals do not allow them to follow the bearing surfaces to be sealed effectively. Therefore metal seals used in systems, due to the effect of pressure and temperature, use other extra heavy structure such as pre-loaded bolts and fasteners to minimize the effect of separation of mating surfaces due to effect of pressure and temperature, which is a precondition for proper sealing.

In some specific applications, this can be eliminated if metal seals are designed as leaf springs using the spring effects to follow the separation of the mating surfaces e.g. sealing metal flanges. This property is very often used as in the case of piston rings which can produce the required contact pressure to form the seal line.

Another good property of metal seals is their inertness to the effect of most hydraulic fluid media. They are also suitable for radioactive environments and therefore most suited in such cases.

(b) Leather In hydraulic systems, leather (mostly chrome-tanned polyurethane impregnated leather) is a commonly used seal material—both for dynamic and static seals. The physical characteristics of leather and its wide availability makes it a common choice for seal manufacturers and fluid power engineers. Leather seals are preferred because of their following physical properties:
1. High resistance to abrasion and wear
2. Low coefficient of friction
3. Inherently pliable
4. Better lubricating property.

The randomly interwoven leather fibers are porous and inherently tough and strong in low tension. This makes leather a good choice as a general sealing medium both for common and specific applications. They are non-abrasive, non-corrosive and provide a good polish to the metal surface with which they come in contact. The coefficient of friction is low and less heat is generated during the sliding action enhancing the seal life considerably. It is, therefore, preferred from the maintenance point of view. The porosity of leather permit it to store lubricant. They are easily impregnated with various substances, e.g. wax. However, this porosity of leather also makes it swell when it is in continuous touch with oil thereby increasing its size and volume which is sometimes undesirable. The open pores of leather seals may allow the fluid being sealed to seep through which is also not very desirable in some cases. The temperature range of a leather seal is $-50°C$ to $+100°C$ but generally it is used up to $+80°C$. They are naturally acidic and operate better in a medium of a pH with 3 to 7, a pH of 8 being the border line. Leather cannot withstand an alkaline condition.

(c) Asbestos   Asbestos is used both for static as well as dynamic seals. They are quite suitable for high temperature applications. However, from the point of view of environmental hygiene and health, a lot of restriction is imposed nowadays in the use of asbestos in modern engineering industries.

(d) Rubber, other elastomeric and plastic seal materials   In hydraulic and other fluid power systems various type of rubbers, elastomers and plastics are found to be commonly used as seal materials due to their excellent resilience, wide temperature range, wide availability and easy fabrication possibility. They are either used individually or in combination with other ingredients or additive materials to suit specific characteristics compatible with the medium used, environment and sealing requirement.

As per the ISO Standard (ISO 1629 & ASTMD-1418-81), all rubber and synthetic materials are classified in seven groups, i.e. M, N, O, R, Q, T, U. These are:

(i) *M-group:* Polyacrylate Rubber—ACM
Chloro-polyethylene—CM
Chloro-sulphonated polyethylene—CSM
Ethylene-propylene—Diene rubber—EPDM
Ethylene-propylene—Rubber—EPM
Fluoro-elastomer-FPM
} Saturated rubber

(ii) *N-group:* Rubber with Nitrogen (N_2) atom in the main polymer chain

(iii) *O-group:* Epichlorohydrin rubber—CO
Epichlorohydrin co-polymer rubber—ECO
Propylene oxide co-polymer—GPO

(iv) *R-group:* Butadiene Rubber—BR
Chloroprene Rubber—CR
Isobutane isoprene rubber—IIR (butyl)
Brom-butyl isoprene rubber—BIIR
Chlorobutyl isoprene rubber—CIIR
Isoprene rubber—IR
Nitrile butadiene rubber—NBR
} Unsaturated rubber

      Natural rubber—NR   ⎤ Unsaturated
      Styrene-butadiene rubber—SBR   ⎦ rubber
(v) *Q-group:* (With silicon as the main substance)
      Fluorosilicone—MFQ ⎤
      Methyl phenyl vinyl silicone—MPVQ   Saturated
      Methyl phenyl silicone—MPQ   rubber
      Methyl silicone—MQ
      Methyl vinyl silicone—MVQ ⎦
(vi) *T-group:* Rubber with sulphur atom (e.g. normal vulcanized rubber)—T
(vii) *U-group:* Polyester-urethene—AU
      Polyether urethene—EU

The common ingredients used with natural rubber are

1. Carbon black (as reinforcing agent)
2. Sulphur (as vulcanising or curing agent)
3. Fillers
4. Activators
5. Plasticizers
6. Accelerators
7. Antioxidants
8. Antiozonants, etc.

An interesting feature of elastomeric seals is that with a particular base polymer various types of additives could be used to form various types of compounds each having specific and distinguishable characteristics from the other. This necessitates compound selection to be done with care and the seal compatibility factor with sealing medium as well as sealing requirement always under maximum consideration. The most common elastomeric seal materials and other compounds are as follows:

1. Butyl
2. Ethylene propylene
3. Fluorocarbon
4. Isoprene
5. Fluorosilicone
6. Neoprene (Chloroprene)
7. Nitrile or Buna N
8. Buna S
9. Natural rubber
10. Plastics
11. Nylon
12. Silicone
13. Polyurethane
14. Polysulphide
15. Polyethylene
16. PTFE (Teflon)

Characteristics of some of these polymers are given below in brief:
(i) *Butyl* It is a common all-petroleum elastomer widely used for making seals. This is prepared by co-polymerizing isobutylene with just enough isoprene in order to provide the desired level of unsaturation necessary for vulcanization. Sometimes brominated and chlorinated butyl elastomers are also used. Butyl is resistant to gas permeation which makes it quite desirable for vacuum applications. Butyl seals are recommended for petroleum oils, phosphate-esters, silicone fluids, grease and diestor base lubricants. They can generally work satisfactorily within the temperature range of $-50°C$ to $107°C$ (up to a maximum of $+180°C$ in some exceptional cases).
(ii) *Ethylene propylene* A compound of ethylene and propylene monomer, (yielding ethylene propylene copolymers) this elastomer was introduced in the elastomer industry first in 1961 and was immediately accepted as a sealing agent because of its excellent resistance to phosphate-ester base hydraulic fluids for which earlier butyl was the only choice. These seals are now widely recommended for
(a) Phosphate-ester fluids
(b) Silicone oils
(c) Automotive brake oils
(d) Grease
(e) Steam (up to $200°C$) and
(f) Water

They are commonly used within a temperature range of $55°C$ to $150°C$, but are not to be used for petroleum oils, diester base lubricants, etc.
(iii) *Fluorocarbon* Fluorocarbon elastomer has been extensively used as an excellent sealing material since the last four decades. These materials have a
(a) Wide spectrum of chemical compatibility
(b) Wide temperature range of operation ($-30°C$ to $205°C$).
(c) Higher tensile strength
(d) Tough and elastic even after long exposure to hot oil and air
(e) Good compression set
(f) Good sealing ability even at high temperatures (up to $316°C$ for a short duration only)
(g) Good sealing ability even at low temperature (up to $-54°C$ in static applications only)

Fluorocarbon elastomeric seals are recommended for:
(a) Petroleum oils
(b) Die-ester base lubricants
(c) Halogenated hydrocarbons
(d) Acids etc.
(e) Silicate ester base lubricants
(f) Silicone fluids
(g) Selected phosphate esters

Seals and Packings 405

Table 13.1 Comparative Picture of Seal Materials

Sl. No.	Seal material	Positive point	Limitation	General range of working temperature in °C	Shore hardness at room temperature
1.	Butyl rubber	Better compatibility with synthetic oil	Poor shear strength	−50 to 150	
2.	Neoprene rubber	Better weather resistance in exposed condition	High friction if not reinforced with fabrics	−50 to 180	
3.	Ethylene propylene rubber	Good weathering	Permeable to higher pressure Not to be exposed to petroleum based oils	−50 to 150	80 to 90 A
4.	Fluorocarbon rubber (Viton)	Better heat resistance	Slow memory and high mould shrinkage	−50 to 200	75A
5.	Leather (Chrome tan polyurethane impregnated)	Low friction, high abrasion resistance, slight polishing action in bore	Useable for low temperature range and for simple seal configuration only	−50 to 80	
6.	Nitrile rubber (Buna N)	Good general purpose use	Prone to damage due to weathering	−50 to 150	70 to 90 A
7.	PTFE (Teflon)	Very wide fluid and material compatibility, less friction	Slow memory	−250 to 400	
8.	Polyurethane	High abrasion resistance, low friction. Most suitable for high pressure, shock load and abrasion contamination	Poor compatibility with synthetics and hot water	−50 to 150	95 to 105 A
9.	Polyurethane	Resistance to abrasion and extrusion	Poor resistance to synthetics and hot water	−20 to 100	
10.	Silicone	Excellent heat resistance		−115 to 370	70 A

Table 13.2 Selection of Seal Material

Fluid	Preferred for		Not to be used at all
	Static application	Dynamic application	
Mineral oil	Sl. No. 2, 4 and 6	Sl. No. 5, 7, 8 and 9	Sl. No. 1, 3 and 10
Phosphate ester	Sl. No. 1	Sl. No. 3 and 7	Sl. No. 2, 6, 8 and 9
Water glycol	Sl. No. 6 and 9	Sl. No. 3, 6 & 7	Sl. No. 5
Water	Sl. No. 2	Sl. No. 3, 6 & 7	Sl. No. 8 and 9
Nitrogenous fluid	Sl. No. 1, 2 and 6	Sl. No. 1, 4, 6, 7 and 9	Sl. No. 3

However these seals are not recommended for most of the phosphate ester base fluids, low molecular weight esters, ethers etc.

(iv) *Fluorosilicone* Fluorosilicones combine the high and low temperature properties of silicone with basic fuel and oil resistance. High strength fluorosilicone seals have also been developed which have high resistance to tearing and good compression set. They are used upto a temperature of 177°C and also in applications where the dry heat resistance of silicone is required. It has generally 70° A shore hardness at room temperature.

(v) *Neoprene* Neoprene seals are synthetic elastomers extensively used in modern fluid power systems and have a good working temperature ranging from 55°C to 150°C. They are homopolymers of chloroprene (chlorobutadiene) and exhibit mild resistance to acids. They are moderate in cost and resistant to deterioration when exposed to oil or oxygen. Neoprene seals are recommended for high aniline point petroleum oil and silicone ester lubricants but are not recommended for phosphate ester hydraulic fluids.

(vi) *Buna N* Buna N is one of the most widely used elastomers in fluid power systems. It is also called by its common name "nitrile". It is made by compounding 18 to 50% acrylonitrile with butadiene and has excellent resistance to petroleum-based oils and works well over a temperature range of 54°C to 121°C. Compared to other seal materials nitrile has better physical properties like better compression set characteristics, higher resistance to tear and abrasion. It has been found that with increase in acrylonitrile, the seal's resistance to petroleum-based oils and hydrocarbon fuels increases, but it becomes unsuitable for low temperature applications. These seals have poor resistance to ozone, sunlight and weathering. Because of this nitrile seals should not be exposed to direct sunlight and should not be stored near equipment which may generate ozone. The seals' suitability to low temperature can be increased by minimizing the acrylonitrile content in the compound. Nitrile or Buna N seals are extensively used for petroleum oils, silicone oils, diester based lubricants, ethylene-glycol based oils, silicone grease, water, etc, but are not suitable for halogenated hydrocarbons, phosphate ester oils, acids, brake oils, nitrobenzene, etc.

(vii) *Polyurethane* This elastomer has excellent mechanical and physical properties such as high tensile strength, excellent resistance to wear, abrasion and extrusion which makes it a good sealing material for general purpose

Seals and Packings   407

Table 13.3 Comparison of Physical Properties of Synthetic Rubber

Compound	N	G	D	B	C	H	E	V	L	I	R	A	T	P	S	Teflon®
Number prefix used by Parker Seal Co	Buna N or Nitrile	Buna S	Butadiene	Butyl	Chloroprene	Chlorosulfonated polyethylene	Ethylene propylene	Fluorocarbon	Fluorosilicone	Isoprene	Natural rubber	Polyacrylic	Polysulfide	Polyurethane	Silicone	Teflon®
Ozone resistance	P	P	P	GE	GE	E	E	E	E	P	P	E	E	E	E	E
Weather resistance	F	F	F	GE	E	E	E	E	E	F	F	E	E	E	E	E
Heat resistance	G	FG	F	GE	G	G	E	E	G	F	F	E	P	F	E	E
Chemical resistance	FG	FG	FG	E	FG	E	E	E	FG	FG	FG	P	G	F	GE	E
Oil resistance	E	P	P	P	FG	F	P	E	G	P	P	E	E	G	PG	E
Impermeability	G	F	F	E	G	G	G	G	P	F	F	E	E	F	P	G
Cold resistance	G	G	G	G	FG	FG	GE	FP	GE	G	G	P	G	G	E	E
Tear resistance	FG	FG	GE	G	FG	G	GE	F	P	GE	GE	FG	P	GE	P	E
Abrasion resistance	G	G	E	FG	G	G	GE	G	P	E	E	G	P	E	P	E
Set resistance	GE	G	G	FG	F	F	GE	G	GE	G	G	F	P	F	GE	P
Dynamic properties	GE	G	G	F	F	F	G	GE	P	F	G	F	P	F	P	P
Acid resistance	F	F	FG	G	FG	G	G	E	FG	FG	FG	P	P	P	P	E
Tensile strength	GE	GE	E	G	G	F	GE	GE	F	E	E	E	F	E	F	E
Electrical properties	F	G	G	G	F	F	G	F	E	G	G	F	F	F	E	E
Water/steam resistance	FG	FG	FG	FG	F	F	E	FG	F	FG	FG	P	F	P	F	E
Flame resistance	P	P	P	P	G	G	P	E	G	P	P	P	P	P	G	G

P—Poor; F—Fair; G—Good; E—Excellent
Teflon® is a registered trademark of E.I. Du Pont De Nemours & Co.(Inc.) *Courtesy*—Parker Seal Co.

hydraulic systems which operate at high pressure, experiences extreme shock-loads, etc. They are resistant to petroleum-based oils, hydrocarbon fuels, oxygen and ozone. Because of its good mechanical property polyurethane can also be used for hydraulic components and systems with higher abrasive contaminants and wide manufacturing tolerances. Though these seals show good resistance to weather and work well over a temperature range of –50 to 95°C, they tend to lose strength at higher temperatures specially above 95°C, they also try to revert towards the monomer and become softer when it comes in contact with steam or hot water. Polyurethane seals are often designed preloaded or activated by more resilient element as they have poor compression set compared to many other elastomers. They are mostly recommended for petroleum-based oils.

(viii) *Silicone* Though silicone seals are not very good as a seal material so far as tensile strength, tear resistance and abrasion resistance are concerned, there are special silicones with high heat and compression set resistance. Silicone has excellent resistance to both low and high temperature from 114°C up to 371°C. For certain high temperature applications they are well suited. However, the maximum recommended temperature for continuous service in dry air is about 230°C as above this range silicone can withstand temperature for a short period only. Silicone which is a compound of O_2, H_2 and C can retain its property much better at high temperatures compared to other elastomers, but its limitation is its low strength and high coefficient of friction making it unsuitable for dynamic application. These seals are not recommended for petroleum-based oils, but can be used for high aniline point oils, chlorinated diphenyls and dry heat.

(ix) *UHMW polyethylene* Ultra high molecular weight polyethylenes are tough high performing engineering plastics having high resistance to abrasion and chemical reaction. They are self-lubricating with a low coefficient of friction. This excellent physical property combined with toughness makes them quite ideal for many engineering applications, e.g. fabrication of cylinder pistons, wear rings, rod seals, scrappers, wear slides, valve and pump parts like valve seats, seals, etc. UHMW polyethylene can tolerate temperatures upto 80°C but the temperature should not go beyond 90°C. They are also quite economical as they can be processed by compression molding and free sintering of powder just like the powder metallurgy technique which suits mass production of small and intricate parts. The coefficient of friction of polyethylene is comparable to PTFE making it ideal as a low friction seal material.

(x) *Polytetrafluoroethylene (PTFE)* Modern sealing techniques have improved a lot due to the advent of TFE (Tetrafluoroethylene resin) plastics. Nowadays these are very widely used in various sealing applications in fluid power systems. The physical and chemical properties of PTFE have made it a most essential seal material even in most critical applications.Their compatibility with most of the oils, their inertness to chemical attack, very low coefficient of friction are some of the important characteristics. The low coefficient of friction enables the PTFE seals to slide over the component surface smoothly without any need for lubricant.they can be easily compounded with elastomers to provide better characteristics.

PTFE can be used over a wide temperature range from –100°C to 260°C in general and –250 to + 400°C sometimes and they usually maintain their tensile strength and toughness between this range. PTFE is non-flammable and possesses non-adhesive characteristics. It also maintains a high degree of flexibility at low temperature and heat resistance. Though PTFE is a widely used seal material, it has certain limitations. PTFE tends to flow under load. However tremendous development has taken place in the recent past to improve this cold flow characteristic by combining PTFE with filler materials, e.g. graphite, fiber glass, asbestos, bronze, etc. The "memory" of the seal, i.e. its response to change of state during operation is also poor. However the chemical resistance of PTFE is excellent and may be used for all hydraulic fluids and most chemicals from cryogenic temperature to over 200°C in general.

(xi) *Plastics* Plastics are generally stiffer and have less spring back characteristics compared to most of the elastomers discussed earlier. They cannot follow relative motion between mating parts and do not fill and follow completely the irregularities in mating surfaces. This poor property of plastics do not make them a good seal material. However certain plastics are used with spring loading to provide the requisite sealing effect.

(xii) *Glass reinforced nylone* These materials are compatible with all common hydraulic oils, water-oil emulsions, glycols, phosphate ester oils, etc. They are free from cold flow even with wide clearances. They are generally tough, stiff, high strength and wear resistant. They are, therefore, widely used to prepare wear rings, adapters, bushing, piston seals, etc. The glass fibers are oriented randomly to provide strength in equal measure in all directions. They are free from abrasion and do not score the cylinder wall.

13.2.2 Seals for Fire Resistant Oil

(i) Oil-in-water emulsion or synthetic polymer solution (HFA) Generally water content in this type of oil may be over 80% and the suggested operating temperature is between 5° and 55°C. Materials like Zn and Al may get affected when in contact with HFA.The seal material suggested is generally nitrile (NBR). The influence of HFA oil on the seal material depends mainly on the water content. Specific polymer chemicals are affected by such oils destroying the seals. As this oil absorbs heat, polyurethane and polyacrylate will be very sensitive and hence unsuitable. Therefore materials like nitrile (NBR), silicon rubber (MVQ) and fluorocarbon (FPM) are used. Also the arrest of swelling of seal material with HFA oil is more unfavourable than with mineral oil. This depends on the art and type of concentrate used. Though the swelling can go up to 20% usually 2 to 5% is acceptable. With NBR the swelling gets more with time and may be as high as 50%. One can use special class FPM which may be more costly.

(ii) Water-in-oil emulsion (HFB) Water content in this oil may be over 40% and ranges of operating temperature 5°C to 60°C. Zn and Al are generally attacked in contact with such oil. The suggested seal material here is nitrile (NBR). The swelling/shrinkage is almost like HFA oil with NBR seal material.

(iii) Water-glycol (HFC) Over 35% water is used in such oils and the operating temperature is between –20°C to 60°C. Zn, Al, Cu and Mg alloys get affected by this oil. The common seal materials suggested are nitrile, ethylene propylene rubber (RPDM) and SBR. Polyurethane and polyacrylate are not used. One has to make sure about the swelling characteristics before selection of the seal.

(iv) Synthetic anhydrous phosphate ester or chlorinated hydrocarbon Water content is less than 0.1% and temperature range is –20°C to 150°C. The seal materials preferred are fluorocarbon (Viton) and EPDM. PTFE may also be used but here also the swelling may be variable.

13.3 SEAL FORM

The evolution of seal form is generally associated with seal material and the constructional method that is best suited for such material apart from many other factors. The selection of a seal form for a specific use may be decided on consideration of the following factors:

1. Possibility of preloading of the seal which is determined by the size of the seal as well as the resistance of the material to deforming. The working pressure decides the seal size and its cross section.
2. Longevity of the seal lip and its capacity to withstand the working pressure without failure, i.e. to what extent the seal hardness or softness will provide the optimum life to a seal.
3. Hard seal with low deforming property is stable and may not extrude into the clearance gap of the mating surface as a soft seal does, but a hard seal may not expand and may be unsuitable at low pressure. As a result there is now a tendency to combine both hard and soft material in one compact seal as a method of preloading as shown in Fig. 13.4 (a). Here the O-spring functions as a preloader to the main seal body when pressure actuates over it as shown in Fig. 13.4(b). The O-ring could be made out of a softer rubber material and the main seal body from a harder material. The O-ring loads the seal lip to provide excellent sealing which increases continuously with increase in pressure thereby maintaining an effective sealing through a combination of a squeeze type and lip type seal.
4. The elasticity of a mounted seal between two fixed surfaces get reduced. This will be more critical if the free volume of space is not enough. More care is to be taken when very small size seals are used and there is no possibility of expansion of the seal to provide a seal-tight condition.

Seals and Packings **411**

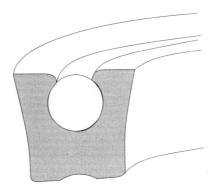

Fig. 13.4 *(a) O-Ring preloaded compact seal*

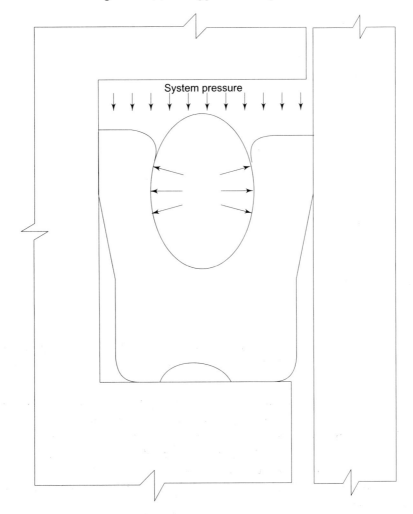

Fig. 13.4 *(b) Action of the O-ring (preload) under oil pressure*
[Courtesy: Parker Seal Co., USA]

13.3.1 Types of Hydraulic Seals and Packings

Hydraulic seals and packings are available in various geometric shapes and designs. A few of these are described here.

(i) Cup seal Mostly used in cylinder pistons, these cups are available in a variety of geometric configurations and made of various materials. While polyurethane and leather piston cups provide abrasion resistance and anti-extrusion at even extreme pressure system (upto 700 bar) for rough cylinder with bore finish (i.e. 250 RMS) and larger than recommended piston-bore clearance for the system pressure, for low pressure applications neoprene cups are used. Cups are also made out of compact "polypack" and are designed as (shown in Fig. 13.5):

(a) Standard flat bottom
(b) Thin-walled
(c) Straight side and
(d) Tapered lip.

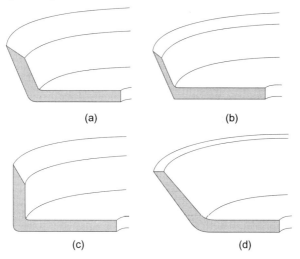

Fig. 13.5 *Piston cup seal (a) Flat bottom (b) Thin walled (c) Straight side (d) Tapered lip [Courtesy: M/s Parker Seal Co., USA]*

(ii) V-packing Fig. 13.6 shows a diagram of the V-packing commonly used in hydraulic system components at various locations. They are made of fluorocarbon reinforced with cotton, asbestos or neoprene reinforced with asbestos for high temperature and exotic fluid application. Sometimes to improve the sealing properly, the use of a homogeneous rubber V along with the fabric reinforced packing is suggested. V-packings are available in split to facilitate assembly and specific sizes.

(iii) U-ring packing U-packing is a common lip type seal most versatile in various applications and used both as ID rod seal or OD piston seal. The diagram

Seals and Packings   413

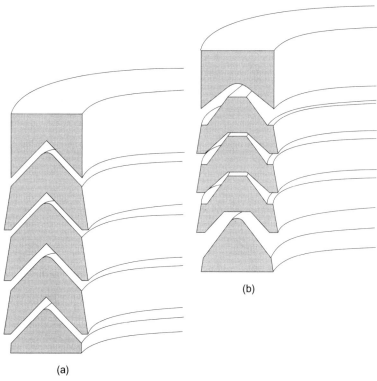

Fig. 13.6   *V-packing: (a) Standard, (b) With adapter*
[Courtesy: M/s Parker Packing Co., USA]

in Fig. 13.7 (a and b) illustrates such a ring. U-packing with long thin lip may generate very low dynamic friction and has the ability to accommodate eccentric hydraulic elements. If the seal has a short heavy lip, it is more suitable for a low pressure system. U-packings are available with rectangular cross section to suit as piston seals and ram packings, or with square cross section used with back up rings. They are also available as square ID lip ring (for mobile equipment where the heavy square ID lip provides stability and anti-roll characteristics), flat trim U-ring or back bevelled symmetrical or non-symmetrical rings.

(iv) O-rings   These are most common and simple seals with a circular cross section as shown in Fig. 13.3 (b). These elastomeric seals are available in various compounds and resilience and are used in seals both dynamically or statically. When choosing an O-ring, hardness must be optimal; hardness being at least 70° A at which O-ring seal swell at low pressure. Further details on O-rings are given subsequently.

(v) Back-up rings   It is a common fact that O-rings of about 70°A shore hardness at high pressure are subject to extrusion in the gap between the sliding components. They are more prone to extrusion than are harder rubber seals. At

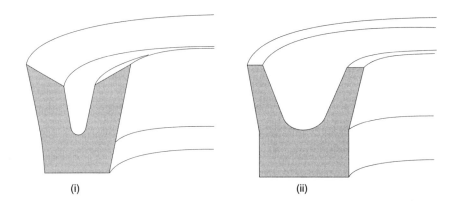

Fig. 13.7 *(a) U-packing seal (i) Rectangular (ii) Square (Height = Width)*

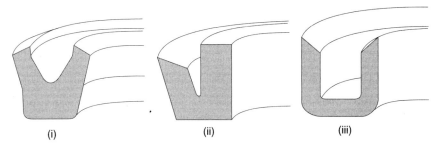

Fig. 13.7 *(b) U-packing seal (i) Square with back bevel (ii) Square with back bevel non-symmetrical (iii) True 'U'*
[Courtesy: Parker Packing Co., USA]

one over O-ring seals exceeding a hardness value of 85°A are seldom successful in dynamic applications as harder seal materials may not follow the irregularities of the sealing surface. Hence harder O-rings are generally prone to leakage. In order to avoid extrusion as well as stopping leakage a back up ring is therefore used to support an O-ring as shown in Fig. 13.8. A back-up ring is not a seal in itself and its main purpose of use is to reduce the clearance gap on the low pressure side of the O-ring. They also help in enhancing the operating pressure range of ordinary O-rings. Other advantages are:

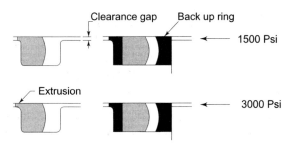

Fig. 13.8 *Use of back up rings*

1. Provides increased lubrication of the O-ring as some oil may get trapped between it and the O-ring seal.
2. Permits higher 'metal-to-metal' clearance between moving parts.
3. Helps overcome sealing difficulties encountered with wide or out-of-tolerance bores and shafts.
4. Enhances life of O-ring seal.
5. Resists higher pressure.
6. With proper groove designs back-up rings do not collapse or cold flow into open groove areas thereby reducing system failure.
7. Helps to keep the O-ring round even under high pressure.

It may be mentioned here that back-up rings are less sensitive to temperature changes than O-rings as they are required only to protect the O-ring from damage due to extrusion. The graphical chart in Fig. 13.9 indicates the clearance and pressure conditions that are likely to cause extrusion of various O-ring materials. From the diagram it is clear that if the diametral clearance exceeds 100 microns, the 70°A material will extrude at around 120 bar.

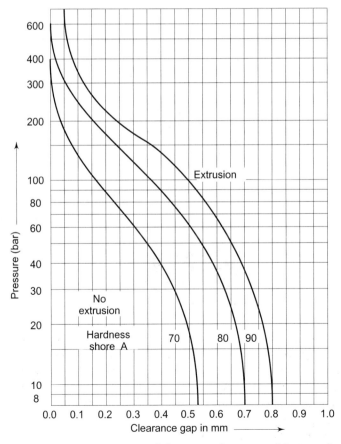

Fig. 13.9 *Pressure, diametral clearance and extrusion of O-ring seals*

(vi) T-seal With the development of the T-seal, a long felt need for a seal which would be as simple to install as the O-ring as well as to seal effectively both in static and dynamic condition, was met. T-seal as an O-ring, functions without rolling, spiral failure and does not extrude and can be used both as rod and piston seals. They function best even with 32 RMS groove surface finish and 16 RMS on dynamic surface. Unlike O-rings, T-seals require separate basic configurations for sealing pistons and rods as shown in Figure 13.10 (a) and (b). Figure 13.10 (c) shows the application of T-seal on a cylinder piston rod.

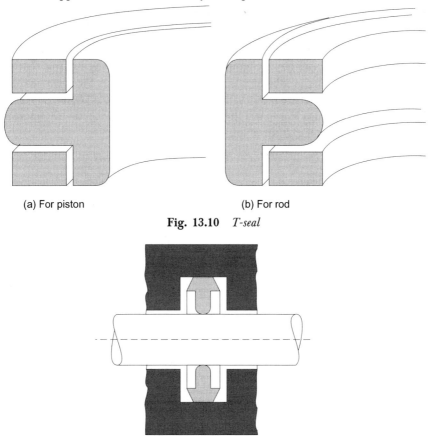

(a) For piston (b) For rod

Fig. 13.10 *T-seal*

Fig. 13.10 *(c) A T-seal used on a piston rod*

(vii) Wiper rings It is important to note that the cylinder rod is a major component responsible in helping outside impurities to enter into the cylinder and thus contaminate the whole system. To protect the hydraulic system from such external contaminants, a wiper seal is used in the cylinder. Wiper seals (also called simply as wipers or scrapers) are available mainly in two categories:
1. Snap-in wiper
2. Press-in-metal encased wipers

Seals and Packings **417**

They are also categorized as lip type and squeeze type as shown in Figs 13.11 (a) and (b). The function of a wiper is to protect the primary seal i.e. the rod seal from damage and also to act as a secondary seal.

Diagrams of various wiper ring seals are shown in Fig. 13.11 and 13.12. Wipers are used to 'wipe out' foreign materials like dirt, abrasive materials or other contaminants from the piston rod. Being continuous rings, deep type of wipers which can compensate for wear and side motion are preferred. According to their geometric design the snap-in-wipers are termed as deep universal wipers. AN style wiper, H type wiper and K type wiper as shown in Figs 13.12 (a), (b), (c) and (d).

In press-in wipers, a metal outer shell or can is molded along with the rubber wiper element as shown in Fig. 13.11(c). These type of wipers are mostly preferred for mobile hydraulic systems and are easily replaceable.

Better and effective wiping action at higher and prolonged seal life is the ultimate goal in using a wiper seal for which proper choice and design of the seal is a must. A very important factor in this regard is the lip geometry selection. In Fig. 13.11 (d) four types of lip geometry are shown with their functional characteristics.

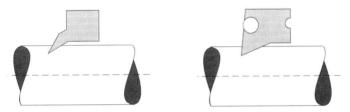

Fig. 13.11 *(a) Lip type wiper*   **Fig. 13.11** *(b) Squeeze type wiper*

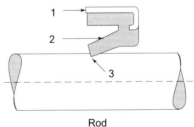

1. Metal shell press fits into wiper gland
2. Elastomer element
3. Wiper lip

Fig. 13.11 *(c) Press-in wiper with metal shell*

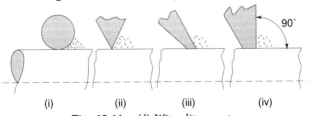

(i)   (ii)   (iii)   (iv)

Fig. 13.11 *(d) Wiper lip geometry*

418 Oil Hydraulic Systems: Principles and Maintenance

(i) Round contact point—Poor in wiping as contaminants tend to collect between the wiper and rod causing rapid wear and abrasion.
(ii) Chamfered contact point (back bevelled)—Best in sealing but worst in wiping and hence not used.
(iii) Chisel contact point—Makes good contact with the rod surface and better wiping action. But easily prone to quick damage.
(iv) Flat contact point—Best contact and desirable wiping action, minimal wear and abrasion.

13.4 COMPACT PACKING

In recent years various types of combination of seals or compact seals made of more than one material or elements combined together are being developed. "Polypak" is one such seal material (patented first by M/s Parker Packing Company, USA) and is being used for preparing various "compact seals" for quite a long time.

It is a unique design combining an automatic lip seal squeeze type preloaded by O-ring made of synthetic rubber to provide an excellent sealing from vacuum

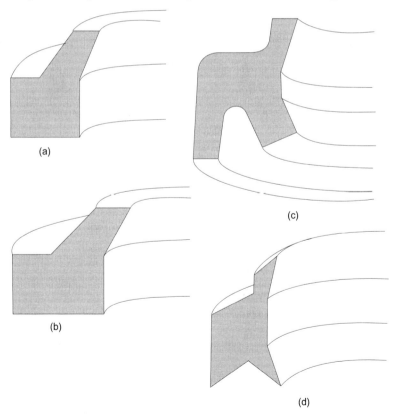

Fig. 13.12 *Wiper seals (a) Deep universal type wiper seal (b) Army-Navy type (AN type) (c) H type (d) K type*

to extreme high pressure. "Molythane polypack"—a seal marketed by M/s Parker Seal Company, USA is a superior blend of polyurethane impregnated with molybdenum disulphide to provide seal lubrication. Figure 13.13 depicts two varieties of the seal. The synthetic rubber O-ring loads the sealing lips providing an excellent sealing under vacuum and low pressure. With increase in system pressure, the pressure is applied on the lips and lip loading is increased automatically to maintain an effective sealing. Some interesting properties of these seals are:

1. They are self-lubricating.
2. They are tough and extrusion resistant.
3. They are ozone and radiation resistant.
4. They provide optimal fluid compatibility except for automative brake fluid, esters, ketones, pure aromatic compounds, strong concentrated acids or bases or very strong oxidizing agents, etc.
5. They possess higher temperature tolerance from $-50°C$ to $+95°C$.
6. They tolebrate high pressure up to about 4000 bar.
7. They have wide adaptability to rough surfaces like 16 RMS for moving surfaces, 32 RMS for static surfaces and 5–10 RMS for maximum sealing.

Use of such "polypacks" is shown in Figs 13.14 and 13.15.

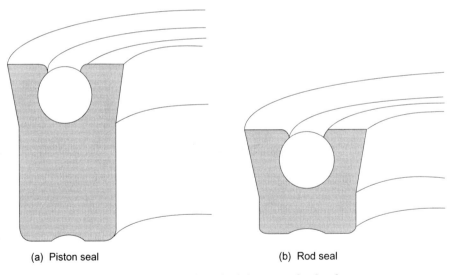

(a) Piston seal (b) Rod seal

Fig. 13.13 *O-ring preloaded piston and rod seal*

13.5 HOW TO FIT AN O-RING

O-rings are the most common seals used in hydraulic system. In case other factors like pressure, temperature, oil and component geometry permit it to be used O-rings become the cheapest seal form. The O-ring is generally positioned in the gland machined on to the moving or fixed component body in three different configurations as shown in Fig. 13.16. These are:

420 *Oil Hydraulic Systems: Principles and Maintenance*

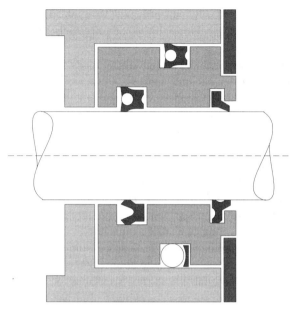

Fig. 13.14 *Polypack compact seal on cylinder rod*

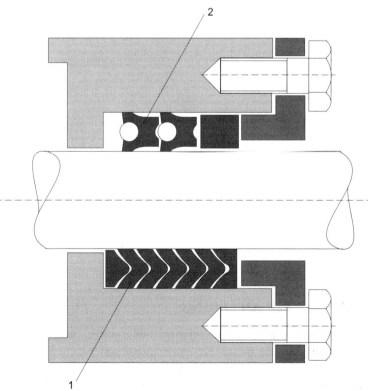

Fig. 13.15 *Replacing V-packing (1) With polypack (2)*

Seals and Packings 421

(a) Face seal gland [Fig. 13.16 (a)]
(b) Piston type seal gland [Fig. 13.16 (b)]
(c) Rod type seal gland [Fig. 13.16 (c)]

The gland is the cavity which receives the O-ring seal. For a face type seal, the gland consists of a groove in a flat surface and the mating surface is flat.

In a piston type seal (which is also termed a male gland), the groove is on the male member whereas the groove is machined in the bore in a rod type seal (female gland).

It may be mentioned here that in case for a given O-ring internal diameter, number of O-ring cross-sections are available with the seal manufacturers. From the point of sealing efficiency, the largest cross sections may be selected as smaller cross section O-rings are generally more sensitive to variation in geometric tolerance, surface finish, foreign particles and extrusion under pressure. Larger diameter O-rings have better compression set and resist extrusion.

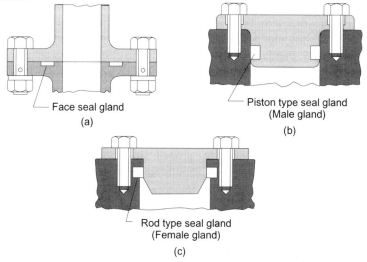

Fig. 13.16 *Groove design for O-ring*
[Courtesy: Hydraulics & Pneumatics, April, 1986 p. 68]

Because of its circular shape the O-ring seal can be squeezed quite readily and uniformly to establish perfect sealing at lesser installation load (3 to 15 kg per linear). Generally groove width is designed for a normal swell of about 10 to 20% of the seal when exposed to oil at high temperature. In piston and rod type of glands the O-ring is moved to the low pressure side of the groove when pressure acts on it and it will remain in the same position unless pressure reversal takes place. This point has to be taken care of while designing a gland for an O-ring.

O-ring seals are assembled with an amount of squeeze (Fig. 13.17) which is the amount of deflection applied to the seal cross section in one direction. The amount of squeeze is dependent on the gland depth and to establish a perfect

422 *Oil Hydraulic Systems: Principles and Maintenance*

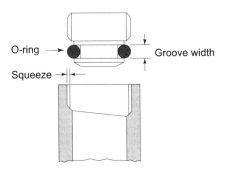

Fig. 13.17 *Squeeze of O-ring*

sealing, a heavy squeeze if preferred which may push the rubber seal into the minutest irregularities of the mating surfaces effecting perfect sealing. The rate of shrinkage of elastomers is 5 to 16 times as much as metal when temperature drops. Hence a lightly squeezed O-ring may shrink away from the seal line creating leakage problem even if the temperature drop does not affect the resilience of the elastomer material. The minimum squeeze should be therefore, at least 10% for any standard O-ring cross-section to eliminate the adverse effect of geometric inaccuracy, compression set, etc. A heavy squeze is easily possible for a face type gland but not easily for the other two types. For a rod or cylinder bore, a long gradual taper is needed to ease the O-ring into the bore which may otherwise pose an assembly problem. For a piston type gland, it is better to select an O-ring with an ID slightly less than the groove diameter so that the ring will be stretched slightly to fit snugly into the groove. But stretch should not be higher than 5% as O-rings tend to age more rapidly under stress and there is resultant reduction of the cross section due to stretching. In case the amount of stretch is more than 5% it is preferred to use the next larger standard size O-ring even if it is bit loose in the groove.

In a rod type gland, the OD of the ring should be the same or slightly larger than the groove dimension so the rod can be inserted without damaging the ring.

It should be kept in mind that a soft resilient seal material can be used to provide a sealing at minimum pressure but a stronger material is needed to avoid extrusion both of which has its plus and minus point which can only be solved by a compromise. Hence seal selection in a hydraulic component is always done keeping in mind the optimum efficiency of sealing taking into consideration all the factors we have discussed.

13.6 DYNAMIC SEALS

Generally soft seals with fibers or wire reinforcement which can seal due to compression are preferred as dynamic seal installed with solid lubricants like graphite, mica, PTFE, etc. Generally they find their application in reciprocating or rotary systems where speeds are not high and good heat dissipation is possible. Dynamic seals come in three basic designs, i.e.

1. Squeeze type, which has a shorter life than the lip type as they are subjected to greater wear
2. Lip type which are used for continuous service
3. Piston ring

Lip seals are generally preferred for continuous service. Generally flange and cup seals have single lips whereas V and U-seals have double lips. All lip type seals are fitted with an interference fit. Seals made of homogeneous materials without any fabric reinforcement are used with back-up reinforcement if the working pressure is more than 100 bar. However polyurethane due to its higher wear resistance and strength can be used even if the pressure is above 200 bar. One must keep in mind that natural homogeneous elastomeric seals may seal the system more tightly than fiber reinforced seals, but such homogeneous seals are more vulnerable to wear and extrusion and hence need replacement very often due to failure. For certain applications, elastomeric materials like nitrile rubber are used. Nitrile is compatible with water, aqueous solutions, emulsions and petroleum base oils upto 120°C. For higher temperature (up to 250°C) elastomers like silicones, fluoroelastomers, PTFE etc are used.

For rotary devices running at slow speed, soft seals may be used. But lip type seals which seal automatically should not be used except for slow speed applications.

13.6.1 Maintenance of Dynamic Seals

A hydraulic engineer and maintenance mechanic should follow the following guidelines in order to maintain a trouble-free life of a dynamic seal.
1. Soft seals should not be extra tightened. The tightness should be adjusted such that a steady leakage rate and a satisfactory lubricating film is maintained all through.
2. Adjustment should be done carefully and slowly about 1/6 of a turn at a time, i.e. one flat of a hexagonal bolt at a time. Adjustment should be run for about 10 minutes so that it can stabilize. The tightening sequence shown in Fig. 13.18 is to be followed always.
3. Ensure that cooling and flushing fluid is running.
4. Look out for undesirable vibration due to mechanical failure e.g. worn out/broken/bent shaft, bearing etc. and take appropriate preventive action in case of any such eventuality.

13.6.2 Static Seal

Compared to dynamic seals, in static seals generally no relative movement of the mating components is allowed. However a perfect static seal is impossible to achieve as over a period of time for all systems there may occur a slow and slight movement due to vibration, shocks, mechanical load, change in working pressure, etc. A very common example of a static seal is a flat gasket held in position by compressive force created by fastening bolts. A more complex static seal is the

U- or V-type lip seal (which is self-energizing), square type seal. O-rings and rectangular rings are also used as static seals. A static seal should have the following properties:
1. Capability to withstand mechanical load.
2. Non-permeable to the fluid medium.
3. Non-contaminating to the hydraulic oil.
4. High chemical stability and resistant to rapid disintegration.
5. Adequate property to deform to distribute the mechanical load.
6. Adequate elasticity or resilience to follow mechanical movement of joints.
7. Strong enough to resist extrusion.
8. Ability to generate enough friction with sealing surface to resist displacement.
9. Easy to install and remove without adhesion, etc.

13.6.3 Installation Procedures of Seals

The following procedure may be followed:
1. Ensure that before installation, the seal and the component on which it is to be fitted have been thoroughly cleaned.
2. Ensure that the right type of seal with size and shape as specified is used.
3. Ensure that the sealing surface are in appropriate surface finish and physical condition.
4. Ensure that the shafts are correctly centered and shaft surfaces are smooth enough not to damage the seals.
5. Ensure that flushing and cooling lines are open.
6. Ensure proper lubrication of seals.
7. Ensure that all burrs and sharp edges are removed so that the seals do not get damaged during installation. This is more important if single rings are used.
8. Seals should not be over tightened as this will increase friction. Remember that with higher friction the seal life is shortened significantly.
9. While installing lip seals, ensure that lips face the pressure oil.
10. Ensure that V-seal rings do not get twisted during installation.
11. Special care is to be taken when joints of cut open or split type V-rings are installed in blind locations. It is better to install and seat them in place one at a time.
12. Stagger the joints of cut open or split rings.

13.6.4 Installation of O-rings and Gaskets

O-ring O-rings or rectangular rings positioned in groove with clearances held to a few microns are capable of sealing up to 125 to 200 bar pressure. These seals should be fitted with an initial squeeze as otherwise they may result in a lot of leakage when oil pressure is applied. Lubricants are commonly used on O-rings and other elastomeric seals, e.g. a O-lube grease or oil to help seat the O-ring properly, quicken assembly work and protect the seal from damage by abrasion

pinching or cutting. During assembly of O-rings, the following points may be observed:
1. O-ring should not be twisted.
2. ID stretch as installed in gland position, should be maintained within 5% otherwise the seal life will get shortened.
3. ID expansion to position the seal should not exceed 100% except in the case of very small diameter rings. In case it cannot be avoided, the gland should be closed after allowing the seal to return to its original condition.
4. The assembly lubricant should be compatible with the system oil and should be capable of forming a thin, strong film of oil and be chemically stable.

Figure 13.16 illustrates the design of groove during O-ring installation.

Gasket Gasket is a non-moving static seal used in hydraulic components where relative movement of the parts is absent. Both rings and flat seals are used as seals. Flat gaskets should be positioned over a flat surface instead of wavy, warped and distorted surfaces. The surface finish for general applications should be 64 to 250 RMS value. The surface should not be too rough or too fine. If the surface is too fine, the gasket may creep or flow too easily. On the other hand a gasket may not fill the valleys in the sealing surface if the surface is too rough. The load on the gasket should be applied uniformly following a staggered bolt tightening procedure as shown in Fig. 13.18 e.g. moving in pairs from one bolt to the one directly opposite across the flange.

A gasket can last longer and can be maintained satisfactorily well if the following procedure is followed:
1. It is better to retorque flat gaskets a day or two after initial installation, in order to overcome the initial loss of stress due to creep-relaxation.
2. Ambient condition should be compatible with the gasket. Avoid destructive/damaging factors like steam line, heater, spills of various liquids close to the gasket.

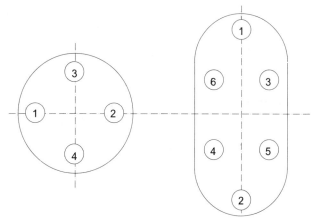

Fig. 13.18 *Proper sequence of bolt tightening for a gasket*

3. All residue from the joint should be removed when a gasket is replaced during preventive or breakdown maintenance.
4. Ensure that undersirable scratches are not made on the surface.
5. Before installing a gasket, the surface has to be thoroughly cleaned.

13.7 FAILURE OF SEAL

For sealing it is needed to maintain perfect contact under pressure between the relatively smooth surface of the hydraulic component. Any clearance developed between the moving parts and the supporting member, e.g. between the piston rod and the rod bushings must be compensated for by the seal if excessive leakage is to be prevented. To prevent out-of-round running, wobbling, vibration and fluid chatter design engineers should avoid excessive clearances.

Failure of seals in a system may result in loss of system fluid, loss of volumetric efficiency and system power, heat generation, etc. It has been observed that seal failure is relative and progressive in nature. With increase in pressure, fluid temperature and drive system speed, leakage tends to increase. As a seal begins to fail due to abrasion or ageing, it becomes progressively worse in its capacity to maintain a leak-proof chamber. It may be noted that pressure is exerted over the entire length of the seal's contact surface with the moving element it is supposed to seal. This pressure tends to separate the two, enlarging the clearance and promoting fluid flow which is actually leakage flow.

It may be also mentioned that even a static seal may not maintain a seal-tight system as even static seals also tend to move a little over time, i.e. they tend to "breathe" with change in operating and environmental conditions, e.g. pressure, temperature, abrasion due to fastening and locking load, etc.

Failure of seals can be better addressed in case of dynamic seals if the design parameters are properly taken care of during—system as well as component design. The enormous variety of applications of seals between various reciprocating surfaces has resulted in development of comparable varieties of the sealing elements and their methods of application. The following design parameters are most important for a reliable seal performance.

(a) Clearance All reciprocating seals are assembled into the mating device with a clearance gap. The clearance may lead to problems of extrusion of seal lip through the gap. This may also cause eccentricity especially for heavy horizontal pistons when supported entirely on the seals or due to lateral shock and vibration or if the piston rod transmits lateral forces back from the load. Seals should never be used to absorb these loads as this may lead to damage both to the seal as well as the friction surface. Self-lubricating modular bearings and wear rings are used to minimize the effects of side loading, absorb high shock loads and act as anti-extrusion devices for the primary seal. If there is excessive clearance, extrusion of the O-ring seal may take place which can be avoided by use of a back-up ring. A T-seal which gets locked in place due to its specific geometry, can provide positive resistance to misorientation.

(b) Friction Friction between the seal and the sealing surface may be affected by various factors, e.g.
1. Lubrication
2. Evenly distributed compression or squeeze of the seal
3. Surface finish of the mating surfaces, etc.

If a seal gets unevenly squeezed, it means uneven friction and may result in twisting of the seal element. Operating pressure, seal form and materials are equally important as may be evident from the friction behaviour of four types of seals shown in Figs 13.19 (a) and (b).

Surface finish It is an important factor to both friction and wear. The resultant effects of heat and temperature are also equally critical as high temperature may swell the elastomer increasing the squeeze force but decreasing hardness whereas low temperature on the other hand may harden it increasing its friction and making it non-resilient.

(c) Lubrication It has been found that running friction is much less than breakout friction in a reciprocating device. The breakout friction may be at least three times the running friction and tends to rise sharply during idle time between the stroke of the piston. This can only be minimized with proper lubrication. Though in a well lubricated sealing arrangement, the rubber seal reciprocates over a film of oil, with passage of time, the rubber may gradually mold into the micro-irregularities of the metal surface displacing the lubricant and increasing the breakout friction. A system designer should therefore provide the right attention to see the optimum availability of lubricants to enhance seal reliability, enhanced life and reduce the friction. If the system works in a high temperature environment, the light ends of the lubricant may get evaporated exposing the adhesive and abrasive surface leading to seal abrasion. Hydraulic components

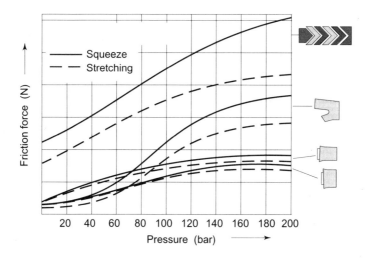

Fig. 13.19 *(a) Seal friction depends on pressure*

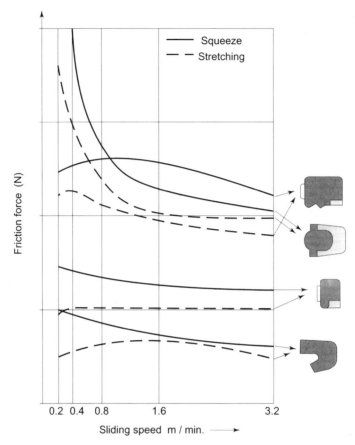

Fig. 13.19 *(b) Seal friction depends on sliding speed*
[Courtesy: Oel Hydraulik U. Pneumatik, Germany]

should be designed so that the surface finish is kept within 32 RMS value in grooves and 6–16 RMS on dynamic surface on which the seal has to move. This may minimize the risk of such abrasion.

(d) Hardness   It is the durometer reading in degrees indicating the hardness of rubber expressed in 'Shore A' Scale. Seal hardness is an important factor for seal selection. It is obvious that softer elastomers have lower friction but may have less wear resistance. Hence appropriate compromise between seal hardness and other seal requirement is the only solution.

Hardness of 35°A is considered soft and 90°A as hard.

(e) Compatibility of the seal material   Compatibility of the elastomer with environment, i.e. fluid medium, working temperature, heat, level of cleanliness, etc. are other important design parameters which a designer and maintenance person should give equal importance.

(f) Seal misorientation It is seen very often that an O-ring used as a dynamic seal may fail due to misorientation during operation (e.g. roll or spiral failure). During installation, accurate positioning of the O-ring seal may minimize the possibility of misorientation. Seal hardness may also help minimize such failure.

13.7.1 Causes for Seal Failure

The following points are noteworthy:
1. Damaged or worn out shafts.
2. Incorrectly installed seals.
3. Incompatibility of the seal material and oil e.g. if hydraulic oil has been changed to synthetic or fire resistant oil but the seals have not been changed.
4. In reciprocating applications, excessive side loads or overloading may be the prime cause of wobbling of piston rod movement change to a lip type of seal from a square type may be preferred as it can follow better shaft movement.
5. All causes of vibration is to be removed e.g. bent shaft, out of balance or damaged impeller, broken vane, misaligned or damaged bearing, defective coupling etc.
6. Soft seals are particularly susceptible to excessive radial movement and are causes of losses due to short life.
7. Seal should not allow any leakage at all and in case of leakage, thorough investigation should be undertaken to detect the cause.
8. Common seal materials can continuously be made to work effectively at high or low temperature within their prescribed limit with occasional exposure to wide variations. But at extremely low temperatures the seal may get brittle or at very high temperature prolonged exposure to excessive heat may cause permanent hardening and damage the seal.
9. Unbalanced speed of stroke and length of stroke (low speed is responsible for spiral failure of seals).
10. Lack of lubrication.
11. Incidence of excessive pressure differential.
12. Squeeze.
13. Groove design and shape.
14. Working temperature.
15. Gland and groove surface fixture.
16. Side loads.
17. D/d ratio of O-ring.
18. Deposit of gummy or contaminants on metal surface.
19. Breathing.
20. Eccentricity of mating surfaces.
21. Lack of back-up ring.
22. Poor installation, etc.

13.7.2 Failure of Dynamic Seals

The common factors for failure are:
(a) lack of cleanliness
(b) extrusion
(c) breathing
(d) surface finish of the metal
(e) hardness of seal material
(f) temperature
(g) side load
(h) direction of pressure
(i) shock load
(j) vibration
(k) squeeze
(l) floating gland
(m) spiral failure

Extrusion Extrustion of dynamic O-rings is one of the most common type of seal failure. However this can be avoided if the gland design and seal hardness are taken into consideration amongst other factors like use of back-up rings. Back-up rings are generally made of leather, teflon (PTFE) and hard rubber.

Hardness Hardness is a vital factor in proper functioning of a seal; it has been seen that majority of applications can be better served by seal hardness of 70°A shore hardness as wear and abrasion resistance is best at this value. Softer or harder seals wear more rapidly. Dynamic seals do not function satisfactorily if hardness is above 85°A shore value as tendency to leak increases.

Friction Friction, both breakout or running can be troublesome due to various reasons like seal hardness, surface finish of metal, length of time the seal is in contact with the metal surface, speed of motion, lubrication, temperature, pressure, etc. Friction can be adjusted by judicious choice of the above factors.

Temperature A hydraulic system is generally designed to function at above 30°–35°C in normal condition. If the temperature is above 38°C the system starts becoming critical as at higher temperature the oil at the light ends of an O-ring gets evaporated resulting in acceleration of wear. At very high temperatures the residual region will char and leave a hard abrasive surface which may abrade the seal. If the seal has been operating at high temperature for sometime, the rubber seal will take a compression set and if the temperature goes down, there may be insufficient resilience to overcome the relatively high coefficient of shrinkage (10 times of steel) at low temperature.

Side loads Side loads on a piston or piston rod may cause the clearance in the gland on one side only. If the clearance is too large extrusion may take place and if adequate squeeze is not applied, leakage will result. Higher side loads may also cause uneven friction causing galling or scoring of the mating surface.

Shock-loads Shock or surge pressure caused due to stopping a high speed cylinder with load may increase the pressure many a times over the designed pressure resulting in excessive shock load on the elements in the system. A seal in such a situation may also give in causing seal-failure. A shock suppressor if used may alleviate such shock load.

Direction of pressure The placement of a groove should be conforming to the direction of pressure. If the friction of the moving metal surface across the O-ring is in the same direction as the direction of pressure, the O-ring will tend to drag into the gap more easily and extrude.

Vibration O-rings can get quickly worn out due to excessive vibration of cylinders caused due to small frequent motion. This is often encountered when the hydraulic system is in transit. A mechanical lock may be used to avoid such vibration.

Squeeze The amount of squeeze is very important with regard to friction. Squeeze is needed during the period of very low pressure as during high pressure the O-ring may squeeze into the clearance gap creating a seal tight condition. Enough squeeze is essential to offset the great difference in the coefficient of shrinkage of rubber and metal.

Floating gland In order to reduce or eliminate high bearing load float glands may be used to allow the piston or rod bearing containing the O-ring groove, to pivot, adjust or float a small amount of setting misalignment. The gland design increases O-ring life and eliminates many unscheduled failures.

Spiral failure Spiral failure of the O-ring may happen in a hydraulic system. It is called spiral failure because when it occurs the seal looks as if it had been cut and damaged halfway through the O-ring cross section in a spiral or cork screw pattern. The conditions causing swelling or torsion of the O-ring is responsible for spiral or torsional failure. This type of failures are most common in reciprocating O-rings.

During any reciprocating stroke the hydraulic pressure generally produces adequate holding force which holds the O-ring inside the groove. This prevents the seal from rolling or twisting. But under certain conditions the segment of the ring is forced to slide or roll simultaneously. Though small amount of twisting may not cause much problem, excessive twisting may induce torsional failure of seal ultimately leading it to rupture when the elastic limit of the rubber is crossed due to rapid stress-ageing. Certain operational, environmental and design factors given as follows may contribute towards torsional or spiral failure:
1. Speed of stroke specially when speed is unbalanced.
2. Length of stroke.
3. Working temperature and range of temperature the system is put to work.
4. Working pressure and change of pressure differential.
5. Improper lubrication and laak of lubrication.
6. Shape and size of groove or split groove.

7. Surface finish of groove and gland.
8. Type of metal rubbing surface and surface finish.
9. Undesirable side loading.
10. Squeeze and stretch of O-ring.
11. Lack of back-up rings.
12. Wrong method of seal installation.

13.8 SEALS ARE AFFECTED BY ADDITIVES

In many hydraulic oils quite a number of chemical additives are used for various purposes. These additives are:
- Antioxidant
- Corrosion inhibitor
- Detergents
- Rust inhibitors
- Pour point depressor
- VI improver
- Antiform agent
- E.P. agent
- Flash point reducer etc.

Many of these additives are mostly essential and hence cannot be avoided. In modern oils. Multipurpose additives are also used, e.g. polymethacrylate, VI improvers and flash point reducers.

Effect of additives on the elastomer The elastomeric rubber materials sometimes may get affected by the additives used in oil. The additive could be classified as
(a) Physically active additives
(b) Chemically active additives

Amongst the physically active additives isobutalene and methacrylate polymer as VI improver and flash point reducer are common as far as physically active additives are concerned because of their high molecular weight. Anti-foam agents, dye and perfume material are used in very insignificant amount and hence need not be considered for any effect on the elastomer.

The major problems are due to chemically activer additive e.g. antioxidants, Extreme pressure (EP), corrosion inhibitor, purifier dispersing agents etc. In the same group of these additives falls vulcanising material which works at high temp. and thus further hardens the rubber material reducing its resilience and increasing its brittleness and breaks up subsequently. The influence of the contact fluid on the elastomer does not limit only to volume change (swelling). Apart from this, the tearing creep failure due to factors like stretching, torsion, hardening, reduction of resilience and tensile strength, cracks may also take place.

13.9 GENERAL GUIDELINES FOR SEAL SELECTION AND INSTALLATION

Besides the working condition, the desired function and working life of a seal depends on the appropriate installation procedures for which the following basic guidelines may be followed.

1. For a specific application, the appropriate and compatible seal material should be selected.
2. For selection of a respective seal form, it is better to select a favorable material for ease of fabrication.
3. The seal contact surface should be fine finished and polished and all sharp corners should be avoided.
4. The clearance between the seal and the sealing surface should be kept small enough in order to prevent the seal from squeezing in between the clearance gap. Sometimes back-up rings are also used in conjunction with the seal to prevent squeezing.
5. Seals used in cylinder piston rods should be capable of wiping out all undesired contaminants from entering the system as it is a fact that such reciprocating rods are one of the main source of oil contamination.
6. While mounting seals, care should be taken that they are not tilted due to the quality of the piston rod in order to prevent their premature failure.
7. While mounting seals, the manufacturer's guidelines and instructions should be meticulously followed.
8. Cleaning of the seal and sealing surface are of paramount importance in ensuring a longer life for the seal.

13.9.1 Influence of Oil Contaminants

It is needless to emphasise how important it is to keep the flow medium free from contaminants. Metallic and other solid contaminants if present in the oil may damage the elastomeric seals used in the system through abrasion. The seal apart from losing its sealing ability may be a source of contaminants of elastomeric materials which ultimately may form gummy material in the oil chamber in association with other oxidized materials.

13.9.2 Influence of Mechanical Features of the Seal

The function of a seal depends essentially on the behavior of the seal material, constructional features like art of seal construction and seal form, preloading, seal size and its cross section, friction, wear, functional life, working safety, etc. The optimum reliability of the seal is ultimately determined by optimizing all the above factors. Seal hardness is also a very important factor. However appropriate compromise between seal hardness and other seal requirements is the only solution regarding seal selection. Hardness of $35°A$ is considered soft and $90°A$ as hard.

13.10 FAULTY FITTING OF SEALS IN CYLINDERS

It has been emphasised that adequate care is to be taken to see that seal selection and installation is done with perfect care and attention. In Fig. 13.20 faulty selection and installation has been illustrated using numbers 1 to 19 which are discussed as follows:

1. Piston rod bearing is too small—it should be proportionate to $L/D = 1$.
2. No spanner space (i.e. flat ends to hold spanner) should be machined on the round piston rod. It may create a sharp corner which may damage seals.
3. Seal assembly may get difficult.
4. O-ring installation false, O-ring may get crushed.
5. Seal under pressure of bearing load. May be undesirable.
6. Excessive gap between the seal and sealing surface, extrusion may take place.
7. Insufficient bearing surface of piston. The seal elements may not withstand the bearing load.
8. The construction/mounting space for the seal is very small. Pressure impact for the seals may be insufficient.
9. Details of seal material not provided.
10. Excessive uncovering of cylinder port cross section may be a constant source of air entrain.
11. Rod bearing too long and without oil groove.
12. Seal cross section not enough.
13. Seal may extrude.
14. Assembly of cover over thread may create large axial deviation; press fit is a better proposition.
15. Working details are inadequate e.g. finishing details, surface roughnen etc.
16. Too sharp chamfer to be avoided. O-ring assembly may be damaged.
17. Length of thread is too small. No reliability against loosening and ultimate leakage.
18. Seal elements against the intermediate spacer not to be axially positioned.
19. O-ring back up missing.

13.11 BURN TEST

A maintenance person may come across many unknown elastomers during maintenance and repair of hydraulically operated machines and components. Correct replacement of such elastomers may turn out to be a nightmarish experience. It will be quite detrimental to the system if the correct seal material is not selected as a replacement. In such a situation one may try the "Burn Test" to identify the elastomers correctly. A simple procedure for burn test is given below.

Seals and Packings 435

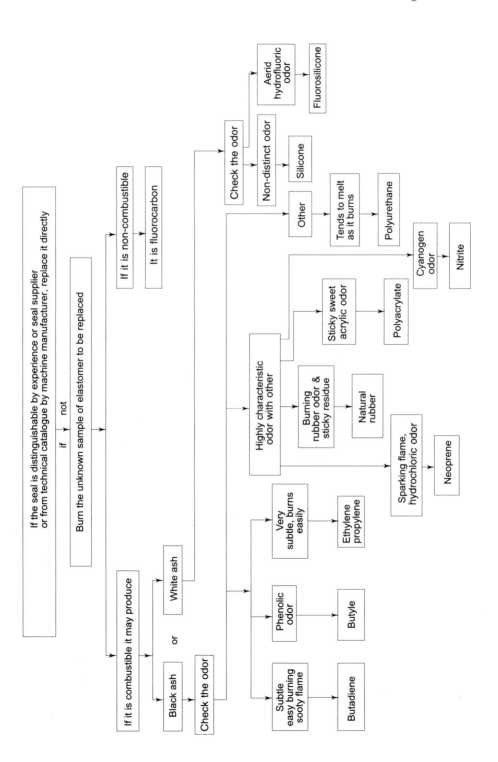

436 Oil Hydraulic Systems: Principles and Maintenance

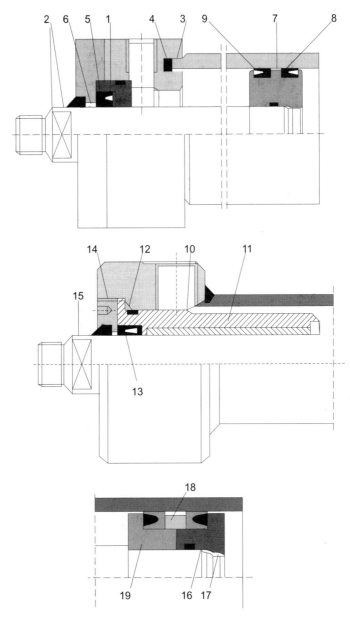

Fig. 13.20 *Faulty fitting of seals in cylinder*
[Courtesy: O and P Oel hydraulik and pneumatik Jahr Buch, 1992–93 p 54]

Review Questions

1. Classify types of seals as per ISO norms.
2. Differentiate between dynamic and static seals. What is the effect of seal materials on friction?

3. Compare the application of leather, metal and rubber as seal materials.
4. What is PTFE? State the specific merits and demerits of PTFE as a seal material.
5. State at least three defects of seals indicating the precautions to be taken to avoid or eliminate them.
6. What is meant by seal pre-loading? Give one example of a pre-loaded hydraulic seal explaining its function. Why pre loading is needed?
7. State various reasons of seal failure.
8. Discuss the effect of oil on hydraulic seal failure.
9. What type of seal material will be preferred for fire resistant oils? Justify.

References

13.1 Stone, J.G. "Properties of seal materials—How they work for you", *Hydraulics & Pneumatics*, Nov. 1977. p. 87, Parker Hannifin Corporation, California, USA.

13.2 Einsatz von Hydraulik Dichtungen, Verhalten, "Konstructions hinweise—Konstruction Jahrbuch", 84/85, Oel-Hydraulik u. Pneumatic, Germany, p. 41.

13.3 Ramsdell, RG, Sr. "A primer of fluid power sealing—Fundamentals of static seals", *Hydraulic and Pneumatics*, USA, April/1986, p. 68.

13.4 Parker O-ring Handbook, Parker Seal Co., Culver city, California, USA, 1971.

13.5 Dr. Ing and Nick Lappiatt, "Hydraulik Dichtungen fuer den Untertageban", *Oel-Hydraulic u. Pneumatik*, Germany, Nov./1987.

13.6 Kleinbreuer,-Ing W, *Requirements on Design and Production of Hydraulic Hoses*, Oel-Hydraulik u. Pneumatik, June/1986, Germany.

13.7 Publicity materials from M/s Martin Markel GmbH, Hamburg, Germany.

13.8 "K. Nagdi Dichtungstechnik", *Oel-Hydraulik u Pneumatik*, Germany Nov. 1986.

13.9 Geisu H and J Stuermer, "Umweltfreundliches Dichtungskonzept", *Oel-Hydraulik u. Pneumatik*, Germany, June 1991.

13.10 Albertson K and RK Pierce, "The Exclusion Solution", *Hydraulics & Pneumatics*, USA, Dec. 1990.

13.11 K. Nagdi, Wirkung von Additiven auf Dichtungswerkstoffe aus Gummi, O + P Oel-hydraulik u. Pneumatik, Nov. 1986, p. 852, Germany.

13.12 O + P Oel-hydraulik u. Pneumatik Jahr Buch, 1992-93, p. 54, Germany.

14

Hydraulic Pipes, Hoses and Fittings

PIPES and tubes are very important parts in a hydraulic system and evolution of the hydraulic piping system is as interesting as other hydraulic components. A pipe can be defined as a functional connection for fluid flow in the fluid power system and the fluid flow efficiency is greatly influenced by the physical characteristics of the piping system. Though the terms pipes, tubes and hoses are generally synonymous, each have their own specific characteristics which need to be understood before using them for a specific application.

14.1 IRON PIPES AND STEEL TUBES

In hydraulic systems iron pipes may be used for low to medium pressure range as they are widely available and economical. But the heavy wall thickness, lack of annealing characteristics and inability to absorb high hydraulic pressure surge are certain basic problems associated with iron pipes. Compared to iron pipes steel tubes are more commonly used because of their advantages. The term pipe is very often used to denote nominal bore (NB) size and tube for precision drawn steel tube OD material. Similarly hose means plastic or synthetic rubber tubes. However these definitions are not necessarily conclusive and are simply academic and may vary due to other factors like method of manufacturing, usage, etc.

In general pipe sizes are specified by inside diameter (ID). The actual ID for pipes of very small diameter is not equal to its designated diameter for any wall thickness of pipe. Outside threads remain the same regardless of changes in wall thickness of the pipe. For various wall thicknesses the OD for that size remains constant but the ID may vary. Hence the nominal ID (NB) and the pipe schedule number which denote the wall thickness are necessary to avoid confusion. However pressure ratings and tube/pipe bore finish and their art of manufacturing are important factors for selection of pipes. Cast iron pipes used by civic bodies for domestic water supply may not have a pressure capacity of more than 2.5 bar. For a common steam line the pressure capacity of fittings varies from 10 to 20 bar but for a hydraulic system if cast iron fittings are to be used, the pressure should be selected with a factor of safety of 4:1, i.e. the minimum burst pressure of such a system, for safety reasons, should not be less than 80 bar. However, cast iron fittings or iron pipes can be used in the hydraulic system in its return line or in application where the pressure does not go above 15 to 20 bar and no surge pressure is generally present.

14.1.1 Pipes and Tubes—Common Materials

The material for the hydraulic piping system is usually steel. Cuprous nickel alloys and stainless steel are also used but while stainless steel is very costly, copper tends to harden and is to be used in low pressure applications or small bore size only. Zinc coated galvanized pipes and copper tubes accelerate the chemical deterioration of certain oils limiting their use specially if the oil is not treated with anti-oxidant additives.

In general, steel tubes are made of annealed quality mild steel or soft ductile carbon steel. They are generally manufactured as cold drawn seamless tubes both in NB or OD sizes. Outside diameter sizes have better external finish and more tightly held tolerances which makes them most desirable.

In order to minimize friction losses as much as possible, the inner bore of the tube should also be highly smooth and appropriately polished, free of scales and should be reworkable without heating which is possible with seamless cold drawn MS tubes. Electric Resistance Welded (ERW) tubes are not desirable for hydraulic systems in general.

One advantage of seamless drawn tubes is that the wall thickness of the tubes can be controlled to any thickness to a high degree of accuracy due to ease of manufacturing which has been perfected over the years. If screw threads are to be cut on the tube wall NB tubes are most preferred due to their high wall thickness. Seamless tubes with high wall thickness, higher tensile strength, better bending quality, etc. are some of the specific properties which make such tubes most suitable for use in many high pressure hydraulic systems. Tubes from 4 to 30 mm diameter OD are generally manufactured with a tolerance of ± 0.1 mm, from 30 to 38 mm OD with a tolerance of ± 0.15 mm and for higher sizes the tolerance level is generally ± 0.2 mm.

Finished tubes are generally given some anti-corrosive treatment like phosphating which gives the tube surface a pore like crystalline outer surface or painted with a hydraulic quality paint compatible with the oil.

14.1.2 Pipe Specification

For hydraulic tubes and pipes the major area of concern is prevention of oil leakage and the ability of the tube to have a long working life without bursting. Wall thickness and the joints play an important role in this respect. Again the wall thickness of tubes and pipes is decided by the system pressure and therefore the factor of safety has to be correctly decided in order to select the appropriate tube wall thickness commensurate with the system pressure. Generally the following norms are preferred for designing the correct thickness.

Table 14.1 Factor of Safety for Tubes

Pressure range in bar	Factor of safety	Points to note
1. 30–70	8:1	For a system pressure of 40 bar the burst pressure should not be less than 320 bar.
2. 70–130	6:1	The burst pressure of tube to work for a hydraulic pressure of 100 bar should be at least 600 bar.
3. >160	4:1	For a hydraulic system working at 250 bar the tubings and fittings should be designed with a burst pressure of at least 1000 bar.

From the above it is clear that the factor of safety, burst pressure and proof pressure are important parameters in designing a hydraulic system tube and its fittings. Their definition therefore must be clear to hydraulic engineers.

Factor of safety This represents the margin of safety between the working pressure and the pressure at which the system may fail.

Working pressure This is the normal operating pressure at which the system has been designed, i.e if a system operates mostly at 100 bar, the working pressure is taken to be 100.

Proof pressure This is defined as the test measurement of the pressure to determine the reliability of the tube material. Generally the ratio of proof pressure to working pressure is taken as 2:1. This means if the proof pressure of tube material is 850 bar, the normal working pressure will be 850/2 = 425 bar.

Burst pressure This is the pressure to which the tube material is subjected to cause it to fail or burst. If it is said that the burst pressure of a tube is 500 bar, it is assumed that the tube should be able to hold any pressure above 500 bar.

Flareless bite type fittings are designed for use with pipes specified by their outside diameters or nominal bores, and are made with the materials described below. In all cases these pipes must

(a) be of seamless construction
(b) have precision tolerances on their outside diameters
(c) have a blemish-free surface
(d) be in a fully annealed condition with hardness not exceeding RB 70.

In certain medium and low pressure applications, cold drawn ERW pipes of nominal bores are also suitable.

Materials Carbon steel pipes conforming to DIN 2391/C or B.S. 3601/3602/1778, fully annealed, descaled and phosphated/oiled for adequate protection against rust in storage and while in use.

Stainless steel pipes conforming to DIN 2391 or B.S. 3605, provided their outside diameter complies with the close tolerance system.

Copper, aluminum, cupro-nickel and other ferrous and non-ferrous metal pipes conforming to the pertinent specifications and subject to the tolerance constraints thereunder.

Pipe wall thickness recommendations The minimum wall thickness specification in various standards for seamless pipes are based on the nominal working pressures, and the same for the popular types are given in Table 14.2. The fittings featured here will be applicable to pipes of other relevant standards on an equivalent basis.

Metric size outside diameter pipes Minimum wall thickness for metric size OD pipes to DIN 2391/C based on nominal working pressure is given in millimetres in Table 14.2.

Table 14.2 Minimum Wall Thickness

Series light		Series medium			Series heavy	
NP 100	NP 250	NP 160	NP 100	NP 630	NP 400	NP 250
4 × 0.5						
6 × 1	6 × 1	—	—	6 × 1.5	—	6 × 1
8 × 1	8 × 1	—	—	8 × 2	8 × 1.5	8 × 1
10 × 1	10 × 1.5	10 × 1	—	10 × 2.5	10 × 2	10 × 1.5
	12 × 1.5	—	—	12 × 3	12 × 2.5	12 × 1.5
	15 × 2	15 × 1.5	—	14 × 3.5	14 × 2.4	14 × 2
		18 × 1.5	—		18 × 3	18 × 2
		22 × 2	22 × 1.5		20 × 3.5	20 × 2.5
					25 × 4.5	25 × 3
			35 × 2			30 × 4
			42 × 3			38 × 5

* Based on a Safety Factor of 4:1.

Tolerance on outside diameters The standards for pipes require precision tolerance on their outside diameters to provide a perfect joint. These are reproduced below for the guidance of the user in Table 14.3

Table 14.3 Pipe O.D. and Tolerance

	Metric OD Pipes		Inch OD Pipes			Inch NB Pipes	
	Pipe OD	Tolerance	Pipe OD	Tolerance	Pipe NB	Outside diameter	
LL	4–8	± 0.07 mm	1/4"–3/4"	± 0.004"	1/4"	0.533"–0.541"	
	6–18	± 0.07 mm	1"–1.1/2"	± 0.005"	3/8"	0.674"–0.682"	
L	22–28	± 0.1 mm	2"	± 0.006"	1/2"	0.842"–0.850"	
	35–42	± 0.11 mm			3/4	1.056"–1.066"	
	6–16	± 0.07 mm			1"	1.332"–1.342"	
S	20–25	± 0.10 mm			1 1/4"	1.673"–1.683"	
	30–38	± 0.11 mm			1 1/2"	1.905"–1.915"	

14.1.3 Standards

Flareless bite type fittings for metric OD pipe sizes are now available in the field of hydraulic, e.g. in Germany these are manufactured to DIN 2353 standard which specifies the dimensions, material of construction and standard of workmanship for individual components to provide a conforming joint. These fittings also conform to B.S. 4368 which corresponds to the DIN standard for metric OD pipe size fittings.

Indian standards for metric OD pipe size fittings have been formulated in great detail to cover all aspects of their design and performance. IS: 8805–1978, for instance, covers "general requirements for ferrule type couplings used in oil hydraulic systems."

There are as many as 16 Indian standards covering virtually all such fittings. The British standard B.S. 4368 for inch OD pipe size fittings is not very stringent and leaves to the manufacturers to decide on the pipe end. The American standard SAE: J 514 has been laid down for the design of individual components and ferrules.

The fittings for inch OD and inch NB pipe sizes featured above are designed basically on the DIN and IS specifications for the pipe end, and are therefore totally interchangeable with the respective standards.

14.1.4 Storage of Steel Tubes

All hydraulic steel tubes should maintain a fine and bright bore which is greatly influenced by the way they are stored. It is therefore most necessary that these tubes are stored in a dry and clean place, the tube lengths being tightly plugged at both ends. The end seals should be fully air tight with hygroscopic material inserted into them. This will prevent the possibility of the pipe breathing during storage. However care should be taken to see that the tubes are subsequently cleaned meticulously when they are used again.

14.1.5 Copper and Other Metals as Hydraulic Tubings

Generally copper tubings are not preferred in hydraulically operated machines except in low pressure systems where there is no possibility of pressure surge or the oil is not compatible with ferrous metals. Sometimes specific alloys of aluminium and titanium are also used in aviation hydraulic systems as tubes.

14.2 PIPE FITTINGS

The end of the metal tube or synthetic hose is attached to the port of the hydraulic elements by means of various adaptors or fittings. These fittings are shown in Fig. 14.1. In order to have a proper fit the adaptors should conform to the port configurations of valves, cylinders etc. Generally four main types of port forms are used in a hydraulic system as given below (see Fig. 14.2).
    1. Female threaded—taper and parallel
    2. Male threaded—taper as well as parallel

Hydraulic Pipes, Hoses and Fittings **443**

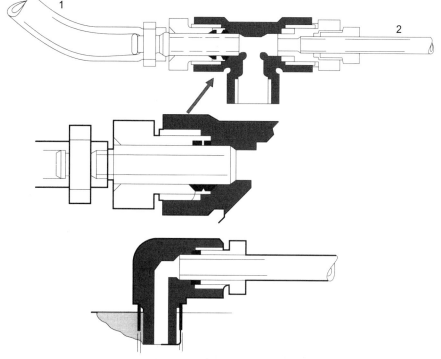

Fig. 14.1 *Typical hydraulic fittings*
[1-hose to tube, 2-tube to tube]

3. Plain female
4. Plain male (shank)

The common method of tube joints are as given below:
  (a) Flareless compression joints
  (b) Flared joints
  (c) Screwed joints
  (d) Formed joints (Formed pipes)
  (e) Brazed joints
  (f) Welded joints
  (g) Flanged couplings
  (h) Clamp couplings

14.2.1 Types of Couplings and Connectors Used in Hydraulic Piplines

Different type of tube fittings are used to connect pipes and tubes in hydraulic systems. Except for welded joints, all tube fittings connect the tubing with a union which lets the tubing remain stationary while the fitting is tightened.

Connectors and end fittings frequently used in hydraulic systems are shown in Fig. 14.3 with their internal construction.

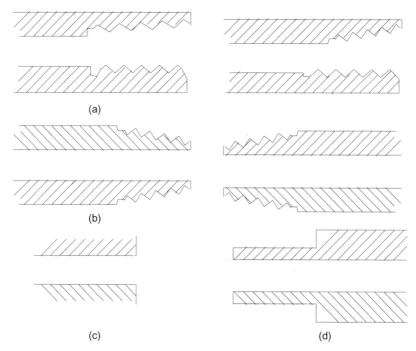

Fig. 14.2 *Forms of hydraulic fittings (a) Female threaded—Parallel and taper (b) Male threaded—Parallel and taper (c) Plain female (d) Plain male*

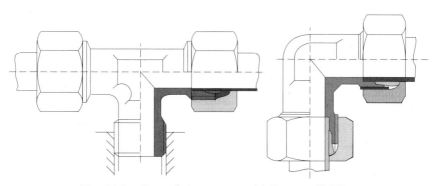

Fig. 14.3 *Types of pipe connectors (a) Tee joint (b) Elbow*

(a) Straight coupling
(b) Tee joint
(c) Bracketed elbow
(d) Bend
(e) Elbow
(f) Tube to tube connector
(g) Reducer
(h) Cross tee

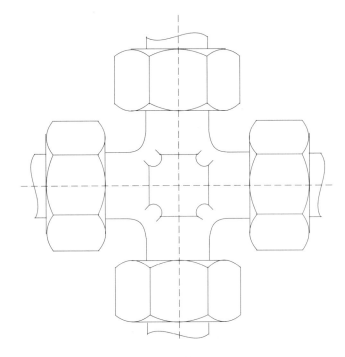

Fig. 14.3 *(c) Cross joint*

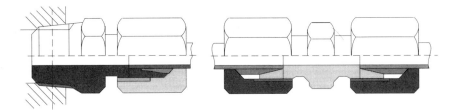

Fig. 14.3 *(d) Straight coupling*

(i) Blanking plug
(j) Nipple
(k) Female swivel
(l) 45° female swivel
(m) Male fixed
(n) Split flange (30°, 45°, 60°, 90°)
(o) NPTF male fixed
(p) Female rigid etc.

14.2.2 Selection of Fittings and Connectors

Types of pipe fittings used in hydraulic pipings may vary with the size of tubes. For example, flanges may be used for 63.5 mm (2-1/2″) and above. Flareless compression fittings are used from 1/8″ to 2″ diameter tube and may or may not be re-usable. Flared type compression fittings may be used on 50 mm (2″) and below. They require the use of a flaring tool and are normally re-usable.

Also compression (ferrule) type fittings using a metal ferrule or olive may be used on 30 mm sizes and above and serrated collet type fittings on 22 mm and below.

Compression type of joints are ideally suited because of their ease of use, compactness, etc. They are also cheaper than flanges. Generally compression type fittings are not available above the size of 50 mm OD.

Though welding joints are used sometimes, it is generally advisable to avoid them, if possible. In case it is unavoidable, only a sleeve type joint should be used instead of a butt joint. This will reduce the risk of internal scaling. However large sized pipes and heavy duty systems may have to be joined through welding. Brazing of fittings are to be restricted to 15 bar pressure only.

Flanged couplings are generally used if the tube sizes are above 2 inches. Clamp couplings may be used above 1 inch tube sizes in general.

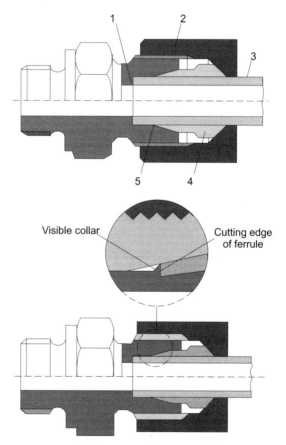

Fig. 14.4 *(a) Ferrule type compression fitting*
[1-Shoulder of internal cone, 2-Coupling nut, 3-Tube, 4-Ferrule, 5-Intenal cone]

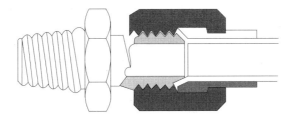

Fig. 14.4 *(b) 37° Flared fitting*

14.2.3 Self-sealing Compression Ring

Compression couplings of this type are shown in Fig. 14.4 (a). They are made in a forming tool on a bench before being connected into the pipe fitting. The sealing is effected by a seal-ring (ferrule or olive) which has to cut its own seat on the pipe wall. Adequate mechanical force is needed to bite into the pipe wall and this is provided by mechanical aids or spanner extensions. One must keep in mind that all compression type fittings are flareless fittings. This can be achieved by hand tightening first and then turn a set amount (often $1\frac{1}{2}$–2 turns) to complete the joint using a spanner.

14.2.4 Elastomeric Seal with Serrated Collet Grip (O-ring Compression Fitting)

Generally for high pressure systems where perfect sealing is needed, this seal is used. The seal becomes tighter as pressure increases. The sealing will depend on the performance of O-rings which will be deformed by pressure to effect the sealing. Mechanical effort is only needed for the pipe to be gripped and hence the effort is comparatively little. This type of coupling is re-useable because the coupling parts can be slipped off the tube when the point is disconnected. Bleeding this type of coupling is hazardous and it is not to slacken it off while under pressure. (The ferrule type compression coupling can be slackened off for bleeding purposes while under pressure.)

14.2.5 Flare Fitting

Though various types of flared fittings are available, only two types of flared fittings are generally used. The 37° flare fitting is the most common fitting for tubing that can be flared. Sealing is achieved by squeezing the flared end of the tube against a seal as the compression nut is tightened. Flaring is also done with 24° and 45°. The 45° flare fitting is used for low pressure hydraulics. It is also made in an inverted design with male threads on the compression nut. For flaring the tube a flaring tool is used. The 37° flare fitting is used for high pressure hydraulic systems. It provides a longer sealing service and less flare distortion and is primarily used in hydraulic steel tubing. A 37° flare fitting is shown in Fig. 14.4(b). Some compression fittings do not bite into the tube such as those used with copper or plastic tubing. However, the large tube contact area of the metal ring, or sleeves, controls tube distortion and holds the fitting in place.

Vibrations in all flareless compression type of fittings against the tube will depend on whether the material bites into the tube or is sealed by a compressed surface area.

The advantage of compression fitting is that the tube is not flared thus eliminating extra work. The tubing can also be bent to a smaller radius. But the disadvantage is that the metal ring (sleeve) is not re-useable when damaged.

14.2.6 Pipe Run

The pipe run should be planned very carefully and one should know that the time and energy spent on it to do it with utmost care are ultimately paid back in most cases in the shape of a tidy, systematic and well arranged system. When fully developed it may provide easy access and clear space with the minimum obstruction. This will result in easy and quick line tracing during trouble shooting and fault diagnosis and will also give the maintenance engineer adequate spanner space to lock and unlock a joint as well as facilitate thorough checking without ambiguity.

14.2.7 Pipe Thread

Pipe thread is another important area to take note of. In earlier times for pressure oil applications two types of threads were being used standardised on the basis of the American National standard—one being American or National Standard Taper (NPT) and the other being the Dryseal American (or National) taper pipe thread Fuel (NPTF).

The difference between the two depends on the method of sealing applied. BSP threads are also more common. In the recent years ISO Metric Threads are also being used in many component manufacturing.

On NPT threads, sealing is by interference contact between the male and female thread flanks. Sealing tape or sealing compounds are used to stop the fluid leaking through the spiral clearance at the thread root. But on the NPTF, the roots and crests of the thread engage before the flanks touch and due to wrench tightening, the crests are crushed creating a perfect seal. Due to historical and technological reasons even now hydraulic pipe threads are mostly influenced by American standards and American JIC hydraulic standards specified that all pipe threads should be NPTF. Though NPT and NPTF threads are interchangeable, to ensure perfect sealing, male and female NPTF threads will be better if used together.

14.2.8 Surface Protection of Fittings

End usage largely determines the choice of materials from which fittings and components are made.

The fittings offered are made of the following materials of construction.

Carbon steel Carbon steel fittings are made from steel to 14C14S14 of IS: 1570 (Part III)–1979. Carbon and carbon manganese free cutting steels are used for straight body and nut. Shaped fitting bodies are made either from 14MIS14 of Schedule III of IS: 1570-1961 for bar stock construction, or Class IA of IS: 1875-1978 for forged construction. Carbon steel fittings are available with 'Parkerized' (manganese-based phosphatizing) finish as standard. Cadmium plated ferrule is also made. Fittings with fully cadmium-plated finish can also be used but are costlier. These fittings are suitable for use with carbon steel, copper, brass, aluminium and stainless steel pipes where line fluids do not affect the material.

Brass Brass fittings are made from free cutting brass for straight body and nut, and forging brass for shaped bodies. The ferrule is usually made of brass for low pressure applications, and of cadmium-plated carbon steel for higher pressures. Brass fittings are available with passivated finish as standard. Nickel-plated finish for special applications in chemical firms e.g. urea plants, etc. is also available.

These fittings are suitable for use with copper, aluminium and cupro-nickel pipes.

Stainless steel Stainless steel fittings are generally available in AISI: 304 grade stainless steel, but can also be offered in AISI: 316 grade stainless steel where the application and line fluids warrant it. The body is made from stainless steel of the quality desired. Nuts in standard fittings are made of cadmium-plated carbon steel since they do not come in contact with line fluids. This choice prevents "galling" of the threads in assembly, especially where the use of thread lubricants is forbidden. The nuts can also be made of stainless steel of the requisite specification, if desired. Stainless steel is not an ideal material for ferrules. Being non-hardening, the primary requirement of a hard outer casing and a soft core is not achievable. So where stainless steel fittings can be used with carbon steel ferrules, excellent results may be obtained. Stainless steel fittings with stainless steel ferrules may, however, be necessary where line fluids are highly corrosive. In such cases, the recommended operating pressure has to be suitably reduced.

Instrumentation fittings For instrumentation requirements in various industries, a complete range of instrumentation fittings may be necessary of small diameter like 6 mm–12 mm OD ($\frac{1}{4}''-\frac{1}{2}''$ OD) copper and PVC coated copper pipes. Standard fittings are available in brass, and special fittings in stainless steel. Brass fittings are suitable for a maximum working pressure of 100 kg/cm^2, while stainless steel fittings can be used for working pressures upto 200 kg/cm^2. However precautions need be taken for brass are as told eartier.

Surface protection The surface protection to be used with various fittings is a matter of choice left largely to the designer, piping engineer or user. The salient points that determine the choice of finish, are detailed below:

For oil hydraulic service Standard carbon steel fittings available with 'parkerized' finish offer the best results. The 'parkerizing' process creates a fine, corrosion-resistant skin with sealant oil entrapped within. This combination not only prolongs the life of the finish against rusting and atmospheric corrosion; more, the presence of the entrapped oil lubricates the individual parts when the fitting is assembled on the pipe.

For service in corrosive atmospheres Where carbon steel is not attacked by the atmosphere or by the fluids to be handled, the use of standard carbon steel fittings with 'parkerized' finish is recommended. Where, however, the presence of the entrapped oil is likely to damage the fluid to be handled, galvanized or cadmium-plated finish can be provided. Cadmium-plating, besides being a superior finish, has the added advantage of "lubricating" the individual parts during assembly because of its softness. Galvanizing, on the other hand, has the advantage of lower cost, though appropriate care needs to be taken to see that it does not affect the oil.

Where carbon steel is attacked by the fluids to be handled or by the atmosphere, brass or stainless steel fittings are preferred. Brass fittings are generally restricted to pneumatic and low pressure applications and are available in either natural, nickel-plated or cadmium-plated finish. Brass fittings are not suitable for use in atmospheres which contain ammonia. Special brass fittings with carbon steel ferrules can be offered for high pressure applications (e.g. in argon gas plants).

14.2.9 Double Ferrule Fittings

The relative disadvantages of stainless steel fittings resulting in downgrading of the pressure rating, have been referred to earlier. An entire range of double ferrule fittings in stainless steel for higher pressure ratings and for use with corrosive liquids and atmospheres are available in all varieties depicted herein for ferrule-type fittings.

Classification Flareless Bite type fittings are available for the following types of pipes:

(a) *Metric size Outside Diameter* pipes to DIN 2391/C or equivalent standards.
(b) *Inch size Outside Diameter* pipes to B.S. 3601/3602/3605 or equivalent standards
(c) *Inch size Nominal Bore* pipes to B.S. 1387-1957 or equivalent standards.

The pressure ratings and general application recommendations in respect of each type are given in Tables 14.4 and 14.5.

Metric size Flareless Bite type fittings for metric OD pipe sizes are classified into three broad categories according to DIN 2353, and are based on the nominal pressure ratings of pipes in common use:

Table 14.4   Metric Pipe O.D. and Nominal Pressure

Series	Pipe OD	Nominal Pressures*	Applications
Light	4–8 mm	100 kg/cm^2	Pneumatic and lubrication systems, synthetic piping and low-pressure applications not exceeding 100 kg/cm^2
Medium	6–15 m 18–22 mm 28–42 mm	250 kg/cm^2 160 kg/cm^2 100 kg/cm^2	For medium pressure range applications up to 250 kg/cm^2 and below
Heavy	6–14 mm 16–25 mm 30–38 mm	630 kg/cm^2 400 kg/cm^2 250 kg/cm^2	Heavy industrial, ship-building, mining and chemical industry pressure piping involving severe mechanical stresses

[*Courtesy:* Hyd-Air Engineering Works, Mumbai, India.]

* These nominal working pressures are based on a Safety Factor of 4:1, and assume uniform load conditions at temperatures up to 120°C. Allowance must, however, be made for ambient conditions involving heavy impact pressure, mechanical strain and vibrations.

Inch size   Flareless Bite type fittings for inch OD and inch nominal bore pipe sizes are designed for applications with the following nominal pressures as shown in Table 14.5 below:

Table 14.5   Inch Pipe O.D. (Bite type fitting) and Nominal Pressure

Pipe size		
Outside diameter	Nominal bore	Nominal working pressure
1″ and under	3/2″ and under	400 bar
1 1/4″ and 1 1/2″	1″, 1 1/4″	200 bar
2″	1 1/2″	125 bar

14.3   ENERGY LOSS

When the hydraulic oil flows through a pipe or fitting, a portion of its energy is lost due to pipe friction, leakage, etc. It has been found that higher the viscosity of oil, greater the loss of energy. Again lesser the oil viscosity, higher the leakage losses. Oil energy is also lost due to presence of various types of valves, pipe fittings, e.g. tees, elbows, etc. The nature and art of flow path and orifice design determines the amount of loss through them.

14.3.1   Loss in Pipes

The resistance loss in a pipe can be calculated by using various empirical formulae developed. Applying the same in the Bernoulli's theorem and continuity equation the pressure drop due to the resistance can be calculated. The head loss in a straight normal pipe is determined by the following equation:

$$h_p = \lambda \cdot \frac{v}{2g} \cdot \frac{l}{d}$$

where, h_p = Head loss, m
λ = Friction coefficient
v = Mean flow velocity of oil, m/s
l = Length of pipe, m
d = Inside pipe diameter, m
g = Acceleration due to gravity = 9.81 m/s.

The frictional resistance λ is dependent on factors like type of fluid flow, inner surface finish of pipe, oil viscosity, etc.

14.3.2 Loss in Pipe Fittings

Head losses in pipe fittings, valves, etc. are proportional to the square of fluid velocity and the formula to calculate the same is shown below:

$$h_f = \frac{K v^2}{2g}$$

where, h_f = head loss in fitting, m, k = is called the K factor for valves and fittings, v = mean flow velocity in m/s, and g = acceleration due to gravity = 9.81 m/s. The values of K factor for some valves and fittings are given below in Table 14.6.

Table 14.6 Values of K Factor

Sl. No.	Fittings	Values of K
1.	Gate valve–quarter open	24
2.	—Do—half open	4.5
3.	—Do—wide open	0.19
4.	Globe valve–half open	12.5
5.	—Do—wide open	10
6.	Tee	1.8
7.	Elbow (standard)	0.9
8.	Elbow–45°	0.42
9.	Elbow–90°	0.75
10.	Check valve–ball type	4

14.3.3 Equivalent Length

The equivalent length of valve or a fitting is defined as that length in a pipe which produces the same head loss as in a valve. The formula given below may be used:

$$l_e = \frac{K \cdot d}{\lambda}$$

where, K = K factor,

l_e = equivalent length, m
d = inside diameter of fitting, m
λ = frictional resistance.

Though hydraulic system designers may calculate head losses in a fitting by applying the empirical formulae, it is more common to use calculated values from Table 14.7.

Table 14.7 Equivalent Length of Pipeline Fitting

Type of fitting	Equivalent length in m/mm (l/d)
1. Check valve	83
2. Valve with poppet disc	26.3
3. 90° bend	38
4. 75° bend	25
5. 60° bend	16.5
6. 40° bend	7.5
7. 90° elbow	19–32
8. 45° elbow	10–16.5
9. Tee	13
10. Square elbow	36
11. 90° Large radius elbow	13

14.3.4 Nature of Fluid Flow

We know that the fluid velocity in a pipe is zero at the fluid layer of adjacent to the pipe wall and is maximum at the center of the pipe. Nature of flow is also not the same always. Flow can be laminar or turbulent. In a laminar flow the fluid flows in smooth layers (laminae), i.e. a particle of fluid in a given layer does not change its line of motion. This means in a laminar flow the fluid flows in a streamlined condition in the same direction.

But with increase in velocity, the streamline condition of the flow is disturbed and the flow particles start fluctuating. The flow particles in this condition move in a random direction perpendicular as well as parallel to the main flow direction creating a turbulence. This turbulent flow causes more resistance to flow effecting more energy losses compared to laminar flow. A scientist Osborn Reynold noticed the nature of flow depends on a dimensionless parameter

$$\text{Re} = \frac{v \cdot d \cdot \rho}{\mu} = \frac{v \cdot d}{v}$$

whack is called Reynold's Number.

where, v = flow velocity, m/s
d = inside diameter of pipe, mm
ρ = Mass density, kg/m^3
v = Kinematic viscosity, m^2/s
μ = Absolute viscosity, Ns/m^2

14.3.5 Critical Values of Re

The losses are also dependent on some other factors:
(i) Coefficient of friction
(ii) Valve coefficient
(iii) Coefficient of conductance
(iv) Reynold's Number.

Flow speed, length of the pipe, etc. are also equally important factors in determining the pressure loss in the system. The losses in the system line are undesirable as it is ultimately instrumental for heat generation in the system and leakage of oil which may result in ultimate energy loss. The following formula my be used to calculate the pressure loss:

$$\Delta P = \frac{\lambda \cdot \rho \cdot v^2}{2g} \cdot \frac{l}{d} + k \cdot \frac{\rho \cdot v^2}{2g}$$

$$= \frac{\rho v^2}{2g} \left[\frac{\lambda \cdot l}{d} + k \right] \frac{\text{kgf}}{\text{cm}^2}$$

where, ΔP = drop in line pressure, kgf/cm^2
λ = friction resistance
k = k-factor dependent on valve form
ρ = specific weight, kg/cm^3
l = length of pipe, cm
d = pipe diameter, cm
v = flow velocity, cm/s.

The friction resistance (λ) is dependent on the Reynolds' Number (Re) and the art of orifice design of the fittings, i.e. whether it is a sharp bend, a fine orifice, a ball valve, etc. The Reynold's Number at a specific critical value determines whether the flow is laminar or turbulent. If Re > Re critical, the flow is turbulent and if Re < Re critical, then the flow is laminar. The friction resistance (λ) is related to Re as shown below:

For a laminar flow through straight polished pipe, $\lambda = \dfrac{64}{\text{Re}}$.

For laminar (non-isothermal) flow, $\lambda = \dfrac{75}{\text{Re}}$

For turbulent flow $\lambda = \dfrac{0.316}{\text{Re}}$

The following critical values of Re are important.
For a round polished pipe Re = 2000 to 2300
For a concentric polished opening Re = 1100
For an eccentric polished opening Re = 1000.
In general the following values of Re may be accepted for design purpose.
For laminar flow, the value of Re is < 1200.
Between laminar and turbulent flow the values of Re lie between 1200–2500.
For turbulent flow Re is > 2500.

The value of critical Re = $v_{crit} \times d/v$ = 2320. At this critical value the laminar flow becomes turbulent.

The flow speed again determines whether a flow is turbulent or laminar. In order to maintain a laminar flow through a pipe, the following guidelines on flow speed may be observed;
Speed in suction line = 1 to 1.5 m/s
Speed in pressure line = 3 to 6 m/s
Speed in valves, etc. = 6 to 10 m/s
Speed through pressure valve = 30 m/s
Speed in return line = 2 to 2.5 m/s

14.4 ESTIMATION OF LINE DIAMETER

Instead of calculating the diameter mathematically, the pipe diameter can also be estimated by using various types of nomograms developed by various experts. If the flow velocity and flow rate (Q) of the system are known, the pipe diameter can be found out.

Example 14.1 The nomogram is shown in Fig.14.5 (a). The flow rate (Q) and flow velocity (v) of a hydraulic system are as follows:
Flow rate (Q) = 100 l/min
Flow velocity = (v) = 5.3 m/s
Determine the pipe diameter
Solution Here Q = 100 l/min and flow velocity v = 5.3 m/s.

In the diagram in Fig. 14.5 (a) let us take the point 5.3 m/s in the oil speed scale and the point 100 l/min. in the flow rate scale and join both of them. The line is then extended up to the nominal diameter scale line which intersects the diameter scale at the point 20 i.e. 20 mm diameter (n.b.), which is the selected nominal bore. Thus the nomogram gives the diameter to be selected for this case as 20 mm. It is to be noted that in the nomogram,
a = speed of oil in pressure line.
b = maximum speed of oil in the return line.
c = probable range of speed of oil in the suction line with oil having viscosity 20–200 cSt.

14.4.1 Estimation of Drop in Pressure in Straight Pipes

Very often engineers and designers instead of calculating the pressure loss using empirical formulae as stated earlier, make use of various nomograms available in Designer's Hand Book. In Fig. 14.5 (b) such a nomogram is shown.

Example 14.2 The following data is given:
Kinematic viscosity of oil = 30 cSt
Specific weight of oil = 0.9
Flow rate = 30 l/min.
Nominal bore of pipe = 25 mm
Length of pipe = 3 m
Find the pressure drop per meter length of pipe.

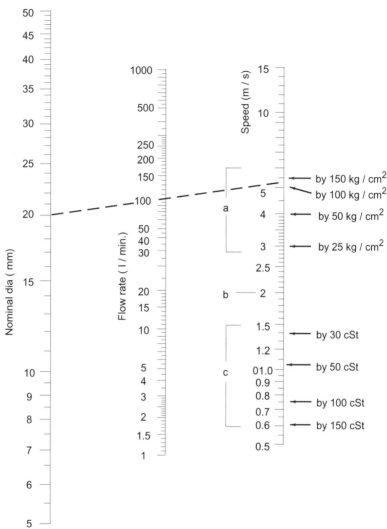

Fig. 14.5 *(a) Nomogram for pipe dia. (n.b.) estimation. Recommended values: (a) For velocity range (oil viscosity 20 to 200 cst). (b) Maximum return line velocity (20-200 cSt). (c) Velocity range for suction line.*

Solution In Fig. 14.5 (b) the line joining the viscosity point of the viscosity line at 30 cSt to specific weight line at 0.9 pt. bisects the ref. line 1 at point P_1. The line joining the point P_1 to the flow rate line at the point 30 l/min, bisects the ref. line 2 at point P_2. If P_2 is now joined by a line to the nominal bore line at the 25 mm point, it bisects the pressure drop line at 0.014 kgf/cm² in point which is the estimated pressure drop.

∴ The pressure drop in a 3 m long pipe = 3 × 0.014 = 0.042 kgf/cm²

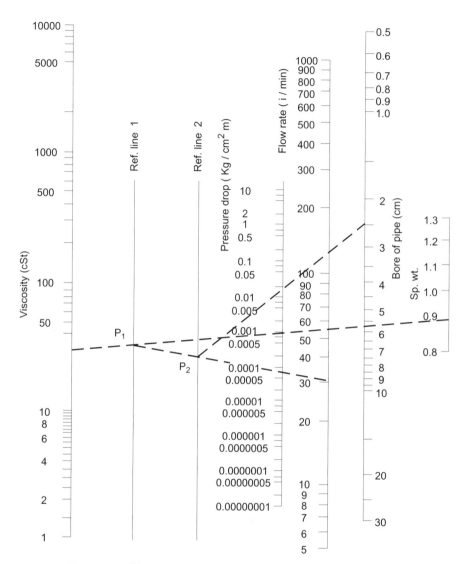

Fig. 14.5 *(b) Nomogram to estimate pressure drop in straight pipe*

14.5 SYNTHETIC HYDRAULIC HOSES

In many hydraulic systems where it is necessary to move the drive unit along with the oil lines, e.g. in a hydro-copying unit a rigid piping system will be difficult to use. In such places hydraulic hoses made of synthetic rubber are preferred. Plastic tubings are also used in hydrualic systems to convey the fluid. The advantages of using a rubber hose are:

1. Rubber hoses are flexible.

2. Their capacity to absorb or withstand shock and vibration is much higher than metal tubes.
3. They are easy to install and dismantle.
4. They can be manufactured in long lengths.
5. They are capable to take high pressure.

However the following disadvantages are also to be noted:
1. They are poor in abrasion resistance.
2. Their initial cost is higher.
3. They are subject to damage due to abrasion and incompatible oil.
4. They are poor in weathering resistance.

14.5.1 Construction of Hydraulic Hoses

The constructional view of various types of medium and high pressure flexible hoses, is shown in Figs 14.6 and 14.7. They are generally made of rubber reinforced with fiber or steel wire braiding. Usually hoses are constructed with an inner tube, reinforcement and an outer protective cover. Wire braids of high tensile strength steel wire may tolerate very high working pressure, permit longer working life, and are more durable. The steel wire may be either spirally woven or cross-woven. Spiral reinforced hoses are stronger but the fittings are to be supplied by hose manufacturers. Cross-woven braids tend to fail by brittle fracture at the place where the wires cross. The advantages of cross-woven braids are that they are reuseable and can be assembled by the users.

A hydraulic hose consists of an inner tube through which the fluid is conveyed and thus it comes into direct contact with the hydraulic fluid. The rubber or other synthetic material used for this must be compatible as well as should be able to withstand the range of working temperature without losing its chemical and physical stability. Depending on the factors of oil-compatibility, abrasion resistance, etc. Various types of synthetic materials could be used for hoses for fluid power system.

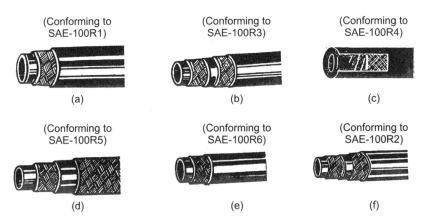

Fig. 14.6 *Types of flexible hoses a, b, c—Medium pressure d, e, f—Medium high pressure*

Hydraulic Pipes, Hoses and Fittings   459

(Triple wire braided hose)

Conforming to SAE-100R9

Super high pressure hose
Hi-impulse
(Parameters to match SAE-100R9)

Heavy duty high impulse hydraulic hose
'Insap-12' conforming to SAE-100R12

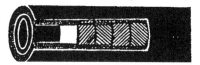

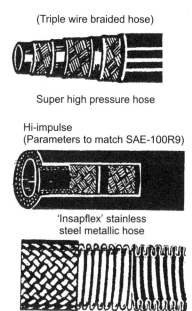

'Insapflex' stainless steel metallic hose

'Insapflex' PTFE teflon hose

Fig. 14.7   *Types of flexible hose—super high and heavy duty type including teflon and metallic hoses*

14.5.2   Hose Materials

1. Plastic   Homogeneous plastics can be used for 7–35 bar pressure only. However they are temperature dependent. The most common plastics used for hydraulic systems are nylon, polyethylene vinyl (polyvinyl chloride), etc. Generally plastic tubings may be used for low pressure hydraulics only. A plastic hose with a seamless nylon inner tube covered by dacron braided tubes is also used. A brief discussion about some of the materials used for hoses is done below.

(i) Nylon   These tubings have good abrasion and impact resistance but cannot be used above 15 bar pressure. They can work satisfactorily within the temperature range of 75° to 105°C. But high temperature and moisture affect their burst pressure rendering them unsafe.

(ii) Braided nylon hose   They can be used for medium to high pressure. They are fatigue-free hoses and apart from dacron, the outer cover could be by other material also. A polyurethane cover is compatible to both mineral and synthetic hydraulic fluids. It lasts longer than rubber and wire types, lighter in weight, cheaper, dimensionally more stable and expands less when pressurized compared to other materials.

(iii) Polyvinyl chloride   It is used for low pressure pneumatic applications only. It is more flexible than nylon and polyethylene and can be produced in longer lengths. However its poor temperature capacity (–10 to + 40°C) is a serious disadvantage. Its corrosion and abrasion resistance are poor rendering a reduced working life for hoses made of this material.

(iv) Textile braided hoses Cotton braided tubes are also used for low pressure hydraulics up to 30 to 35 bar pressure only.

(v) Thermoplasts Generally they can work satisfactorily within a temperature range of – 40 to 110°C for various types of oil like phosphate esters, petroleum, water-based oils, etc. Polyester yarn with polyurethane covers are also used.

(vi) Teflon Poly tetra fluoroethylene (DTFE) which is commonly called Teflon is used as a hydraulic tube material in a number of cases. Teflon can be used effectively within a temperature range of –50 to 204°C. It is available in the market in bending radius of 38 to 400 mm and working pressure as low as 10 to 105 bar in general. However very high pressure (up to 600 bar) Teflon tubes are also being marketed by some firms. These tubes are braided with steel reinforcement, can withstand high oscillation, vibration and pulsating force. They can be used with all types of oil except water-glycols and some petroleum-based oils. They are non-flammable, have inertness to chemical action, high mechanical and chemical resistance and very low coefficient of friction. But they are not readily used as in binding, lamination, painting or metallisation.

The rigidity of teflon and its tendency to cold-flow may limit its application.

(vii) Chloro sulfonated polyethylene (Hypalon) Its tensile strength is very good; tear and abrasion resistance fair to good and has excellent compatibility to water glycol emulsions. Its ozone resistance is also very good but it is poor in low temperature characteristics.

(viii) Ethylene propylene diene (EPDM) A hose made out of this material will have excellent weathering, ozone and heat resistance but its compatibility with petroleum-based oil is poor and may be avoided for such oils as well as water-oil emulsions. It is also compatible with water glycol though its resistance to flame is poor. Its low temperature resistance is good.

(ix) Chlorinated polyethylene This is quite excellent for water-glycol and water-oil emulsions as well as petroleum-based oils and very good for phosphate ester. The other properties including ozone resistance, permeability, etc. are also good which make it an ideal hose material for general purpose hydraulic system.

2. *Homogeneous Synthetic Rubber* Various types of synthetic elastomers are used for medium to high pressure hydraulics system as inner tubes as given below.

(i) Buna N It is also called nitrile rubber . It is most commonly used in hydraulic hoses within a temperature range of –50° to + 150°C. It is a synthetic rubber and can be used with petroleum-based oils but not with fire resistant phospate ester based fluids.

This material has the greatest resistance to swelling when used with petroleum oils and does not absorb oil. Hence there is no possibility of restriction of the passage to obstruct the flow of oil. Its abrasion resistance is poor but as inner tube it is quite excellent.

(ii) Neoprene It is a synthetic rubber compound which is also termed as chloroprene, has good oil resistance but is not resistant to aromatic hydrocarbons. Its physical characteristics are excellent having very good abrasion resistance. When used with oil, moderate swelling may occur with passage of time. It makes an excellent outer cover where oily condition exists on the outside. Therefore in many hydraulic system having such conditions it is used.

(iii) Natural rubber Natural rubber or GRS is mostly used in lines only. It can also be used satisfactorily with hydraulic system using water glycol as the fluids. Generally it is not oil resistant and will swell to over 20% of its original size when used with petroleum based fluids though it is very good in its abrasion resistance property too.

(iv) Butyl It is a synthetic compound suitable for use within the temperature range of $-50°$ to $150°C$. But its oil resistance is very poor and deteriorates rapidly if it comes into contact with petroleum based liquids. However it is extremely compatible with phosphate ester oils.

14.5.3 Hose Reinforcement

As has already been mentioned the inner tube of a hose needs reinforcement. Therefore the inner tube is covered with single or more number of outer layers as a source of reinforcement. These layers provide the hose its structural strength to withstand the hydraulic pressure applied on it. The prolayers may be made out of:
1. Cotton or textile
2. Nylon
3. Synthetic elastomers
4. Rayon
5. Steel wires
6. Synthetic yarn, etc.

A brief description of their use and application is given below.

(a) Rayon It is a standard material used for low pressure hydraulic systems only. The burst pressure for this material varies generally from 20 to 70 bar in ID sizes of $3/4''$ downward to $3/16''$. The flexibility of the material is its advantage. Also the minimum bending radius is smaller than that of other hoses. Some high burst capacity high tensile closely woven rayon braided hoses are also available that can withstand higher pressure.

(b) Synthetic yarn It is a loosely woven fabric which is used for low pressure hydraulic systems. It can be used from 7 to 35 bar and is temperature dependent. Their characteristics are similar to rayon braided hoses. However, the hose fitting

can be pushed into it and the hose material can grip it, because of which it is called a press-on hose.

(c) Steel wire

(i) Single wire braid is the common reinforcement used in hoses where high working pressures are involved. Wire has a higher tensile strength than rayon braids or synthetic yarns. Wire braids are used with either one or two layer construction. The single wire braid hose is commonly used in systems operating at 100 bar when flow volume permits a hose ID of 5/8 inch or less (19 mm). The two layer wire braids boost the working pressure of the hoses by as much as 150 bar in the smaller 1/4 inch (6 mm) and 3/16 inch (4.5 mm) ID sizes. The two layer braided hose is a high burst pressure hose.

(ii) Spiral wire In this type of reinforcement wires are laid out in layers on the inner tube and are not braided. This design affords high pressure for enhanced performance under repeated flexing or pressure surges. They provide higher flexibility, and better impulse life. Spiral wire braid construction is generally found in the larger ID sizes ranging from 19 to 50 mm. The normal pressure rating is 350 bar.

Nylon braided Nylon braided hose can be used from low to medium working pressure upto 100 bar as compared to the rayon braided hose. Nylon is used in smaller ID sizes and especially with light mobile equipment or in power line applications where a non-conductor of electricity is required. It does not have the scissoring effect that wire braids generally have.

14.5.4 Outer Protective Cover and Devices

The reinforcement of the hose is protected by the outer cover from corrosion, abrasion, and other damage from within or accidents, etc. Various types of synthetic elastomers as has been discussed earlier are used for this purpose. Neoprene or synthetic GRS rubber are commonly used for this cover because they have excellent abrasion and weather resistant characteristics. Cotton yarn may also be used as a protective cover in the case of one wire braid hydraulic hose and for synthetic yarn push-on hose. Cotton protection cover has an advantage in putting on a re-usable hydraulic hose fitting in the field. This cover does not have to be cut down to the wire (skinned) to provide a holding grip for the coupling.

Sometimes to protect the hose from damage such as from external rubbing, abrasion, etc. various protective devices are used. These are discussed below and are schematically shown in Fig. 14.8.

(i) *Nylon sleeves*—Apart from protecting the hoses from abrasion these sleeves allow bundling of hose lines.
(ii) *Plastic coil sleeves*—This is a light weight plastic sleeve generally unaffected by the hydraulic fluid. This coil can also be used for group bundling of hoses.

Hydraulic Pipes, Hoses and Fittings **463**

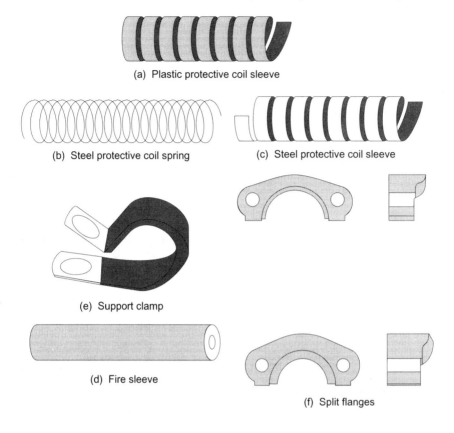

Fig. 14.8 *Hose protective devices*

(iii) *Steel protective coils*—Made of anti-rust steel wire they are capable of protecting the hose cover and reinforcement from damage, abrasion, etc.

(iv) *Steel protective coil sleeves*—They are recommended for use when the hoselines are subjected to excessive abrasion, kinking or accidental damage. These are made of rust-proof spring steel.

(v) *Fire sleeves*—These sleeves are capable of protecting the hose from direct flame. They are made of a uniform single layer of braided fiber glass tubing impregnated with fire resistant rubbraided fiber glass tubing impregnated with fire resistant rubber.

14.6 COMPATIBILITY OF FIRE RESISTANT OILS WITH THE HOSE MATERIAL

As per ISO classification (ISO DIS 6071) fire resistant oils are categorized in the following groups and the compatible hose material is mentioned against them.

Table 14.8 Fire Resistant Oils and their Compatible Hose Materials

Type and nomenclature	Compatible hose material
HFA–Oil-in-water emulsion or synthetic	NBR
HFB–Water-in-oil emulsion	NBR
HFC–Water poly-glycol solution	NBR, EPDM, SBR
HFD–Synthetic anhydrous phosphate ester or chlorinated hydrocarbon	FPM (Viton), EPDM

It should be kept in mind that materials like, Zn, Cd, Mn, Al, etc. are affected by oils of group HFA, HFB and HFC.

Table 14.9 Chemical and Commercial Name of the Hose Materials

Chemical name	Commercial name
1. NBR–Nitrile butadiene rubber	Perbunan
2. FPM–Fluoroprene rubber	Viton
3. EPDM–Ethylene propylene diene rubber	Nordel
4. CR–Chloroprene	Neoprene
5. YBPO–Polythene polyester elastomer	Hytrel
6. PTFE-Poly tetra fluoro ethylene	Teflon
7. PA–Polyamide	PA12, Nylon

14.7 INSTALLATION OF HOSES, TUBES AND PIPES

To avoid frequent maintenance and to facilitate easy working and to increase the service life of the hoses, the following points are of utmost importance to the machine builders and maintenance mechanics.

1. While connecting two elements through a hose, it will be better if excessive taut in the connected hose is avoided.
2. Some slackness is to be permitted to avoid strain to the hose as it is noted that when pressure is applied to hose, the hose tends to bulge and decrease in length. Elastomeric hoses fastened between two fixed points are found to have about 5% change in length.
3. Similarly enough slackness is also helpful to avoid kinking of the hose. It should be noted that only the hose and not the fitting is flexible. Hence the hose run should be long enough to bend during the machine movement.
4. Hoses should never be installed with a twisting as too much twisting may weaken the hose structurally and may also cause to loosen the fittings.
5. Sometimes to avoid long loops, a number of fittings may have to be used. However one should not forget that more number of fittings may be detrimental to the system performance as lot of heat may be generated due to excessive number of line fittings.

6. High temperature may damage the hose cover and may shorten its useful life. Hence under hot working condition it is better if the hoses are protected through proper heat insulation by using fire sleeves or only heat resistant hose material.
7. Design and installation of the hose run should be such that rubbing and abrasion with the metal members of the system are avoided. Clamps are very often used to group the hoses and keep them away from the moving parts. Hose guards are also used for this purpose.
8. During installation of hoses in a hydraulic system it is important to take into consideration the manufacturer's instructions and recommendations. The hose manufacture's guidance and specific instruction should be given maximum importance with regard to bend radius, environment and oil compatibility, pressure capability, etc.

14.7.1 Correct Hose Installation Procedure

The installation of hoses should be made correctly as shown in Figs 14.9(a) to (i). Most hoses have a longitudinal bonding mark on its outer cover as shown in Fig. 14.9 (a). The hose should never be turned or twisted along this line. It will be always better if instead of a straight connection between two fixed points the hose is given an amount of sag at angle $\alpha = 15°- 45°$ as per need a sharp bend or curve should be avoided while installing a hose as shown in Figs 14.9 (b) and (c). The hose should not be kinked and snapped as shown in Fig. 14.9 (d). The hose should be installed such as to have the least change in flow directions of the fluid as shown in Fig. 14.9 (e). The length of the hose (L) as shown in Fig. 14.10 should also be measured accurately for which the following formula can be used.

$$L = 2S + \pi R + H, \text{mm}$$

where L = length of hose, mm
  S = straight length of the hose end including the length of the hose fitting. The value of S for various nominal diameter of hose is given below in Table 14.10.
  R = the minimum bend radius of the hose, mm
  H = amount of hose travel, mm.

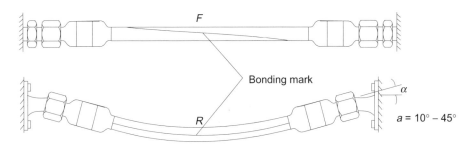

Fig. 14.9 *(a) Right method of hose installation*

466 Oil Hydraulic Systems: Principles and Maintenance

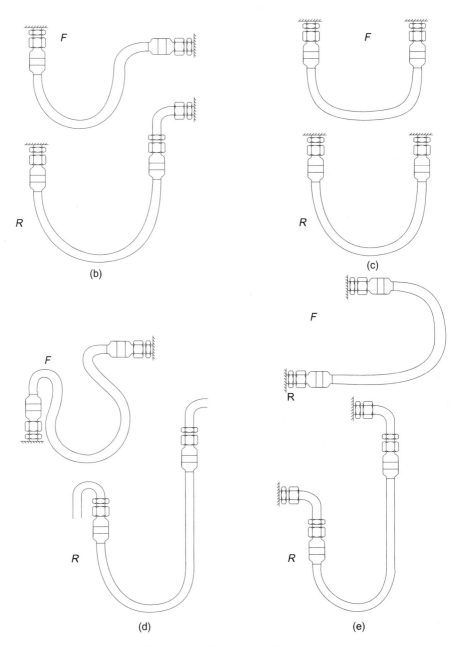

Fig. 14.9(b to e) *Right method of hose installation*
R = Right α = 15° – 45° F = False or incorrect

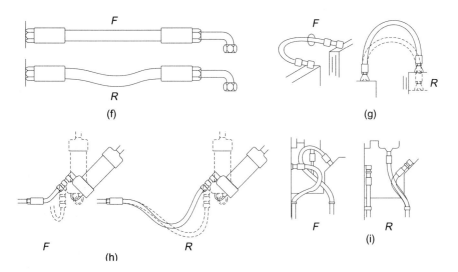

Fig. 14.9(f to i) *Right method of hose installation*

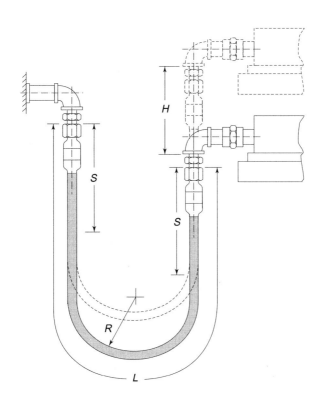

Fig. 14.10 *Calculation of hose length (L)*
*[Courtesy: Hydraulische Anlagen, Mr. Leiber Wolfgang, Germany,
C.H. Verlag, Munich, p. 33]*

Table 14.10

Value of Nom.-dia in mm	4	6	8	10	13	16	20	25	32	40	50	60
S in mm	150	175	200	230	260	300	340	380	430	480	530	580

14.7.2 Making a Ferrule Assembly with a Pipe or Tube

The correct assembly procedure of a ferrule is shown in Fig. 14.11. As shown in Fig. 14.11 (a) the pipe has to be cut at right angles and the inside and the outside have to be deburred lightly. The wall edge should not be cut bevelled.

Pipe cutters will bevel the edges. In such a case the edge will have to be filed to remove the level. The nut to be fitted to the pipe as well as the cone of the body has to be lubricated thoroughly with oil as shown in Fig. 14.11 (b). Grease should not be used. The nut and then the ferrule should be slided over the pipe as shown in Fig. 14.11(c). It should be ensured that the ferrule is not fitted backwards—if so, the ferrule will not make the necessary seal. The biting edge of the ferrule must face towards the interior cone of the body.

If it is impossible or very difficult to push the ferrule over the pipe end, the ferrule should not be opened out. This will ruin the joint.

Smaller diameter pipes may be fitted direct into the couplings if these are screwed into their ports. First tighten the nut by hand until one feels it binding on the ferrule i.e. finger tight as shown in Fig. 14.11 (d). Now press the pipe against the stop in the inner cone and tighten the nut with a spanner as shown in Fig. 14.11(e) by about 3/4 turn make sure that the pipe is not turned round with it. Once the ferrule grips the pipe firmly it is no longer necessary to press the pipe. Finally the nut is further tightened by another 3/4 turn—the ferrule now bites perceptibly into the pipe to the desired depth and pushes up a visible collar in front of the cutting edge as shown earlier in Fig. 14.4(a).

14.7.3 Bending Radius of Steel Tube

While installing a steel tube, the bending radius should be given an adequate amount in order to facilitate easy flow of the oil. In Fig. 14.12, three cases of tube bending radius are illustrated. Between the tube fitting and start of the bending, it is better to have the tube to be straight enough otherwise it will be difficult to install the tube perfectly with the fitting free of leakage. The tube should be bent into a perfect right angle but not with a very small radius [as in Fig. 14.12 (b)]. The bending radius shown in Fig. 14.12 (a) is very large, not at right angle and so it is difficult to have proper connection. In Fig. 14.12 (b), the radius is very small which will result in a higher pressure drop. There is also danger of pipe burst with such a sharp bend. The bending radius shown in Fig. 14.12 (c) is most ideal. The straight portion of the pipe before the start of the bend, should be 2 times the height of the connecting nut as shown in Fig. 14.12(d). The bend radius should be

Hydraulic Pipes, Hoses and Fittings 469

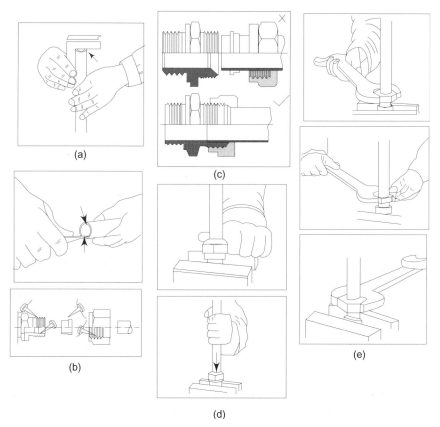

Fig. 14.11 *Correct procedure of tube/pipe assembly.*

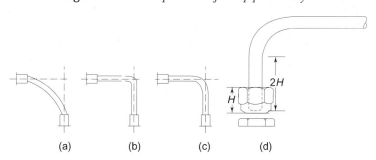

Fig. 14.12 *Bend radius of steel tube*

at least 3 times the pipe/tube outside diameter. In Figs 14.13 (a), (b) the correct method of pipe connection has been illustrated.

14.7.4 Hose Assembly Procedure

Hose assembly is generally made in two ways:
  (a) Hose fitting that is permanently attached to the hose.
  (b) Hose fitting that can be taken off the hose and re-used, i.e. push-on type.

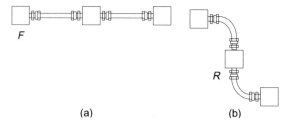

Fig. 14.13 *Correct way of pipe connectin R—right, F—false*

All hose fittings (except push-on type) have three basic components, i.e. an insert, an outer shell and a coupling end. The outer shell pushed down on the insert, clamps the hose between itself and the insert. For higher pressures the outer shell and the insert will have higher wall thickness and length to provide longer larger area.

But the reusable type consists of two components, i.e. an insert or a stem and a threaded sleeve or socket. The assembly is done by first inserting the hose into the socket with a right hand thread provided on the stem. The stem OD being larger in size compared to the hose ID results in enlargement of the hose ID compressing it against the wall of the socket. This type of assembly is not suitable for critical applications as it is not tamper proof and may cut the inner tube while assembling the same.

14.7.5 Permanent Assembly

A hose with its end fitting permanently attached is shown in Fig. 14.14 (a). These assemblies may be made in the shop floor or in the field site by using a crimping or swaging machine. This machine can apply a crimping force through a fluid power operated machine containing a die to form the assembly. A crimping machine is either available, as a portable mode or as a floor type installation. A simple, fast and easy field assembly of hoses in shown in Fig. 14.14(b).

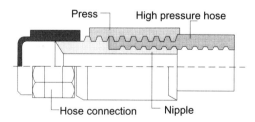

Fig. 14.14 *(a) Permanent hose fitting*

14.7.6 Method of Permanent Assembly

Crimping While making an assembly with a crimping machine, the following points may be observed. A press with eight dies that circle around the coupling, forces the die onto the shell. The press pushes the shell down with a predetermined

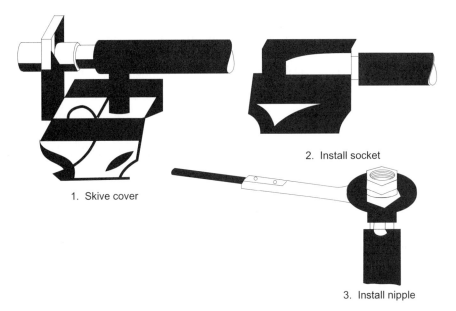

Fig. 14.14 *(b) Fast and easy field assembly of reusable hose*
[Courtesy: M/s Aeroquip, USA]

force on the shell to enable it to embed firmly the serrations into the wire. For critical applications with possibility of shock load and surge pressure this method is more dependable from the point of reliability. Even with variation in hose OD the shell can be pushed down to the hose reinforcement to have the maximum holding capacity. A few points may be noted here

1. At the beginning an exact length of the hose is to be cut and it is ensured that the hose ends are cut in a perfect square.
2. Then clean the hose of all dust, dirt and other components.
3. Ensure that the hose, the fitting and the ferrule are correctly positioned in the die.
4. The fitting is then inserted into the hose. Before inserting the fitting it is better to lubricate the inside of the hose end and stem of the fitting. In some cases a ferrule is to be placed over the hose end.
5. The assembly is to be placed into a correct size collet and crimping pressure is then applied.
6. The assembly is now ready and is inspected for its proper dimension and defects if any.
7. The assembly is then tested at proof pressure and used in the system.

14.7.7 Swaging

Another method of hose assembly is also used which is termed swaging. Here a fitting, which has been onto the hose, is forced through an orifice between two dies. With the hydraulic pressure the shell is forced down to the reinforcement wire of the hose. However this method is not very reliable as due to variation in hose tolerance throughout its length, the coupling either will be pushed down too far and cut the wire if the hose OD is on the high side or in case of low side hose OD, the coupling may not be forced enough to have the maximum holding ability. Hence hose assemblies for low pressure applications only could be made using this technique.

A provision may be given for locking the site, to the sleeve after crimping or swaging so that the insert does not blow off under pressure. Certain types of hoses having thick outer cover have to be skived before inserting the sleeve for better grip.

14.8 DESIGN OF END FITTINGS

Design and manufacture of end fitting and hose assemblies should be given due attention as it is very often seen that excellent hoses fail miserably due to faulty design, inefficient crimping or poor swaging. The following features are most important for a perfect hose assembly:
1. The assembly should be able to provide positive lock with the reinforcement.
2. Full flow characteristics must be attainable.
3. Connection with the machine components must be leakproof.
4. Should resist atmospheric corrosion.

Types of end couplings The coupling end is that part of the fitting which is attached to the component.

There are innumerable types of end couplings available in the hydraulic system but certain broad varieties are mostly used.
They are:
(a) Female swivel
(b) Male swivel
(c) Male rigid
(d) Flange connection.

Suitable bend nipples are also provided to match individual equipment design. The common screw threads used are
(a) BSP
(b) UN or UNF
(c) Metric and
(d) NPT

SAE in its series J 506 gives the complete details of end fittings including flanges for its various types of hoses.

In Section 14.2.7 we have discussed certain salient points on pipe threads. Let us discuss here a few more points. We know that pipe threads are identified by the size of the pipe on which they are machined whereas pipe sizes are denoted by the nominal bore and hence there is no direct correlation between pipe size and the tapering dimension of those pipe threads (inside diameter of pipe size). The standard thread for hydraulic fittings are pipe threads which is cut on a taper of 3/4″ (19 mm) per ft. It is also the standard taper for the dry seal taper pipe thread. But tapered pipe threads were never intended for high pressure hydraulics with pulsation and pressure surge. The tapered pipe thread is a one time connection and is incapable of reducing leakage. The NPT (National Pipe Thread) used for common pipe work needs the use of a sealant. Here the tops of the threads are truncated and the grooves are cut in a very sharp V. Sealing is effected by packing sealants into the gaps between the mating threads. In the NPTF (National Pipe Thread Fuel) the thread tops are cut very sharp and sealing is ensured by crushing of the mating threads. Teflon tape is very often applied over the threads as a sealant. Teflon has a lubricating effect which ensures to overtorque the mating threads to the point of cracking female parts.

14.8.1 Hose Fittings

A variety of hose fittings, adapters, and couplings are very often used in a hydraulic system. A maintenance mechanic or a machine builder must have sound knowledge about their various engineering or application aspect in order to enhance the line efficiency with regard to leakage losses, heat generation, ease of maintenance and line servicing, etc. It may be mentioned that for ease of connecting or dismantling a line quick connecting couplings are very often used in hydraulic as well as pneumatic systems. Let us discuss here the important fittings and other related matters used in conjunction with hydraulic hoses. Some hose fittings are shown in Fig. 14.15.

Coupling end The coupling end is that part of the fitting which is attached to the hydraulic component, a pump, valve, or a cylinder port.

There are many varieties of coupling ends used.

Male pipe coupling end Male pipe coupling with NPTF (National Pipe Thread Fuel) dry seal threads may be very effectively used for low pressure systems to install the hose assembly into the element port, the hose or component needs be turned into the port and tightened using a spanner. A swivel union adapter may be used where the component parts cannot be turned. In mobile hydraulic systems the hose uses a swivel female NPTF adapter.

JIC female swivel Most commonly found on all hydraulic applications this type is a 37° coupling end which is furnished with straight ends, and with 90° and 45° bends. The 37° female end can be turned into a 37° adapter without turning the entire hose. An important advantage is that it has a metal-to-metal flared seat which, when tightened, will provide a leakproof seal. These are most suitable when longer and heavier hoses are needed to make connections.

45 deg. SAE swivel This swivel is the same as the 37° flare JIC swivel, except that the flare set is at an angle of 45° instead of 37°. This fitting is used primarily on automotive applications, the threads are interchangeable and a 45° flare fitting

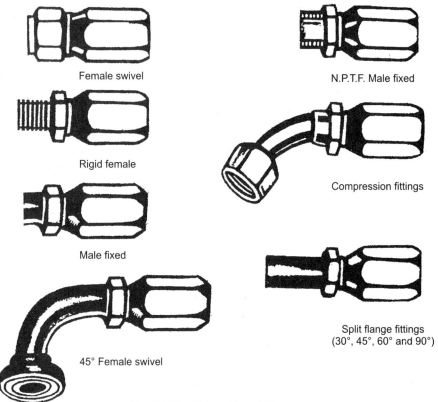

Fig. 14.15   *Types of hose fittings*

can be used with a 37° adapter or in cases of extreme emergency, a 37° fitting can be used with a 45° adapter.

Flange connection   This fitting seals with an O-ring, which acts as a gasket under pressure with the four bolt flanges threaded into the port. They are capable of providing high pressure leakproof connection and can be installed in a confined area having limited access to the components. While making the connection one can position the flange ends and bring the split halves up to it, using a socket wrench to install the bolts. They are available with straight ends, or bent ends to any required degree.

Male-O-ring straight-thread fitting   Known also as an O-ring boss fitting, it is available with an O-ring boss gasket but it is not necessary to use flanged halves and bolts to attach the end of the straight-thread fitting to a port, as done with a flanged fitting. The fitting has straight threads which thread into the port, and are available mostly in straight, 90°, 45° bends.

Inverted flare fittings   These are used for low pressure and are found primarily on automotive equipment. Generally they are used in conjunction with copper tubing.

Male swivel fitting This type is similar to the male pipe end, with the exception that the fitting will swivel into the port so that the hose assembly or component part does not have to be turned. Sealing is done in two parts—on the thread ends and with an O-ring instead of the fitting.

14.9 QUICK COUPLING

Couplings are precision components, engineered for specific uses with exact dimensions and close tolerances. A better understanding of their design, construction, and operation would give a greater appreciation of their advantages. There are a variety of applications in modern industrial plants for quick connect (QC) couplings both for pneumatically operated tools such as nut runners, etc. as well as other fluid power equipments which can be connected rapidly to their power source to permit wide versatility for various production needs, for instance, in connecting or disconnecting a tractor and its hydraulically actuated agricultural component.

QCs make changes simple, do not require additional hand tools, take little time, and do not require the help of an additional trade or skill. They are devices which permit the rapid connection or disconnection of fluid conductors, generally without the use of tools. QCs have two halves: a female half, often called a coupler, socket, or body, and a male half, commonly referred to as a nipple, plug, or tip. Basically, QCs differ in their valving arrangements, locking mechanisms and sealing principles. A schematic diagram of various quick connecting couplings is shown in Figs. 14.16 (a)–(d).

14.9.1 QC Valving

Most QC couplings are manufactured with three valving arrangements straight through, single, and double shut-off. The straight through type has no valving in the plug or the socket and hence this is normally used with external valving to block the fluid flow before the QC is disconnected. Because there are no elements to restrict internal flow passages, pressure drop in these QCs is also minimal. However, lack of valving limits their use in fluid power applications.

Single shut-off QC couplings, are often used with applications where fluid spillage is not important. This type of QC has one shut-off valve, usually in the socket half. The socket then is installed in the upstream or pressure side of the connection to permit shut-off when the coupling is disconnected. This type normally is used for pneumatic applications and should be avoided for hydraulic systems.

Double shut-off QC couplings, have valves in each coupling half. These QCs are used where fluid spillage is to be avoided, as in hydraulic systems.

14.9.2 Type of Valves

QC valves are designed to block fluid flow when the coupling assembly is disconnected. This stoppage is done manually or automatically, depending on

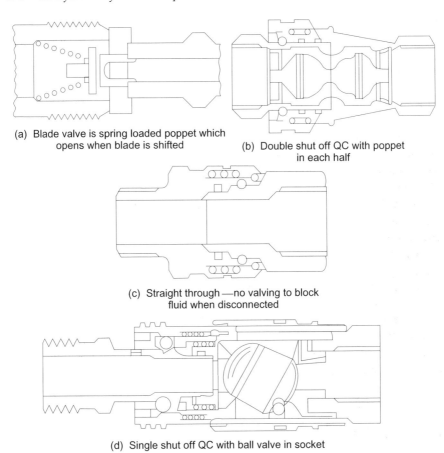

(a) Blade valve is spring loaded poppet which opens when blade is shifted

(b) Double shut off QC with poppet in each half

(c) Straight through —no valving to block fluid when disconnected

(d) Single shut off QC with ball valve in socket

Fig. 14.16 *Types of quick connecting couplings*

valve design. Manual shut-off often uses a rotating lever that cams the valve closed, or manual movement of a sleeve operated valve.

A variation is the type which releases the plug when the springloaded sleeve is moved in one, or in some cases, either direction. Other connecting designs include sleeves which twist to lock, sleeves which must align with pins before they can retract for disconnect, and flanged plugs pinned in place.

14.9.3 Seals

QCs usually have elastomer or metal seals which may be dynamic or static. These seals must be compatible with the fluid media and with anticipated media temperatures. The seal's basic function is to prevent leakage between the two mating halves of the coupling, between a valve and its seat and between component parts of the QC.

14.9.4 QC Materials

QC couplings are most often made of carbon steel, stainless steel, brass or aluminum. Carbon steel is the least expensive and has the highest pressure capabilities. Stainless steel is used for corrosive media and has good pressure capabilities. Brass couplings are for water and air applications and rank third on the scale of pressure capabilities. Aluminum couplings are used mostly in aerospace and air vehicle applications because of their light weight.

14.10 RIGHT SIZE OF HOSE

The selection of hose diameter is a complex subject. The efficiency of hydraulic power transmission depends on two important parameters, i.e. flow rate and fluid pressure. If pressure is a limiting factor in a hydraulic system, more power can be transmitted if the flowrate is increased which means that the fluid conductor diameter should be adequate to keep pressure low and heat generation to a minimum. We know that recommended flow velocity of oil in the pressure line is generally 2 m/s to 4 m/s. We also know that at lower velocity of oil, the flow is laminar. With higher velocity, the flow starts becoming turbulent resulting in loss of pressure, generation of heat and loss of efficiency. Another important fact is that a low viscosity oil tends to become turbulent. We must also keep in mind that viscosity of oil decreases with the heat generated by turbulent flow. This can only be avoided if a high viscosity index oil is used. Hence when one wishes to select a correct size of hose, varieties of other factors have to be considered, e.g. smaller fluid lines cost less and provide compactness of the instalment. Function of the hydraulic circuit in which the hose is to be used is another factor. Suppose the line which conveys the pump output to the tank through open center direction control valve, there may be every possibility of pressure loss and heat generation may result and a designer may have to go for a hose diameter which will give less fluid velocity and less pressure loss. Again the velocity will be decided on the natural exchange of heat and oil cooler used for heat dissipation in the system.

If a portion of a hydraulic system is not operational on a continuous basis, a higher flow velocity can sometimes be chosen by the designer as the overall heat generation may not be of much significance. In case of intake hose diameter, the heat generation is not a big problem as the fluid velocity here is taken very low say 0.5 m/s only. If the pressure of the pump intake is below atmospheric pressure, there is possibility of cavitation of pump resulting in pump wear which is undesirable. On the other hand if a pressurized hydraulic tank is used in the system, much higher fluid velocity could be used at the pump inlet. Nomograms could be used to determine the hose diameter if flow rate and flow velocity are known as mentioned earlier.

Sizes for flexible hoses are usually decided on the recommended flow rate velocity as mentioned below:
Pressure line–2–4.5 m/s.
Pump intake line–0.5–1.2 m/s.

Return line–0.5–1.2 m/s.
The hose bore size can be estimated from the following formula:

$$d = 1.127\sqrt{\frac{Q}{v}}$$

where Q = flow rate
 v = recommended fluid velocity.
 Q_1 v and d are in consistent unit.

14.11 HOSE SELECTION CRITERIA

The following factors are generally to be considered while selecting a hose for a specific purpose.

Table 14.11 Criteria for Hose Selection

Factors	Points to note
1. Pressure	Surge pressure is to be given due consideration otherwise hose life may get shortened.
2. Temperature	Hose material should match the system as well as ambient temperature.
3. Fluid compatibility	Selection of hose material should be done taking into consideration fluid compalibility.
4. Flow capacity	Pipe diameter, losses etc to be considered.
5. Size	Size of hose to be selected such that pressure loss is bare minimum and physical damage to hose due to turbulence and heat generation is avoided.
6. Length of hose	Possibility of sag or over stretching to be avoided.
7. Type of end fittings	Line losses, art of fitting etc.
8. Environment	Certain hose material may get affected due to exposure to salty water, ozone chemicals, heat and ultraviolet light, etc. Hence these factors are to be taken care of while selecting hose material. Eco friendliness of the material is also to be given due consideration.
9. Degree of vibration	Generally hoses are capable to absorb shock.
10. Abrasion resistance	Due protection has to be provided to the hose against failure due to abrasion, erosion, snagging and cutting of hose cover.
11. Flexibility	If a hydraulic hose is to be as flexible as possible, the thickness must be minimum and the material of which it is made must permit bending to a very small bending radius.
12. Shock and mechanical load	Excessive flexing twist, kinking, tensile and side loads, vibration must be avoided.
13. Safety factor	Generally bursting to working pressure the safety factor is 4:1 where shock load is expected, for service it is 2.5:1.

14.11.1 Other Factors

Surge pressure Most hydraulic hoses have the ability to expand and shrink with rise and fall of internal pressure. This tends to create pressure surges and water hammer which is very often heard in high pressure hydraulic systems with rigid plumbing. This tends to damage the pipes. A synthetic hose will be able to dissipate the pressure surge and minimize the water hammer. So if a length of pipe or tubing fails occasionally from either hydraulic shock or vibration, replacing it with a hose may be the correct solution. The bend radius and the type of flexion will have a bearing on hose life.

Hose material The material of the hose must be compatible with the system fluid.

Protective cover Suitable arrangements must be made to guide or protect the hose while it is moving, so as to prevent it from being damaged or kinked.

14.11.2 Speed, Pressure, Temperature and Hose Longevity

It is always to be kept in mind that in general a flexible hose is weaker than rigid tubing. However recent technological development has made it possible to use them at a higher pressure and temperature condition. However for transfer of hydraulic power apart from the system pressure, a definite amount of flow rate is necessary to carry within a specific timeframe.

The cross-sectional area of the inner tube determines the flow velocity. Due to friction and contraction of the inner tube there is possibility of pressure loss. The resultant heat along with high flow speed may damage the inner tube material. To avoid such damage in the hydraulic power transmission system, the flow speed should be limited to 6 m/s and the extreme limit of oil speed should never be above 9 m/s. The working temperature influences the longevity of a synthetic elastomeric hose. Though the permissible range of temperature for such hose materials is between 40°C to + 100°C, for earth moving machineries using high pressure hydraulic system with fire resistant oils of group HFA and HFC, i.e. oil in water emulsion and water-glycol exposed to high pressure for a longer duration of time the range of temperature will be better if it limits to 10°C and 60°C. The longevity will be much unfavorable if this is associated with high pressure peaks and high temperature which is schematically explained in Fig. 14.17.

14.12 RELIABILITY TEST FOR HOSES

The hydraulic hose is an important element in the system which needs to be checked for its reliability and durability during its manufacture. The International Standard Organization advocates the following tests to ensure this:

1. *Visual examination*—To check any visual defects.
2. *Dimensional check*—to ensure correctness in length, concentricity in lining, accuracy in linear and diametrical dimensions.

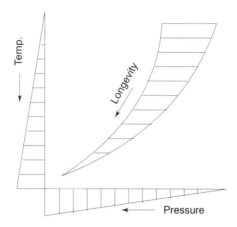

Fig. 14.17 *Influence of pressure and temperature on hose-longevity*
[Courtesy-Oel Hydraulik U. Pneumatik, 1st January, 1987, Germany]

3. *Leakage test*—To ensure perfect hose and end fitting assembly to avoid any leakage.
4. *Cold flexing test*—To maintain high level of the hose material at very low temperature and cold condition.
5. *Oil absorption test*—To ensure that the material of the hose cover and linings are not chemically affected by oil.
6. *Ozone resistance test*—To determine the level of ozone resistance of the hose cover.
7. *Proof test*—To estimate the capability of the hose to withstand sustained system pressure the hose is subjected to.
8. *Burst test*—To ensure that the hose is capable of withstanding the minimum stipulated burst pressure without fail.
9. *Impulse test*—To ensure that the hose is capable of withstanding the impulse/oscillating load or surge pressure which very often is encountered in high pressure hydraulic system.

For this test a specified length of hose is bent with a minimum bend radius. The angle of bend is 90° or 180° for hoses up to NB 25 mm and 90 for size above 25 mm. The test fluid should flow at a specified temperature. The surge pressure is created at a rate 30 to 100 cpm. The hose should withstand 200,000 cycles without failure or leakage.

The impulse/oscillation load or surge pressure is very often encountered in high pressure hydraulic system. This is one of the most critical tests for hose and is carried out under specific conditions as per ISO–1436: 1978 and SAE J343 d. A schematic test curve is shown in Fig. 14.18. The impulse pressure should fall within the area shown by the broken lines and should conform as closely as possible to the curve shown. The pressure rate in bar/s or kpa/s is calculated and the impulse estimated between two points at 15% and 85% of the gradient.

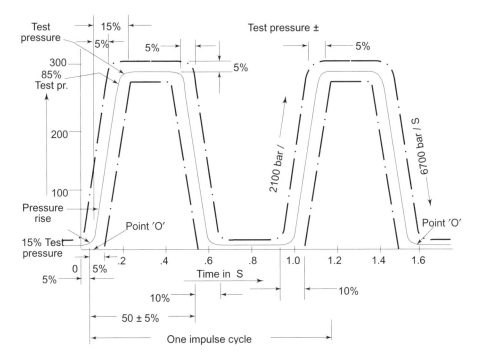

Fig. 14.18 *Impulse test curve of hose*
[Courtesy: M/s Dunlop India Ltd., India.]

From such experiments it has been seen that the rate of pressure fluctuation could be between 2000 and 7000 bar/s. Hoses for high pressure system should therefore be selected after such dynamic behaviour test of hoses only.

14.13 GUIDELINES FOR PIPE AND HOSE MAINTENANCE

Maintenance Even with proper selection and installation, hose life may be significantly reduced without a continuing maintenance program. Frequency should be determined by the severity of the application and risk potential. A maintenance programme should include the following as a minimum.

1. Pre-installation inspection Prior to installation, a careful examination of the pipes, hose, all fittings must be performed. All components must be checked for correct style, size and length. In addition, the hose must be examined for cleanliness, ID obstructions, blisters, loose cover, or any other visible defects.

2. Minimum bend radius Pipes and hoses should not be installed at less than minimum bend radius as this may significantly reduce their life and efficiency. Particular attention must be given to preclude sharp bending at the hose/fitting juncture.

3. Twist angle and orientation Hose installation must be such that relative motion of machine components produces bending of the hose rather than twisting.

In many applications it may be necessary to restrain, protect or guide the hose to protect it from damage by unnecessary flexing pressure surges and contact with other mechanical components. Care must be taken to ensure such restraints, do not introduce additional stress or wear points.

4. Proper connection of ports Proper physical installation of the hose requires a correctly installed port connection while ensuring that no twist or torque is put into the hose.

Proper installation is not complete without ensuring that tensile loads, side loads, kinking, flattening, potential abrasion, thread damage or damage to sealing surfaces are corrected or eliminated.

5. System check out After completing the installation, all air entrapment must be eliminated and the system pressurized to the maximum system pressure and checked for proper function and freedom from leaks.
Note: Avoid potential hazardous areas while testing.

6. Visual inspection Any of the following conditions require replacement of the hose.
  (a) Leaks at fitting or in the hose (leaking fluid is a fire hazard)
  (b) Damaged, cut or abraded cover (any reinforcement exposed)
  (c) Kinked, crushed, flattened or twisted hose
  (d) Hard, stiff, heat cracked or charred hose
  (e) Blistered, soft, degraded or loose cover
  (f) Cracked, damaged, or badly corroded fittings
  (g) Fitting slippage on hose.

7. Periodic tightening of hose (pipe fittings) The following items must be checked, tightened, repaired, or replaced as required from time to time for which a preventive maintenance schedule has to be proposed.
  (a) Leaking port conditions.
  (b) Clamps, guards, shields.
  (c) Remove excessive dirt build-up.
  (d) System fluid type, and any air entrapment.

8. Running test Operate the system at maximum operating pressure and check for possible malfunctions and freedom from leaks as well as vibrations.
Note: Avoid potential hazardous areas while testing.

9. Replacement intervals Specific replacement intervals must be considered based on previous service life, government or industry recommendations or when failures could result in unacceptable downtime, damage or injury risk.

10. Hose storage Hose products in storage can be affected adversely by temperature, humidity, ozone, sunlight, oil, solvent, corrosive liquid and fumes, insects, rodents and radioactive materials. Storage areas should be relatively cool and dark and free of dust, dirt, dampness and humidity.

14.14 PIPE/TUBE PREPARATION FOR INSTALLATION

When installing various iron and steel pipes, tubes and line-fittings of a hydrosystem, it is necessary that they should be absolutely clean, free from scale and all other foreign matters. The following steps may be useful

1. Pipes, tubes and fittings are to be brushed with suitable tube wire brush or cleaned with other pipe cleaning apparatus. The inner edge of the tubes and pipes should be bevelled and reamed after cutting to remove burrs.
2. Rust and scale may be removed from shorter pipes and tubes by sand blasting. But necessary care should be taken so that the sand particles do not stick to blind pockets after flushing. Larger pipes are not to be sand blasted for this reason.
3. The longer pipes and tubes are generally to be pickled in a suitable solution until all rust and scale are removed. Preparation for pickling requires thorough de-greasing in trichloroethylene or other de-greasing solution.
4. The pickling solution is to be neutralized.
5. The parts are to be rinsed and prepared for storage with end sealing, etc.
6. Tubes must not be welded, brazed or soldered after assembly as proper cleaning will be difficult. It must be accurately bent and fitted so that it will not be necessary to spring it into place.
7. In case of flange connections, flanges must be squarely fitted on the mounting faces and be secured with screws of correct length.
8. Care must be taken to ensure that no metallic particles stick to the threaded fittings and also metal silvers from the threaded fittings should not enter the hydraulic system.
9. While filling the reservoir with oil use the proper filtering screen. The system must be operated for a short time to eliminate air in the line. Add hydraulic fluid if necessary.
10. Chemicals used in the cleaning and pickling operations are to be kept only in the proper containers and handled with extreme care to avoid any danger.
11. Before unplugging the stored length of tube, ensure that the seal is in good order and clean round this before removing the same. Remove the seals from both and examine its bore to ensure that it is bright and clean.

484 *Oil Hydraulic Systems: Principles and Maintenance*

Remove hygroscopic paper if present. Remember that it is easier to clean a tube before it is bent than afterwards. If necessary, pull a leather through the bore to remove any dirt particles.

12. No hot work is to be done to avoid scale formation.

14.15 USE OF TEFLON TAPE

The purpose of the dry-seal taper pipe thread is to keep hydraulic systems clean by removing the need for pipe dope which invariably finds its way into the fluid. Sometimes Teflon tapes are wrapped over the threaded pipe or fitting as shown in Fig. 14.19(a).

14.15.1 Pipe Anchoring

Hydraulic tubings may be subjected to vibration or heavy shock loads due to pressure pulsation. When pressurized the material of a free length of a pipe is in tension and hence anchors are necessary for about every 1 m of tubes. A common clamp as shown in Fig. 14.19 (b) may be used (made either of wood or metal).

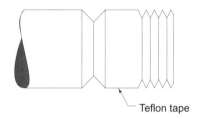

Fig. 14.19 (a) *Use of teflon tape*

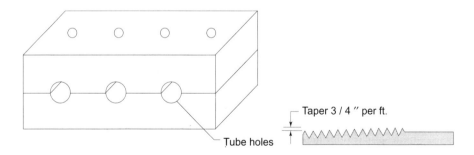

Fig. 14.19 (b) *Pipe anchor*   **Fig. 14.19 (c)** *Standard taper dry seal*

14.16 EFFECT OF FRICTION ON PIPES/TUBES/HOSES

One source of energy loss arises from friction of flow against the containing walls or from friction within the fluid due to kinetic effects. Friction increases as the speed or velocity of the hydraulic fluid flow increases. Friction is reduced in a hydraulic circuit when the work of moving the liquid from the pump through the control valves to the actuators and return to the tank is made easier.

The higher the flow rate higher is the friction loss. It is worth noting that flow can be laminar or turbulent. Laminar flow takes place at lower velocity and turbulent flow at higher velocity. Laminar flow is preferable for control applications. Friction increases rapidly when the flow becomes turbulent and the power used to overcome friction is wasted as heat. Near the critical velocity (the velocity range in which laminar flow changes to turbulent flow) laminar flow can only be obtained when flowing in straight, smooth bore pipes. If pipes are subject to bends or changes in flow area, laminar flow is hard to achieve and friction losses are high. With laminar flow in a given length of pipe the effect of doubling the flow rate is to double the friction, hence pressure loss takes place. If the flow is turbulent the friction varies roughly as the square of the velocity and the loss would be nearly 4 times as great the loss in case of laminar flow over the same length.

It is common in hydraulic circuit design to use a maximum flow velocity of 3 m/s to 4.5 m/s. For common types of hydraulic mineral liquids and at normal operating temperature flow at these velocities will usually be laminar. The frictional resistance to oil flow may be reduced with shorter lines, fewer fittings, longer radius bends and good plumbing.

14.17 STANDARD THREADS ON TUBES

The standard thread for hydraulic fittings are pipe threads which is cut on a taper [shown in Fig. 14.19(c)] of 3/4″ per ft. It is also the standard taper for the dry seal taper pipe thread.

Though hydraulic installations work at fairly ambient conditions, some allowance may have to be made for expansion (specially for long pipes—say 1 to 1.5 m) due to temperature changes. Steel pipings are mechanically strong and can be used as its own structural support. Hence hydraulic valves can be mounted in the pipe line without being attached to a fixed frame work in many cases. Care should be taken so that connecting threads, particularly taper threads, are not subjected to pipe deflection forces which will eventually cause a leak. Hydraulic pipe work however should never be used for supporting a heavy load.

It is generally best to clamp hydraulic pipes individually to a fixed point. As a general guide, a pipe will need support every 50 to 60 diameters of its length (i.e. a 25 mm pipe for every 1.25–1.5 m)

14.18 EASY SCREW-TOGETHER REUSABLE ASSEMBLY

Screw-together reusable hose fittings for spiral wire hose can be assembled by hand with a few simple tools or with one of different reusable hose assembly machines. A machine from M/s Aeroquip is shown here.

14.18.1 FT 1013 Portable Hose Assembly Machine

The Aeroquip FT1013 machine is designed to speed low production assembly of Aeroquip reusable fittings. It can efficiently handle the assembly of – 4 (1/4″) to – 40 (2–1/2″) single-wire braid hose and – 4 (1/4″) to – 32 (2″) double-wire braid and 4-spiral-wire hose. All styles of Aeroquip fittings can be assembled including elbow fittings.

The machine is ideal for shop use when bench mounted. When used with its detachable leg supports, this light-weight, compact machine becomes very portable.

Since the FT1013 is simple to operate, any employee should be able to run it after a short period of instruction. For additional information, request IEB 177A.

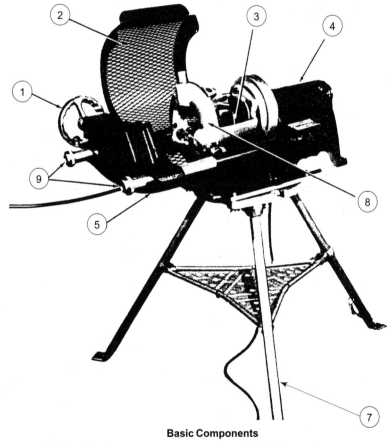

Basic Components

1. Hand wheel
2. Safety guard
3. Pot chuck
4. Power unit
5. Vise
6. Foot switch
7. Stand
8. Jaw control mechanism
9. FT1281 socketing attachment

Fig. 14.20 *Hose assembly machine*

Features

- The machine is supplied with a 110V or 220V, AC-DC 1/2 HP reversible universal motor; as the speed decreases, the torque increases.
- A three-way reversing snap-type safety switch. The switch cannot be turned on unless the safety guard is down.
- A quick-acting wrench-type rotating chuck with center hole large enough to chuck a spiral socket.
- Chuck runs at 36 rpm.
- The machine can be used to disassemble hose and fittings by simply reversing the assembly process.
- It is 45" high, 56" long and 46" wide, and weighs 170 lbs (approx. 75 kg).

Review Questions

1. (a) State the common materials used to manufacture pipes and tubes for hydraulic system.
   (b) How do you specify a hydraulic tube?
2. Give your comments on copper as a high pressure hydraulic tube material.
3. What is a flare fitting? Why is flaring needed and how is it done?
4. What is NPTF thread? Can it be used on hydraulic tubes? What is 1S0 metric thread?
5. (a) What is meant by "equivalent length"?
   (b) State the reasons of energy loss in a pipe.
   (c) How does one estimate such losses in a hydraulic pipe?
6. What is Reynolds number? What is its critical value? What happens when the Re is above or below the critical value?

References

14.1 Cooke HM and PL Martin, "Hydraulic Hose Fittings: Methods of Connection" *Hydraulics & Pneumatics*, M/s Penton IPC, USA (Sept. 1977), p. 94.
14.2 *Oil Hydraulik in Theorie und Anwendung*, F and D Findeisen, Schweizer Verlagshaus A.G. Zurich, Switzerland.
14.3 Leiber Wolfgang, Hydraulische Anlagen Carl Hanser Verlag, Munich, 1966 (Germany), p. 33.
14.4 Nagraj, N, "Loss of Energy of Fluids During their Flow in Pipes & Pipe Fittings", *Fluid Power*, FPSI, Bangalore, India June, 1988.
14.5 Technical Catalogue of Hyd.-Air Engg. Pvt. Ltd., Plot C7 St.No. 22, Marol Industrial Area, Mumbai, p.11.
14.6 Technical Catalogue of Insap Flexibles & Engineers Pvt. Ltd., 134, A.N. Street, Chennai, India.
14.7 "Fluid Power Directory & Hand Book 89", FPSI, Bangalore, India, p. 177.

14.8 Szabo, M, "Do it yourself permanent hose assemblies" (Ed), Hydraulics und Pneumatics, Penton IPC, USA, Oct. 1972, p. 90
14.9 Dipl. Ing. J. Reichel, Essen, "Auswahl und Einsatz Von Hydraulik schlauchleitungen und Leitungs Verbindungen in der Leistungshydraulic", Oel-hydraulik. Penumatik, January, 1987 Post fach 2760 D-6500 Mainz, Germany.
14.10 Singaperumal, M. (Ed.), "Proceedings of the Short Terms Course on Industrial Hydraulics," IIT, Madras, India.
14.11 Catalogue of ENOTS, Lichfield, UK.
14.12 Catalogue of Aeroquip. 1225W, Main street, Van wert, USA.
14.13 Catalogue on Pulselife High Pressure Hydraulic Hose, Dunlop India, Kolkata.
14.14 Hildebrandt Hann, H, "Annforderungen an Konstruktion und Fertigung von Hydraulik schlauchen und Schlau chteitungen aus der sicht des Herstellers", *Oel hydraulik und Pneumatik*, 1986, Germany, p. 423.
14.15 Paire, W.W., "How to select the right size hydraulic hose", *Hydraulics & Pneumatics*, USA, Dec. 1977. p. 55.

15
Hydraulic System Maintenance, Repair and Reconditioning

MODERN hydraulic systems have become highly sophisticated and increasingly complex operating at much higher power and speeds with greater accuracy. They find application in various fields of modern engineering. The failure of these systems therefore may not be desirable at all. When the system is out of order there is non-utilization of machines and resultant loss of earnings. In order to enhance the operational reliability and maintainability of the hydraulic system, shop floor managers have to take up various maintenance and repair measures which may ensure trouble-free functioning of the system. Any system may develop various types of faults and defects which need periodic and routine maintenance. An older system may even need reconditioning.

In a hydraulic system curing a fault alone is not enough. Fault diagnosis is important as this establishes the cause which could reoccur if left uncorrected during the rebuilding process. Replacement of a faulty pump with a new one may apply correction, but the new pump may also fail if the cause of the initial fault is not treated. This may either be relatively simple where the fault is a fairly obvious one, or may need careful analysis and study of the system as a whole to isolate the primary cause from secondary effects. The repair or maintenance technician should be able to understand the constructional and functional features of each individual element, its application in the hydraulic circuit and should be able to relate to the overall behavior of the system as a whole. A thorough knowledge of the hydraulic circuit diagram is an added tool. This is an important asset to the management in these days of ever increasing hike in oil prices and its effect on the maintenance, trouble-shooting, repairing and reconditioning of the oil hydraulic system.

To solve problems in a hydraulic system one should be conversant with the constructional details and functional features of the system. It is also equally important to have an adequate knowledge of the intricacies involved in the design and manufacture of individual components as well as the design of the complete system. Equally important is the basic knowledge of the hydraulic circuit diagram along with its control system. It is important to be analytical in the method of diagnosis in case of a system malfunction or a component failure. While simple malfunctioning can be corrected through slight adjustment of the component or control parts, for major repairs the system needs to be stopped for inspection, fault diagnosis and repair. Reconditioning and overhauling may also be needed if the machine is quite old and the faults can not be rectified through routine repairs. At this stage it may be necessary to decide whether one has to go for reconditioning and when.

Reconditioning of machine tools and equipment is defined as a planned and systematic engineering activity to restore the equipment to its original condition. To recondition a hydraulic system, it is essential to have adequate knowledge of the principles of design and functional performance of the various elements and components of the system. The decision for taking up a hydraulic system for reconditioning is arrived at after studying the following parameters:

(a) Performance of the machine tool utilizing the hydraulic power.
(b) Performance of individual hydraulic components.
(c) Break down and repair/replacement analysis of the system.
(d) Economics of reconditioning versus procurement of a new machine or elements. Reconditioning cost (inclusive of all cost factors) of up to 25% to 40% of the price of a new system is generally accepted
(e) Availability of critical spares and parts of the system.

It has been found that in hydraulic systems less repair is required for control elements and oil distribution mechanism than for the power generation unit. A majority of the hydraulic systems as well as hydraulically operated machines in our country are still important and it may be necessary that while repairing or reconditioning, many firms and organizations may have to prepare the parts by themselves, as parts from original suppliers may not be readily available.

15.1 COMMON FAULTS IN A HYDRAULIC SYSTEM

Before going for repair or reconditioning of a hydraulic system it is important to know the type and nature of defects commonly found in these systems. The most common type of defects are as follows:

(i) Reduced speed of travel of machine tool elements
(ii) Sharp noise in the system
(iii) Steep rise in the oil temperature

(iv) Non-uniform or jerky movement of tables, carriages, etc. especially at low feed rates
(v) Slow response to control
(vi) Excessive leakage in the system
(vii) Excessive loss of system pressure
(viii) Cavitation of pump
(ix) No supply or less supply from pump
(x) High rate of seal failure
(xi) High degree of contamination level of system media and poor oil life.

Figure 15.1 shows comparative failure rates of some of the common hydraulic elements.

It is important to note that the above faults may be caused by many factors that are interrelated. Once the causes are established, remedial action becomes easy during breakdown maintenance or during the system reconditioning period. The decision to recondition a system will depend on the individual condition of the unit members, their interrelated behavioural patterns, and hydraulic performance of the system as a whole. Reconditioning of the system should be considered when the machine tool using hydraulic power is no longer capable of functioning at the rated capacity. Apart from the conditions mentioned earlier, the prime consideration for rebuilding of the system should be centered around:

(i) Whether the hydraulic power generated is effective in transmitting the energy at its related capacity.
(ii) Whether the right type of spare parts or components/elements are available in the plant or in the market.

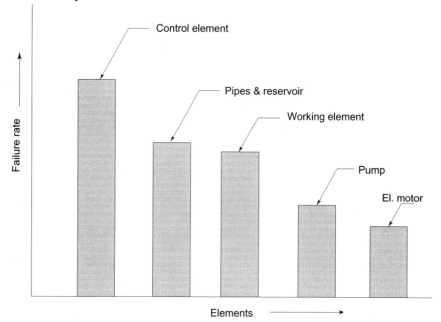

Fig. 15.1 *General failure rates of common hydraulic elements*

(iii) Whether necessary super finishing or polishing techniques like honing, lapping, etc. are available with the plant.
(iv) Whether the hydraulic test rig is available in the plant to check the performance of the individual characterisitics of the system after or before reconditioning.
(v) Whether trained fluid power personnel like technicians and supervisors are available to carry out the reconditioning work.

It is always advisable to replace a hydraulic element rather than rebuild it if the degree of damage is higher. But care must be taken to procure the right type of element compatible with the system. To take a clear-cut decision of procurement of correct spares like control valves, cylinders, seals and packings, filters, etc. it is essential to have a detailed list of spares and the hydraulic circuit diagram. It will be helpful if the functional diagram is also enclosed. Many a times it happens that the machine tool manufacturer may not be in a position to supply the correct item due to change in design or other reasons. It is necessary in such cases to obtain the essential spares or information from the hydraulic component manufacturer. It is advisable to maintain a stock of all the necessary hardware and critical spares like seals, rings, filters, etc. It should be ready before the hydraulic system is taken up overhauling or reconditioning. A properly organized routine or preventive maintenance system will certainly be advantageous while rebuilding the system as most of the components can be reused with or without modification. In reconditioning, as a rule, all bearings, O-rings and other seals, filters, etc. are to be discarded and replaced with items procured from standard suppliers as per the original specifications.

15.2 PROCEDURE FOR REPAIR

As already discussed, the hydraulic system consists of certain basic items connected together by pipes, hoses and other line fittings. While repairing or reconditioning, each item may require individual care and attention but the maximum attention is probably centred around the pumping system. The pumping unit comprises of the pump, electrical motors, couplings, oil reservoir, heat exchanger, etc.

15.2.1 Reservoir

While rebuilding a tank care must be taken to retain the shape and size of the tank (generally 2 to 3 times the pump flow rate per minute) and material thickness (generally not below 3 mm thick for minimum structural rigidity), vent holes, drain plugs, baffles, etc. After fabrication, both the inside and outside of the tank should be painted with a good quality oil compatible paint. The choice of a suitable paint may be strictly limited in case the reservoir is to be used with synthetic fluids. Oil resistant paint would be quite unsuitable in such cases. For better cooling, de-aeration and contamination settling, baffles, and magnetic filters are to be placed in the proper position. Correct pipe line geometry and positioning will help to have better suction.

During routine maintenance an oil reservoir should be checked for contamination and cleaned out if need be. While cleaning, no wool or cotton waste should be used. Only pre-filtered oil as per system design grade should be used. While refilling the tank care is to be taken to see that the receptacles or gadgets used for filling oil are clean. The fluid should be examined for the presence of water, undesirable chemicals and particulate foreign materials. While filling the tank after cleaning, the oil level should be brought to its maximum limit and free passage of air must be granted. While refilling, one may use a portable pump unit as shown in Fig. 15.2.

Fig. 15.2 *Use of a portable pump unit to draw oil from the drum*

In practice it is better to flush the system with a flushing oil which may be removed after flushing is over. The system must be flushed for at least half an hour to one hour with the pump operating at optimum speed. The reservoir is then filled with the specified oil and flushed again by running the pump set at a slower speed. Once flushing is over the oil level of the reservoir should be checked again and topped up if necessary.

15.2.2 Inspection of Oil

The oil should be periodically checked in a hydraulic system by drawing an appropriate oil sample from the drain plug just after the system is shut down and preferably until the oil is at its operating temperature. The oil sample can be kept in a test tube for sometime during which time any water present in the oil will settle at the bottom. If the oil is found to be turbid, it implies that it contains water. If a drop of oil containing water is put on a hot plate the water evaporates with a popping sound or crackle indicating the presence of water in-excess of 0.1 to 0.2%. The color of the sample can also be compared with the original color of the oil kept in a separate test tube by placing both side by side. The presence of foreign particles and oxidation of oil may produce slight darkening of the fluid. If a drop of such contaminated oil is placed on a piece of blotting paper, it will produce a dark patch at the middle due to presence of sludge whereas fresh oil will show only a bright yellow patch. Further tests may identify sludge materials.

Oil viscosity While checking the oil, its viscosity and viscosity index (VI) number should be checked from time to time. The preferred operating range of viscosity (kinematic) of a normal petroleum-based oil should be 17 to 40 cSt at 50°C. However the viscosity range may change from system to system depending on the pump and system compatibility.

Change of oil The system oil should be changed after 15,000 to 20,000 hours of working or one year, whichever is earlier.

15.2.3 Suction and Return Line

It is better to arrange return and suction lines as far away as possible. The lines should terminate below the oil level. Suction line joints above oil level should not permit intake of air. Air leaks into the system can be detected by air bubbles in the reservoir which may induce aeration of pump and resultant pump failure due to cavitation. Excessive aeration may result in noise.

15.2.4 Oil Property

To minimize water evaporation, water based fluids should not be used above a temperature of 65°C. Water loss may affect viscosity. Emulsions experience a decrease in viscosity whereas viscosity of water-glycols increases with reduced water content. Low temperature operation may be also a problem for water-based oils as they may freeze at low temperatures.

Phosphate ester oils can be used at slightly higher temperatures than petroleum oils but their low temperature properties are poor. Sometimes low capacity heaters of 0.7 to 1.4 watts/sq.cm may be used in the reservoir to maintain the temperature.

Selection of a proper fluid is quite important for ensuring quick response and sensitivity of the system. Apart from transmitting power, the oil should have:
1. Adequate film strength
2. Sealing capacity to close clearances between moving parts against leakage
3. Minimization of wear and friction
4. Adequate lubricity to protect sliding surfaces

5. Resistance to chemical change
6. Prevention of rust and corrosion
7. Prevention of oxidation, etc.

During the working life it is to ensure that oil retains its property all through.

15.2.5 Simple Visual Checks of Oil

The following characteristics are likely to change during service life of oils:
1. Visual appearance
2. Smell
3. Water content
4. Solid content
5. Foaming
6. Acidity
7. Viscosity
8. Tardiness
9. Oxidation

Readers may note that in modern hydraulic systems, continuous monitoring of oil is a must if one wishes to maintain a failsafe system. However such monitoring may be quite costly. It is therefore advisable that the following visual checks may at least be undertaken during routine maintenance of hydraulic systems in order to determine the oil health.

15.2.6 Tests of Hydraulic Oil

The initial tests to be run on hydraulic oil or any other oil are sensory tests like:
1. Appearance
2. Color
3. Odor

Further tests are carried out when one finds that the above physical parameters start changing after a considerable period of working life of the oil. At such a point of time the analytical test of oil should be done as follows.

Table 15.1 Simple Visual Checks for Oil in Hydraulic Systems

Appearance	Possible cause	Recommendation
1. Clear oil		No action needed
2. Oil hazy or opaque	Possible contamination of oil	Oil needs change and find out cause of contamination
3. Separated water layer inside tank	Excessive water contamination above emulsification limit.	Permit system to settle, drain off water settled at the bottom of tank.
4. Oil darkened	Oxidation of oil may be the major cause	Check acidity and viscosity and change oil
5. Separated deposits	Dirt, worn out particles or oxidation products	Filter to examine quantify of deposit. Check for oxidation

(*Contd.*)

Appearance	Possible cause	Recommendation
6. Bad odor or smelly oil	Oil may be oxidized and acids might have formed	Check for oxidation and change oil if needed
7. Foam in oil	Poor suction, entrapment of air etc.	Check suction line and oil level in tank

(a) Appearance The oil may be inspected visually to see if it is clear and bright or hazy and cloudy which may indicate water contamination, suspended material or foam, etc.

A frothy or milky oil indicates that it is aerated and contains more air than the maximum of 8% by volume. In such a situation, the system may be kept idle—long enough to allow the oil to settle inside the reservoir. If even after this, it is not found to be useful, one may try to homogenize the water in the fluid. An oil sample can be collected and boiled. If not much change is observed, there is no alternative but to change the entire oil from the reservoir.

(b) Colour Though color is a relative matter one can inspect the oil visually to see if the oil color is same as the original product or not. Oxidation may be suspected if the oil is darkening in color. A dark color may also indicate excessive contamination. However physical parameters of oil like appearance and color may be dependent on the type of service to which the oil is subjected.

(c) Odour This is another comparative test of oil. Used oils normally may have a bland or oily odor. Oils that have been oxidized have 'burnt' or pungent odours.

While testing the oil the following precautions should be observed:
(i) Ensure that the sample to be tested is the correct sample only.
(ii) Ensure that the non-representative oil is not tested.
(iii) Ensure that the sample does not get contaminated further before the test is conducted (e.g. due to dirty container).
(iv) The test should be run properly.

15.2.7 Oil Life and Fluid Temperature

In Fig. 15.3, a graphical representation of the effect of temperature on system performance as well as oil life expectancy has been illustrated. It may be mentioned here that reservoir oil temperature should never be allowed to cross 40.0°C for optimum system performance. However the operating temperature is generally 20–45°C. It is noticed that with every 1°C rise in temperature above 48.8°C there is perceptible change in oil viscosity and resultant loss in life expectancy of oil. It may be noted that the temperature of the system is one of the primary factors which influence the fluid viscosity. The acceleration of oxidation may change with more rapidity with abrupt changes in working temperature resulting in undesirable development in oil degradation as the oxidation inhibitors are used up. This naturally affects oil life to a great extent as may be observed from the table below.

Table 15.2 Ideal Working Temperature and Oil Life

Type of system (1)	Working condition (2)	Working temperature (3)	Oil life (4)
1. Industrial situation	(a) Ideal	Under 40°C	Very long—over 20,000 working hrs.
	(b) Normal	Between 20°–45°C	Moderate life from 15,000 to 20,000 working hrs.
	(c) Acceptable	55°–65°C	Reduced oil life
	(d) Acceptable with inter-cooling	65°–80°C	-do-
	(e) Maximum limit without inter-cooling	95° C	Very much reduced life. Straight mineral oil not suitable.
2. Very high temperature	—	Up to 140°C	Short life. Straight mineral oil not suitable.
3. Hydrostatic	normal	Up to 80°–90°C	Better to use oil viscosity within 10 to 30 cSt at working temperature

15.2.8 Measurement of Temperature

The most acceptable and accurate test would be to measure the temperature using proper instruments. If no gauge or thermometer is available, one can feel the temperature by touching the surface of the pipe or tube by hand.

48 to 50°C (usually safe temperature)—The pipe will be uncomfortably warm.

55°C or above—One may not be able to grip the piping firmly for more than a few seconds.

Some common sources of heat are:
1. Partially full reservoir
2. Inadequate cooling
3. Heat exchangers not working properly
4. Restrictions/obstructions in pipes and fittings
5. Improper oil viscosity
6. Relief valve setting too high
7. Relief valve may be sticking
8. Control valves in 'off' position
9. Worn out pump, etc.

More details are discussed later.

498 *Oil Hydraulic Systems: Principles and Maintenance*

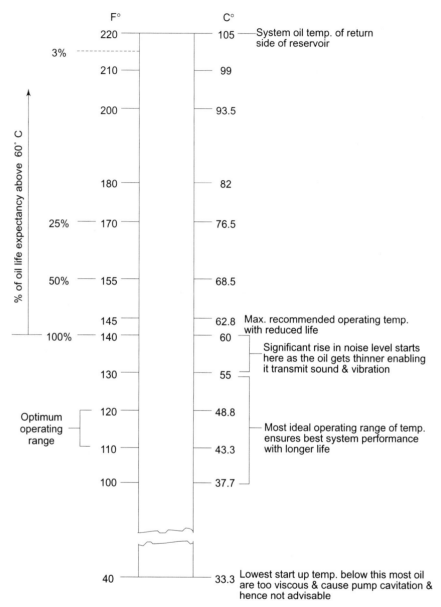

Fig. 15.3 *Graphic representation of effect of temp. on oil*
[Courtesy: Fluid Power Maintenance by Dr. Ned R. Stull, Hydraulics and Pneumatics, USA]

15.3 CONTAMINATION

We have discussed earlier the various effects of contaminants. In order to maintain a healthy system, the hydraulic system needs to be kept contaminant free as far as possible. It is equally important to minimize chances of entry of contaminants.

15.3.1 Causes of Contamination

Contaminants can enter or be generated in a system in a number of ways:
1. They will be present in a new system because of insufficient cleaning of component parts, and because of contaminants generated during manufacturing and/or assembly.
2. Fluid used to fill or replenish a hydraulic system may be contaminated.
3. Wear particles are constantly being generated by an operating system.
4. Contaminants are formed by chemical breakdown of the hydraulic fluid, and the interaction between some hydraulic fluids and water.
5. Contaminants are generated whenever cavitation occurs in a hydraulic system.
6. Accidental presence of water in the system may corrode steel piping and other components. The resultant rust particles can flake off and enter the compressed air system.
7. The system may ingest airborne contaminants particularly through air breathers and cylinder rod seals.
8. Contaminants are introduced when a system is opened and closed during servicing and/or maintenance, or during the installation and replacement of parts and components.

Apart from this, contaminants can enter also into a hydraulic system through
1. Reservoir cover for suction and drain line
2. Breather or oil filling opening
3. Leaks in the pump suction line
4. Piston rod packing of cylinder
5. Cleaning solvents and cutting fluids
6. Air borne dust and dirt
7. Sand particles from component casting
8. Grease, lint, metal chips
9. Fragmental gasket material

A hydraulic system should be thoroughly cleaned before start-up with optimum filters as per the manufacturer's recommendations. Magnetic plugs, filters, screens need to be examined frequently during the first few days of starting up the system.

15.3.2 Problems Caused by Contamination

Contaminants in the fluid can:
1. Accelerate component wear, decreasing system performance and service life.
2. Result in sluggish operation and cause moving parts to seize.
3. Score finely finished surfaces and damage seals resulting in leakage.
4. Prevent valves from sealing properly, again causing leakage and loss of control.

5. Act as a catalyst to accelerate hydraulic fluid oxidation and breakdown thereby shortening fluid life and reducing the useful operating temperature range of the fluid.
6. Oxidation of oil which can also cause corrosion throughout the system with change of fluid viscosity and formation of gums, sludges, and varnishes.

15.3.3 Contamination Controls

There are many ways to help reduce the entry and detrimental effects of contaminants in a system.
1. Plumb the system with pipes, tubing, and fittings that are reasonably free from rust, scale, dirt, and other foreign matter.
2. Flush the entire hydraulic system, preferably with the same type of fluid to be used, before normal system operation is begun. Ideally, flush with hot fluid at as high a velocity as possible.
3. Filter all hydraulics before using, to minimize introducing contaminants into the system.
4. Provide continuous filtration matching with the contaminant sensitivity of the system to remove products of wear and corrosion as well as sludge generated during operation. Locate filters where they will provide the most protection.
5. Provide continuous protection from airborne contamination by sealing the hydraulic system, or installing an air filter/breather on unpressurized reservoirs.
6. Clean or replace filter elements on a routine basis, or when filter change indicators show that the element needs cleaning or replacing.
7. Maintain fluid viscosity and pH level within fluid suppliers' recommendations.
8. Minimize sources of water entry into the hydraulic system.
9. Select cylinder rod seals which have the best ability to exclude external contaminants.
10. Consider use of fittings that avoid generating contaminants into the fluid such as the SAE4 bolt flange.
11. Avoid introducing thread sealants into the fluid stream.
12. Consider using non-corroding piping.

15.4 COMPONENT FITTINGS AND FAILURE DUE TO CONTAMINANTS

It has been observed that the level of maintenance of hydraulic oils in service is quite inadequate. The following factors as listed here may be responsible.
1. Inadequate awareness of cleanliness and failure to set the right quality standard
2. Poor housekeeping, wrong storage and inappropriate handling of oil
3. Poor knowledge of filter and filtration techniques
4. Inappropriate preventive maintenance.

It has been indentified that a hydraulic system may lose its usefulness due to obsolescence (15%), accidents (15%) and surface degradation (70%). The major reason for surface degradation is again corrosion (20%) and mechanical wear (50%). Mechanical wear takes place mostly due to abrasion, fatigue and adhesion. Abrasive wear is generally found to take place due to particles between two moving surfaces. The particles generated as a result of abrasive wear are work hardened. They are found to be harder than the parent material and if not removed by proper filtration may cause additional wear resulting in premature system failure. Erosion wear is caused mainly due to high velocity impinging on a component surface. This is generally found to be a common cause for the damage of flow control valves used for speed control of hydraulic system. Smaller particles are found to be trapped between two sliding surfaces causing failures due to spall, dents, cracks, etc. Silt produced due to such effects may cause failures due to:

1. Increased internal leakage and undue heat generation.
2. Slower cycle time resulting in loss of productivity.
3. Lack of adequate and quick response.
4. Sticky or silted valves such as servo-controlled direction control valve.

The dynamic clearance between mating surfaces may have lot of influence on the mechanism of system failures. Some typical dynamic clearances are shown for common hydraulic components.

Table 15.3 Clearances for Common Hydraulic Components

Component	Clearance in μm
1. Gear pump	
Tooth to side plate	0.5 to 5
Tooth tip to case	0.5 to 5
2. Vane pump	
Vane sides	5 to 13
Vane tip	0.5 to 1
3. Piston pump	
Piston to bore	5 to 40
Valve plate to cylinder	0.5 to 5
4. Servo valve	
Orifice	130 to 450
Flapper wall	18 to 63
Spool to sleeve	1 to 4
5. Actuator	50 to 250

Contamination is a seed which breeds further contamination and forms the nucleus of a subsequent chain reaction which may create problems that are catastrophic in the long run. Hence service engineers have to be very conscious on this issue and be careful to attack contamination related failures at the very beginning. The present trend is towards development of finer filters to arrest finer elements, ensure trouble-free and extended life between two consecutive servicings and permit appropriate replacement of elements before failure. In case of high production machines it is of paramount importance to change the filter element in time so that down time can be kept to a minimum. Indicators on filters are also useful for detecting the contamination level and help in taking timely action. Magnetic traps should be used to trap very fine particles inside the reservoir.

15.4.1 Methods of Measurement of Contamination Levels

For the measurement of particulate contamination levels the following techniques may be used:
1. *Visual methods* such as patch test as described earlier.
2. *Silting index* which is a measure of decay of the contaminated fluid flow through a 0.8 micron membrane and hence in the particulate contamination range 0.5. SAE ARP 788 furnishes more details.
3. *Gravimetric analysis* which is a measure of fluid contamination by weight evaluation in a unit volume of the fluid.
4. *Particulate counting technique* with an electronic liquid particle counter which is very fast and accurate.
5. *Spectographic analysis* which can be used to identify airborne dirt, repair debris, worn out metals, corrosion product, coolant water solution, etc. from their ash residue.
6. *X-ray defraction* which can be used to determine the chemical nature of crystalline sediment, deposit, etc. after separating the solids from the oil.
7. *Infra-red analysis.*
8. *Ferrography.*

15.4.2 The Mechanism of Oxidation

Mineral-based hydraulic fluids tend to produce sludge or resin deposits due to oxidation if the system is subjected to high temperatures coupled with excessive agitation, oil-turbulence and infected with a conglomeration of various particulate and other contaminants. The process of oxidation gets accelerated in the presence of air, moisture and certain metals like zinc, copper, etc. and is also dependent on time. Entrained air has been found to generate localized over-heating and cause an unevenness in the thermal condition of the system which may aggravate the situation further. The moisture content of the air forms an emulsion which may circulate all through and come in contact with the metallic parts of the components and may interact with the fine film of metaloxides on their surface thus producing hydrates of these oxides. These hydrates react with the oil-borne organic acids and produce soap which acts as a catalyst in the entire process of oxidation to form gum or resins.

15.4.3 Effects of Oxidation

(i) The deposits tend to settle on valve spools or bore walls thereby generating a sticky or sluggish spool movement.
(ii) The sluggish spool movement may impair the system efficiency in the form of poor valve response or induce time lag in the valve sequencing.
(iii) The resins may deposit on or around smaller or finer bores/holes reducing oil flow and sometimes blocking it completely.
(iv) Oxidation affects the emulsifying property of the oil.
(v) De-aeration of pressurized air from the oil may also be affected.
(vi) Sludge or resin may clog small diameter, slender and longer pipings, specially those below 2 mm and thus pose a regular problem for maintenance.
(vii) Certain by-products of oxidation may induce corrosive property in the oil.
(viii) Oil viscosity may get affected.
(ix) Cooling characteristics of the oil may get hampered.
(x) Oil life expectancy is greatly lowered enhancing maintenance and replacement cost.

15.4.4 Preventive Actions

This problem can best be tackled if some rules are observed:
1. Avoid or minimize heat generation in the system like heat due to pump, pressure relief valve, throttling, etc. and also eliminate other possibility of localized heating in the system or reservoir.
2. Stop all passage of air entering into the system by arresting undesirable leakage points through glands, fittings, pipings, suction lines, etc.
3. Avoid use of copper and zinc coated pipes, tubes and other components specially for high temperature, high pressure systems.
4. Ensure that the seals, component materials and oils are compatible with one another.
5. In case of resinous deposits on the oil passage clear the passage by imparting axial oscillation to either of the mating parts at the passage at moderate frequency and amplitude. Similarly mild vibration may be induced in the pipings and components to clear the blockage followed by an appropriate flushing operation.
6. Resinous deposits on valve spools or similar items should never be scratched for fear of surface damage.
7. The best precaution is to replace the entire oxidized oil well in advance to save the system components as soon as one suspects oxidation.

15.4.5 Problems in Contamination Control and Contamination Monitoring

We have seen that contamination is a major problem in the effective operation of a hydraulic system. Three basic models of contamination related failures have been identified for hydraulic components.

(a) Transient
(b) Catastrophic
(c) Degradation

The total loss is enormous in terms of money and time due to fluid contamination and resultant wear. Periodic monitoring of contamination levels in the fluid at critical stages is of utmost importance to prevent degradation or failure of the components. Various types of electronically operated hydraulic contamination monitors are available nowadays in the market which can be easily used. These monitors fitted with sensors and samplers can be very effectively and conveniently used to count or measure the concentration of contaminants. The counter has provision to compare results with standard reference contamination data and provides necessary alarm when the concentration limit is crossed.

Various computer controlled electronic contamination monitors are available in the market. The portable particle counter sold in India sometime earlier, is one such gadget. With this gadget it is possible to do online dynamic sampling and actual particle counting is possible at 420 bar pressure as per ISO/NAS Standards. The equipment is generally computer controlled having RS 232 interface connection with scrolling memory capacity. Apart from the contamination monitor, condition monitor equipments of various make are also available for simultaneous measurement of flow, pressure, temperature and peak pressure. This type of equipment is handy and compact with finger tip control for online monitoring with high repeatability.

15.5 FILTER AND FILTER MAINTENANCE

The filter used in a hydraulic system should be subjected to periodic and routine cleaning and maintenance. For doing so it is better to draw a weekly, monthly and annual schedule taking into account filter cleaning, change of filter element, etc.

15.5.1 Filter Life

The life of a filter in a hydraulic system depends primarily on the system pressure, level of contamination and nature of contaminants. The following points are essential to change the life cycle of the system.
  (i) Estimate filter element life on an hourly basis and change filters when found necessary.
  (ii) Use pressure gauges across the filter to measure the differential pressure and indicate a change period.
  (iii) Coupling a clogging indicator to give a visual or audible warning when the element becomes clogged and the by pass is opened.

Filter life decreases with oil degradation. Any marked reduction in filter life should call for an oil analysis. The life of oil depends on the grade employed and the system characteristics. Any high quality hydraulic oil properly matched with the hydraulic system should have a life of more than 15000 hrs. In any oil sample cloudiness will indicate possible water contamination. Workshop test for oil may include hot-plate test, patch test, etc. Dark oil colors indicate oxidation and contamination while lightening of the color may indicate water contamination. Oil viscosity test should be a regular routine check.

In order to enhance filter life, the following action may be carried out on a regular basis.
1. Inspect the filter for damages, tear-ups, collapsing, etc.
2. Frequency of filter cleaning and filter inspection needs to be decided.
3. Ensure good house keeping.
4. Throwaway type of filter elements like filter paper should be meticulously replaced as per a pre-determined time frequency.
5. When filter elements are washed out or replaced, the filter bowl should also be cleaned.
6. Filter element needs to be cleaned initially after 50 hours of operation and later quarterly or earlier as per the schedule specifically drawn.
7. Contamination indicators fitted with a filtration unit should be checked on a regular basis.
8. Clean dispensing type of gadgets and tools should be used as far as possible, while handling oil.
9. Ensure the quality of oil and contamination level before putting fresh oil in case the oil drum is a fresh supply from the store.
10. For topping up of oil reservoir it is better to use a portable pump unit as shown in Fig. 15.3.
11. Arrange appropriate training for the maintenance crew to equip them with good servicing skills and habits.

15.5.2 Centrifuge and Electrostatic Liquid Purification

During the last few decades the world has seen tremendous amount of improvement in various branches of science and technology including the science and art of mechanical filters and filtration. In the days of "Zero defect product system" appropriate and quick filtration may play an important role in enhancing cost effectiveness and longivity of a hydraulic system as well as the hydraulic oil. One has, however, to find out how optimization of filtration could be ensured with higher accuracy and rapidity and also whether the conventional filters will be able to cope with such demands these filters having certain inherent basic difficulties. Hence other methods of filtration techniques are being applied by many to overcome problems of normal filters and filtration.

Difficulties in conventional filtration The basic principle of a conventional filter is to separate out the hydraulic oil and worn out and other contaminating particles from the contaminated oil by allowing the same to pass through a porous filter medium. This is a very quick and most effective method of cleaning any liquid preferably the low viscosity oil. However it has been noticed that it is very difficult and also costly to separate very fine particles from a viscous fluid effectively by such filters as reduction of pore size of such filter medium may cause higher oil friction with enhanced tendency to obstruct oil movement through the filter. Moreover the pore sizes in a conventional filter as used for hydraulic oil filtration may not always remain the same in size throughout its working cycle due to progressive blockage of the filtering pores over a period of time. When a filter with higher pore size is installed such as an in-line filter in a hydraulic circuit, it may allow a selectively higher flow rate with less friction initially. But at a later stage due to its working cycle features like unequal pore size, there is every possibility that the pressure oil may generate undesirable drop in pressure and flow as well as pressure pulsation depending on number of factors, e.g. pore size, contaminating particle size, quality of oil and oil viscosity, replacement cycle of filter, working life element, etc. We know that a hydraulic system operates best when the pressure drop in the system is the minimum. Therefore an in-line conventional filter is in essence a self-inductive flow disturbing element in the circuit but which cannot be avoided. The other problem associated with such mechanical elements in the circuit is the possibility of entraining of air in the system during periodic changing of filter element. So naturally the question arises what is the alternative to conventional filter for contamination control?

Though there is no absolute answer to such a question, people have resorted to go for other forms of filtration such as centrifuge and electrostatic oil filtration. These methods have come as a help for better and easy contamination control in the recent times in certain spheres of hydraulic and lubricating filtration.

15.5.3 Centrifuge and Electrostatic Liquid Purifiers

1. Centrifuge In industries people resort to a wide use of centrifugal separation of solid-liquid mixtures. The apparatus used for centrifugal separation consists of a cylindrical vessel made of strong permeable material such as a metallic gauge adequately supported and capable of rotation on a vertical axis. When rotated, the vessel will experience a centrifugal force with the result that the spinning fluid (with solids) while rotating along the vessel will get separated, the solid being swept to the periphery of the vessel and the liquid at the center. Thus the liquid will pass through the gauge and the solids will cling to the bowl wall. The principle centers around the fact that all particles in the fluid will have the same angular velocity as the bowl and the centrifugal force on the particles varies with distance from the center of axis of rotation. It is also to be noted that while separation occurs, the heavier substances will move towards the outside of the bowl displacing the lighter materials. Under a similar analogy we can assume that during centrifuging of two liquids, the heavier liquid will be concentrated at the bowl wall and the liquids will form interfaces separating each other. It is possible to achieve a degree of filtration upto 5–7 microns using this method. There are two types of centrifuges:

(a) Disc type
(b) Tubular type.

In the disc type, the bowl consists of pack of conical discs which divides the liquid into a number of thin layers and thereby increase the separation efficiency.

The action of separation during centrifuging is shown in a De Lavel type of disc centrifuge (Fig. 15.4 a).

The tubular type centrifuge is relatively longer and smaller in diameter (length = 6 × diameter), The working principle is that dirty oil entering at the bottom strikes a horizontal baffle plate and deflects outward due to centrifugal force. In order to achieve the desired centrifugal force at the bowl wall, the tubular centrifuge operates at very high speed say 15000 to 17000 rpm. Such a centrifuge can be used to separate oil, water and solids. The solid contaminants get tightly packed against the cylinder wall which can be removed at convenient intervals. Just before the heavier liquids leave the top of the bowl, they pass through a special ring where the clearances may be varied to suit the difference in specific gravity of the liquids. This arrangement will serve the purpose of a gravity disc in a De Lavel centrifuge.

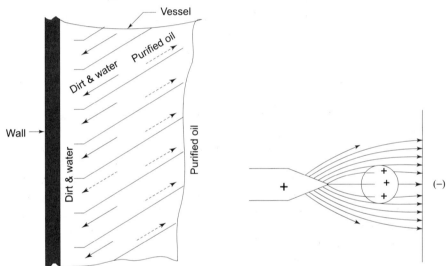

Fig. 15.4 *(a) Action of discs in a De Lavel centrifuge* **Fig. 15.4** *(b) Positively charged particles get attracted to negative pole*

The effective capacity depends upon a number of factors such as oil viscosity, liquid temperature, quantity, size and weight of impurities.

2. Electrostatic liquid purifier/cleaner (ELC) Type and nature of contaminants
We have already seen contaminants in varieties of shape, size, materials, etc. They can be categorized:
(a) According to material e.g. organic or inorganic
(b) According to form e.g. long or short, round or polygon, blunt or sharp, thick or thin, etc.
(c) According to size e.g. large or small from 100 µm to 0.5 µm
(d) According to hardness e.g. hard or soft
(e) According to chemical property e.g. soluble in oil or insoluble under normal condition and with changing temperature
(f) According to electrical property e.g. positive, negative or neutral

With the availability of adequate knowledge and reliable method of estimation of oil contaminants, it is possible today to determine the type and size of contaminants through automatic particle counter. These probes help the maintenance personnel screen off contaminants above the desired value. However it may not be enough to screen off the solid contaminants alone as various laboratory experiments have shown that major problems are created not only by solid metallic contaminants but also by the polymerized products in the oil due to oxidation. However major problem lies in separating the minutest solid particles from viscous fluid wherein electrostatic oil cleaner (ELC) has come as a choice of oil engineers today.

It has been noticed that many of the solid and other contaminants including oxidised products in oil may be very minute in size sometimes as small as less than 5 μm or even less. We know when electric current is passed through a solution having various materials, the particles may be charged as positive, negative or neutral ions. Similarly when the contaminated oil is made to pass through an electrical field, the positively charged particle will be attracted towards the negative pole as shown in the schematic diagram shown in Fig. 15.4(b). In contrast the neutral particles get attracted towards that side of the field where the field intensity is more intensive. The basic theory of electrostatic cleaning is quite old.

Around 1900, for removal of fine particles from fluids (both gaseous and liquids) people started using electrostatic liquid cleaners. Here separation is achieved by means of electric forces established on a practical basis by Cotrell on a theory—"Science of electrostatics and experiments" by Hohlfled.

Electrostatic precipitation is the most effective, economical and technically significant method. Here the electric forces are large and applied solely to the solid particles and not to the oil as being done in typical centrifuges. The electric force for separation of 1 micron diameter particle from a typical hydraulic oil may be as much as 40000 times that of gravity.

In an electrostatic cleaning system contaminated oil is allowed to enter the bottom of a cylindrical vessel. The oil passes upward through a treatment zone which consists of a series of alternate high voltage electrodes and grounded metal discs or collecting electrodes. The suspended particles are removed from the oil by the electric field between the two electrodes and the clean oil emerges at the top. Particles are electrically charged to a high degree by contact with the high voltage electrode and by ion bombardment. The charged particles are attracted to the grounded electrodes which are covered with layers of suitable dielectric, porous matrix such as polyurethane foam, sintered ceramic material, glass wool having as much as 90% void by volume to trap and retain the charged particles. The system is energized from a high voltage DC supply (15 to 30 kV). The particles are removed by manual or other methods and one can achieve filtration of particles ranging from 0 to 100 μm with 90 to 99% collection efficiency.

During electrostatic liquid cleaning additives are not removed from the oil and it does not provide absolute value. The only major restriction is in its application for water in oil. Water reduces the strength of the dielectric and hence ELC should not to be used when the amount of water in oil is more than 500 ppm.

Electrostatic oil purifier removes mainly not only the hard and solid contaminants, but also the soft dirty adhesive products created out of oxidation of oil which is not very easy to separate out by normal filtration process. One need not take out the oil from the machine while cleaning it and the electric charge has no influence on the additive on the oil. This enhances the working life of the machine and reduces the cost of oil radically. Energy conservation and higher machine reliability may be easily achieved through Electrostatic oil purification. A diagram of an electrostatic oil cleaner is shown in Fig. 15.4(c).

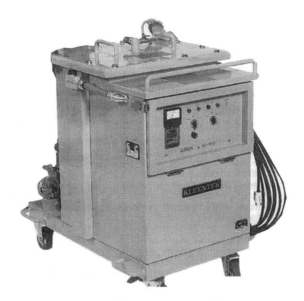

Fig. 15.4(c) *Electrostatic oil cleaner*
(Courtesy: Gadelins GmbH, 6308 Butzbach, Germany)

15.6 PUMP MAINTENANCE

Pump malfunction is one of the major causes for system failure. The general causes of pump malfunction may arise from low oil level in the tank, restriction in suction and delivery lines, air leaks in the suction side, too low or very high pump speed, use of improper oil (e.g. oil viscosity too high), misalignment of pump and motor (e.g. running alignment may differ from static alignment), etc. The decision to recondition a pump is a complicated issue and hence a detailed analysis of its original and present working performance is of vital importance. Analysing pump performance in the system essentially determines the pump-system compatibility. A technician may test a system pump independently to measure volumetric efficiency and pump wear after the pump has been in operation in a system and may compute parameters like noise, vibration, cavitation, etc. at the rated rpm and pressure.

Manufacturers of hydraulic machines very rarely give information in the form of graphical representation of the pump efficiencies at various pressures or rpm.

This may be available from hydraulic pump manufacturers or may be computed in the plant itself. Figure 15.5 illustrates the performance curve of a pump. If plotted properly, this will help to analyse the physical condition of the pump in order to take appropriate action at any point of time for bringing it back to normal working efficiency if it is found to deviate from its normal operating range. The graph given in the above figure shows the characteristics of volumetric, mechanical and overall efficiencies of a pump and the operating range. The pump should be taken up for repairing or overhauling before the efficiency goes down alarmingly at the rated capacity and speed. The operating range shown in the diagram is the best performance zone and any deviation from this should be viewed with caution.

A gear pump may be considered serviceable provided the backlash between the gears does not exceed 0.30 mm, the tooth tip clearances in relation to the housing bore do not exceed 0.08 mm and the side clearances between the gear and end plates do not exceed a total of 0.05 mm. In gear pumps, one-sided wear takes place due to unbalanced pressure subjected to it. The overall efficiency of a worn out gear pump decreases very rapidly resulting in drop of pressure and flow rate. Reconditioning of pump body is worthwhile only when spares or replacement pumps are difficult to obtain. Gears are to be made from case hardened steel of Re 52–58. The non-parallelism between teeth and the hole axis should be within 0.03 mm. The total permissible axial and radial clearance should be within 30–50 microns. Commonly 20°–28° involute teeth gears are used with rolling action only. No sliding action as in transmission gears.

Wear in vane pumps affects mainly the rotor, vanes, discs, seals, etc. The total clearance of the vane in the slots should be within 0.05 mm or less. Worn stator rings may be replaced with hardness of material of nearly Re 60–65 hardness. Manufacturing of new rotors involves considerable difficulties. It is

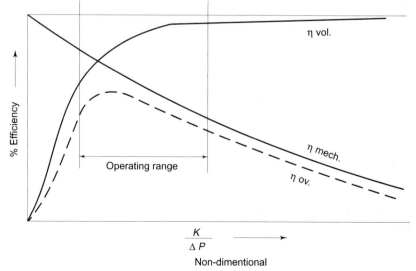

Fig. 15.5 *Performance curves of a pump showing its effective operating range*

advisable to reuse the old one by restoring the parallelism of the slot walls and the worn out journals and end faces. The maximum permissible non-parallelism of slot walls is 0.02 mm. Badly worn out slot walls are machined with a thin abrasive disc and finished by hand. The concentricity of the journals should be maintained within 0.02 mm. Badly worn out vanes are replaced by new ones manufactured/procured to specification.

In piston pumps the overall clearance between the piston and bore on the cylinder block should not exceed 0.01 to 0.02 mm per cm of piston diameter. The finish of the piston should be as smooth as possible. Necessary care should be taken to maintain the parallelism of the piston bore to the cylinder axis. Excessively worn out parts should be replaced. Causes of troubles usually include dirty check valves, stuck pistons, etc. All the internal parts require careful cleaning before reuse. In the case of radial piston pumps, the eccentricity should not be allowed to increase by an amount of 5% of the original eccentricity. Additional eccentricity indicates wear and hence needs immediate repairs.

15.6.1 Pump Cavitation

Cavitation is a common problem found in hydraulic systems which is caused when suction pressure, i.e. the pump inlet pressure is reduced to the vapour pressure of the oil. In such an eventuality, the liquid gets vaporized and a number of cavities may be formed filled with the vapor. If the oil sucked into the pump chamber is not adequate enough to completely fill up the space created by the pump elements, i.e. the rotational elements of the pump, cavitation may take place. The other possibility may be due to admission of air into the system through joints, e.g. pipe joints, pump-shaft sealing point etc. which may result from poor preventive maintenance of the system. Very often pipe failures may be attributed to cavitation which results in:
1. Vibration
2. Damage to bearing due to poor lubrication
3. Over-heating
4. Mechanical damage to internal components of pumps.

15.6.2 Pump Alignment

Mechanical alignment of the pump is to be given a lot of importance to avoid problems relating to misalignment. Pump and motor shaft should be aligned with utmost care and it will be better if the amount of misalignment can be maintained within 175 to 200 microns. Maximum care is also required to use a good quality coupling and mating half of flexible coupling should not be forced on to the pump shaft.

15.6.3 Pump Priming

Proper priming of pump is also equally important. Initial priming may be achieved by jogging a few times. Even a system with a positive suction head needs priming. After initial priming, system oil should be allowed to warm up before the system is run at a specified speed and pressure.

It may be mentioned here that pumps should not be allowed to generate higher speed and pressure from cold start itself as it may load the pump to seize affecting its longevity.

15.6.4 Hydraulic Cylinder

In a fluid power system, the cylinder provides the rectilinear motion of the piston and piston rod assembly. Cylinders are designed as per load, speed and other operating conditions necessary for the system. While reconditioning, the cylinder should be carefully inspected for wear, scoring and pitting. Generally it is seen that the principal wear will be found in the middle as normally only a few jobs require the complete cylinder traverse. As per the permissible wear, the ovality of the cylinder bore must not exceed 0.01 to 0.015 mm per cm of the diameter. Any taper in the bore must not exceed 0.050 mm per cm of the diameter per 300 mm of stroke length. The permissible wear out of roundness and taper of the connecting rod are mostly limited to 0.01 to 0.02 mm. A highly damaged cylinder with scoring and pitting is rebored, honed and lapped using abrasive paste. While reconditioning, the seals and rings are to be replaced as a rule. The tolerance on the piston seal groove diameter is ± 1.5 mm. The tolerance on the width of the piston ring groove is + 0.5 mm approximately. For a honed cylinder a new piston must be matched. To increase the working life of the cylinder, the bore or piston may be chrome-plated. The surface finish of the rod should be within 0.25 μm (micrometer) to 0.4 μm (micrometer). Thin piston rods (below 10 mm dia) are to be replaced. Many fast acting cylinders may be designed with a cushion chamber (if the speed is above 0.3 m/s). The most likely fault with this is leakage developing between the piston hose and cushion collar. A majority of industrial hydraulic cylinders are cold drawn from low carbon steel (up to 0.2% C) with strength of the order of 6000 bar. Drawn cylinder tubes are produced in aluminium, brass and stainless steel. Hot drawn or hot rolled tube is seldom used since this method of production results in insecured mechanical property, low dimensional accuracy and a very poor bore. For large high pressure cylinders, casting is very useful. For cylinders working in a corrosive atmosphere, the outside of the cylinder tube may be protected either by surface coating or by jacketting in a corrosion resistant material.

15.6.5 Valves

All valves in a hydraulic system are manufactured to very close geometrical tolerances. Hence a thorough inspection must be carried out to maintain the same amount of clearances in a reconditioned valve. All seals and balls should be replaced and all springs should be tested and replaced if necessary. The pressure relief valve is most important and faults in a system due to valves are generally attributed to these valves. Worn out valve seatings may be reclaimed by relapping with carborundum paste. Slight scoring of the valve spool may be removed by polishing and occasionally a slightly worn out spool can be reclaimed with a small deposit of chromium plating. Valve spool bores can be corrected by hand reaming, boring out and lapping. In any case it is always advisable to replace the valve if spares are easily available.

15.7 HYDRAULIC SYSTEM MAINTENANCE

Though the hydraulic filter plays a vital role in maintaining the good health of hydraulic system, the following points are worth mentioning here for providing important guidelines to maintenance personnel for proper upkeep of the system and protect it from failure due to contamination.

(i) Ensure that all protective devices, helpful in preventing contamination, in a hydraulic system are maintained in good working condition.

(ii) System designer's and manufacturers' instruction regarding schedule and frequency of preventive maintenance and replacement of devices like filters, seals, breathers, indicators should be adhered to.

(iii) Always use oils which are pre-filtered while filling up the reservoir during starting the system and also during topping up operation.

(iv) Only specified oil recommended by system designer and manufacturer is to be used.

(v) Never use two different oils in the same reservoir.

(vi) Hydraulic oil or the system components are never to be exposed to a dirty atmosphere.

(vii) Flush the system with clean oil only.

(viii) Change of oil and topping up should be carried out as per a pre-determined schedule.

(ix) Oil sample for inspection of oil should be collected from the middle of the oil reservoir and at least 15 minutes after the system is on. Disposable syring may be used for collecting the sample.

(x) Ensure that all components, pipes, fittings, etc. dismantled from the system during repairs properly capped by clean plugs and kept in a neat and clean place so that the chance of dust/dirt entering the system is minimum.

(xi) Leakage of oil from the system should be stopped as quickly as possible.

(xii) It is better to fill the reservoir from an oil drum by means of a portable pump as shown in Fig 15.2. In case containers are used, ensure they are clean.

(xiii) Cotton and jute waste, rags, etc. should never be used for cleaning the hydraulic system and components like spool-bore, spool, threaded parts, oil reservoir interior surface, etc. Similarly hard and soft metallic items should not be used for cleaning.

(xiv) Before painting inside the oil-tank necessary precaution is to be taken to see that the oil is compatible with the paint.

(xv) Before any hydraulic components are dismantled from a system the exterior of the system and the surrounding should be cleaned.

(xvi) After a major component failure it is advisable to service the entire system in order to identify the cause of such failures for future guidance.
(xvii) During servicing it is better to replace all seals. However if the system is under condition monitoring, decision may be taken on the basis of seal condition at that time.
(xviii) Disposable filters should be replaced as per guidelines provided by system designers/manufacturers.
(xix) During shut down one should drain the entire system while the oil is still at the operating temperature. If this is not done there is possibility that solid contaminants may settle out and during the process of cooling certain soluble oxidation products may become insoluable and precipitate.
(xx) Before operating the new equipment under load, the interior of the reservoir should be examined and cleaned manually.

15.8 MAINTENANCE AND PERFORMANCE MONITORING

In any servicing and manufacturing set-up, an effective and timely maintenance of machines and equipment has acquired an important and vital role in minimizing machine downtime. Maintenance is a combination of various activities shown as follows.

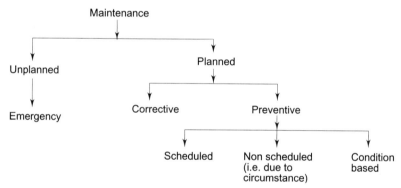

It has been observed over the years that Condition Based Monitoring (CBM) can play a better role over scheduled maintenance in case of a complex and sophisticated system where machine downtime should be maintained at the minimal due to reasons of safety, security and cost. In a hydraulic system in many specific areas CBM can be employed effectively in place of scheduled maintenance which may be time specific. But to accomplish this one needs to monitor the condition of the system which involves four important steps. These steps are:
  (a) Fault detection, i.e. monitoring any adverse change in operating condition.
  (b) Fault diagnosis, i.e. determining the cause of such change.
  (c) Prognostic estimation, i.e. estimation of future state of the system over a specific time frame.
  (d) Fault correction, i.e. repairing the fault at the appropriate time.

In order to have a scientific and realistic condition monitoring it is necessary to have good knowledge of the failure statistics of a machine and its subsystems over a periodic life cycle. In the life cycle of a machine shown in Fig. 15.6, the three stages of maintenance can be classified as:
(1) Wear-in period where failure based maintenance (FBM) is carried out for failures at the early stage.
(2) Stable period where use based maintenance (UBM) is resorted to for failures during the effective machine life period.
(3) Wear-out period where conditioned based maintenance (CBM) is applied.

CBM is based on a realistic mode as compared to other modes as during condition based monitoring one can monitor the actual condition of the machine performance through use of various diagnostic tools and aids developed over the years. These aids help one to ascertain the off-line and on-line monitoring of the fluid power system through various physical signals.

Few such signals are:
1. Temperature rise and heat generated during operation
2. Flow rate through the lines and specific fluid power components
3. Pressure developed
4. Power produced
5. Torque produced
6. Speed of the machine member
7. Acceleration of the actuators
8. Machine efficiency
9. Positional accuracy achieved etc.

Variation in any of the parameters as stated above between the input and output may provide the necessary signal indicating the actual condition of the system at that point of time. For example, speed reduction of the hydraulic motor may be indicative of the losses in the rated flow to the motor or cylinder.

In a hydraulic system two important characteristic parameters are:
1. Dynamic signals such as noise, vibration, etc.
2. Tribo signals such as solid and liquid contamination, mechanical wear, etc.

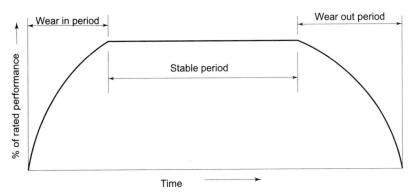

Fig. 15.6 *M/C life cycle and failure vs. time*

These signals are quite effective and easy to pick up electronically with a variety of gadgets available nowadays. They are quite useful for machine maintenance. For example, modern aviation hydraulic systems are provided with a ferrograph to indicate metallic wear debris in the oil. CBM can be effectively used to monitor the working condition of all the important hydraulic elements like valves, pipes, heat exchangers, pumps, seals, bearings, etc. Performance monitoring of seals is an important example.

Generally, during preventive maintenance or routine check up or even during minor repairs, it is customary to replace all associated seals in the hope of eliminating any chance of unscheduled machine breakdown.

"Performance monitoring" of all critical seals in a hydraulic system could be an effective alternative today to ascertain their actual condition while the machine is in operation. Seals can be monitored either continuously or at short intervals so that seal failure between two consecutive repairs or maintenance can be avoided. Moreover, having such monitoring facilities, appropriate decisions can be taken. Seals which are in good condition need not be replaced thus reducing seal replacement cost resulting in tremendous economy in maintenance activities.

Designing an ideal seal to provide minimum leakage with no wear to the sealing surface is quite a difficult task, because of the problems in defining the leakage path and in maintaining a fixed dynamic clearance between the mating surfaces.

Diagnosis of the cause of machine failure generally determines the procedures that establish maintenance intervals, or indicates areas of machine re-design and improvement. Performance or condition monitoring makes it easier to determine the cause of failure particularly where there is interactive failure that can obscure the primary cause.

15.8.1 Condition Monitoring vs Preventive Maintenance

A fundamental question is often raised about the quantum of benefit one may achieve while the machines in a production shop are subjected to regular preventive maintenance and routine checks vis-a-vis attending to maintenance work at the time of actual need as very often it is expressed that preventive maintenance as per schedule may sometimes lead to over or under maintenance. Hence a new school of thought has gained worldwide attention, which prescribed preventive maintenance to be carried out on the basis of results attained by continuously monitoring the system or machine condition for which various types of inspection and measuring gadgets—mostly electronically operated are being used. This means that maintenance activities should be carried out only when needed. This has been found to be most effective as the machine components are kept under constant vigil thereby saving maintenance costs by a considerable amount in terms of avoiding unnecessary maintenance and repairing cost by reducing the downtime. In any modern machine shop, condition monitoring has come to stay as a major tool for machine maintenance and to upkeep the production process to its maximum efficiency.

15.8.2 Methods of Condition Monitoring

The basic methods of condition monitoring that are used in industry today, may be classified as:
(i) Visual monitoring of machine tool parameters and component damages
(ii) Vibration monitoring of various vibratory elements of the machine tool
(iii) Performance monitoring of the machine tool operation
(iv) Wear debris analysis to analyse the quality of debris produced by the machine elements.

Most of these tests are carried out by designing and incorporating suitable instrumentation along with the machine control. Instruments like microscopes, boroscopes and stroboscopes are found to be helpful in estimating the visual analysis of damages caused to machine tool elements by direct or indirect process depending on the suitability of application.

15.8.3 Vibration and Signature Analysis

A signature can be defined and compared to the voltage level indicator of any analog signal given by any electro-mechanical moving system. For example, vibration of any mechanical system can be converted to an electrical wave and measured. To troubleshoot such an analog system, the voltage produced by the moving member is measured at various locations until the measured value deviates from the specified reading thus isolating the trouble. The signature analyzer to measure vibration is used in the same way.

Vibration monitoring thus has become an important aspect of condition monitoring for inspection of bearings, pumps, motors, shafts, bearing housings or other rotating parts. Some newer generation of sophisticated vibro-analyzers have been pressed into service in modern machine tools. A simplified method which finds use for inspection of vibration parameters is to use a vibrometer which will read the amplitude of vibration in a graduated circular dial calibrated in micron/mm values. Another very simple method which can be used for detecting this, is to feel the vibration by touching the running machine structure or part of it at some convenient and safe location. However, any rotating part should not be touched for the same.

This particular method does not require any instrumentation of a sophisticated nature, but is dependent on the machine tool inspector and hence may be to some extent unreliable. Another shop floor method is to place a drop of water on any flat surface of the machine body where vibration is suspected and then to observe presence of the oscillation of the water drop when the machine is run at its expected speed. This oscillation will indicate presence of vibration in the machine tool.

But the surest and the most reliable method is the use of the signature analyzer (as it is mostly termed), which compares the wave patterns of the vibration of the machine tool structure or its hydraulic components like motors. The vibration spectrum of a new machine will vary significantly from the machine which may have worn out due to long usage of the running parts.This process of comparison will help to predict the expected operating life of the part and thus helps to pinpoint when a machine should be taken up for servicing or replacement of parts. Thus this method will help to avoid unscheduled breakdown of the machine to a greater extent. Interesting technical developments have taken place in the field of vibration monitoring techniques and their application. Vibration monitoring equipment and methods which are commonly used nowadays include shock pulse monitor, kurtosis analyser, etc. The process of kurtosis analysis encompasses statistical method to denote the peak distribution of vibration amplitude. We are aware that vibration of machine tools is due to the presence of an excitation force defect mostly developed from an inherent design effect or defects which might have formed at a later date such as pitting marks on surfaces after a certain time of useful life of the part. The shock pulse meter which is having an accelerometer which clips into the mounting stud on the bearing housing records the shock wave or pulse when the resonance frequency is excited.

Wear debris analysis is conducted to analyse the quality of the debris present in gear boxes, hydraulic reservoirs, etc. This method of analysis may be suitable in identifying the nature of contaminants present in the lubricating oils in the machine tool or in the hydraulic oil of the hydraulic reservoir. The particle size of the contaminants present may indicate the cause of the damage already inflicted or any possible future damage to the system. Hence this type of analysis, if properly carried out may help to take prior preventive action to save the machine from future damage, which may sometimes turn out to be quite major in nature.

Though condition monitoring is a new and expensive mode of inspection, this has been found to provide valuable information regarding machine health. It helps in taking precautionary measures well in advance and thus saves a lot in relation to machine downtime enhancing the productivity rate tremendously. Condition monitoring is a new but most promising science to be used extensively in future machine tools.

15.9 ESTIMATION OF SEAL FAILURE

A seal is said to fail when the leakage rate past the seal exceeds a preset maximum flow. The condition of the seal prior to this point is more difficult to estimate. For a face seal the leakage rate may remain reasonably constant even if the sealing surfaces wear away. But at the point of complete wear, the seal fails and there will be sudden increase in leakage. The leakage rate prior to failure may have been satisfactory, with no observable change from its initial performance.

Seal failures can be classified as: degradation and catastrophic failure. While degradation failure is a gradual process catastrophic failure is sudden. It is quite difficult to predict seal failure until one monitors the seal performance through

CBM. The following two methods of oil analysis may be used for monitoring seal performance and predicting seal failure: wear debris detection, and surface/subsurface crack detection. Though there are number of techniques available for measuring the amount of wear debris in fluids, these two nuclear methods are frequently used for performance monitoring of seals: Neutron activation analysis and X-ray fluorescence analysis.

With these techniques one can quantitatively measure minute concentrations of wear debris in the oil using the characteristic radiation from the debris though it may not be possible to identify if the seals are wearing or not.

15.9.1 Neutron Activation Analysis

In this technique a small sample of the fluid is collected periodically and irradiated with neutrons from either a neutron generator or a nuclear reactor. Some of the nuclei in the sample absorb electrons producing a radioactive isotope with its own characteristic radiation. The radiation identifies the debris, and its intensity may indicate the derbis concentration.

A major disadvantage of neutron activation analysis is the time lag between collecting the sample and determining the wear rate. No immediate indication of wear (not even approximate) is available at the test site.

15.9.2 X-ray Fluorescence Analysis

In this technique the sample is irradiated with electrons, rather than neutrons. After irradiation there is a characteristic radiation unique to the debris constituents and their concentration. Because electrons are much easier to produce than neutrons, the irradiation can be performed at the test site, providing immediate indication of the debris concentration.

An online detector may be connected with the hydraulic system for continuous reading but the system becomes more complex including possibility of settling of the debris in the fluid line. Therefore collecting the sample of oil and analysing the same is preferred.

15.9.3 Common Types of Seal Failures

Seal failure may be due to lack of compatibility of the seal material with the oil. Some of the most common causes of seal failure are listed below:
  (i) *Extrusion*—caused by excessive pressure, higher clearance, lack of back-up plate. Seals should be used within the recommended limits of pressure with due regard to the intermittent peak pressure.
  (ii) *Cracking*—may be due to age-hardening, physical failure or thermal hardening at very low temperature, excessive heating due to high friction or abrasive wear. Age hardening may occur during long idle periods at low ambient temperature. Abrasive wear may take place due to tough surface finish which for a good seal life should be less than 0.4 micrometer.
  (iii) *Spiral twisting*—normally limited to O-rings.

(iv) *Surface damage*— abrasion or wear caused by a rough mating surface or sharp edge on grooves or backing ring.
(v) *Swelling*—normally an indication of incompatibility with the oil or contaminated oil.
(vi) *Shrinkage*—may be caused by incompatibility with the oil or due to drying out in an idlle system.
(vii) *Compression set*—normally caused by either excessive loading or exessive temperature.

15.9.4 Leakage

The most common cause of leaks is faulty joints. If a compression type of fitting cannot be sealed adequately by tightening, then ferrule must be replaced. In most cases, leaks developing at joints are due to bad initial fitting such as overtightening of joints. Leakage from components is usually due to damaged seal and packings. When a seal has failed necessitating disassembly of the component, the condition of the rubbing surfaces on the component should also be examined. The cylinder bore of a hydraulic cylinder could have been damaged as a result of seal failure or perhaps been the primary cause of seal failure due to rusting or corrosion. Massive leaks are generally recognised by the oil-spray under pressure, or a considerable pool of oil collecting under the system. Internal leakage should be tackled logically taking one component at a time. Leakage through hydraulic components may be greatly influenced by the seal-bore clearance. The maximum clearance between various seals and their mechanical counterparts should not exceed.

Some values given in Table 15.4 may be considered.

Table 15.4 Clearance between Seal and Mating Parts

Type of seal	*Light duty (microns)*	*Normal duty (microns)*	*Heavy duty (microns)*
O-ring	0.4	0.4	0.4
Molded rubber ring	0.4	0.4	0.4
Semi-reinforced ring	0.4	0.4	0.4
Rubber impregnated fabrics	0.8	0.4	0.3
Fabric packings	—	0.4 to 0.8	0.8–1.2
Leather ring	1.2	0.8	1.3

15.10 MAINTENANCE OF LINE-FITTINGS

15.10.1 Bite Type Fittings and Thin-walled Tubes

Inside the body of the fitting while rotating the nut a few turns, a point comes where one can feel that the rotation of the nut needs more torque to turn it. For a thin-walled tube problems may arise as the hoop strength is not high enough to allow the ferrule to bite into it and the tube may collapse.

To ensure proper tight fitting, proper tools as well as an adequate number of turns of the thread should be used. Lubricating the threads helps to reduce the torque by a significant amount.

Care has to be taken to see that tubings are adequate in length. For a short tube, the ferrule may not even bite into the tubing and even if it does, the tube may be overstretched or stressed.

15.10.2 Reassembling of Tube Fittings

On reassembling of a fitting on the tubing, the nut has to be tightened to the original point which can be felt due to the sudden increase of torque and then the nut should be turned another 1/2 to 3/4 turn more to have tight seal. This is to be carried out every time a fitting assembly is dismantled and reassembled. Ensure that the ferrule seats evenly on both the OD against the body of the fitting and ID against the tubing both of which are possible paths for future leakage.

15.10.3 Handling of Compression Type Fittings

Compression fittings are usually made of non-ferrous materials like brass and are intended for low pressure applications having minimum vibrations. The sealing is effected through a tight compression fit between the OD of the sleeve and the fitting body and on the ID of the sleeve and tubing. They are susceptible to failure due to vibration.

15.10.4 Handling of Tubes and Pipes

Three types of tubes are used in a hydraulic system—rigid, semi-rigid and flexible. Rigid tubes are mostly JIC steel, stainless steel, titanium, hard drawn copper, etc. Semi-rigid tubes includes Al, copper, brass, thin-walled steel, etc. Some of the thin-walled tubes may have seams. Some of them are copper-coated or tin-coated. From a working point of view it is preferable to have dead-soft tubes (fully annealed)—as it yields better and easily. But over a long period of time, copper tubing may get hardened. Brass also may have similar problem and thus may create problems in future. Hence for high pressure-applications it is better to use seamless steel tubings—hardened and treated. Steel tubes after heating should be cooled gradually and not rapidly. Tubes and pipes should be cut by a tube cutter and burrs should be cleaned before using. Tubes and pipes which are used to carry return or pressure oil to or from the tank should be cut as shown in Fig. 15.7.

15.10.5 Some General Tips

(i) Cleanliness should not be compromised.
(ii) Cylinder, pump, hose, pipes, fitting, valves should be capped until the system is put to use.

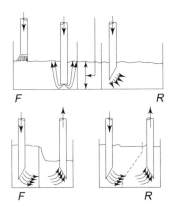

Fig. 15.7 *Use of tubes inside reservoir*
F–False. R–Right.

- (iii) Teflon tape or compound should not be used without ensuring their compatibility and other factors.
- (iv) Grinding and welding operations should be avoided in places where hydraulic components are assembled.
- (v) Pipes should never be allowed to vibrate.
- (vi) To check vibration, one can place a drop of water on the pipe or tube. If it falls off, it indicates vibration.
- (vii) It is understood that amount of air in the oil can be reduced if one can maintain a back pressure of 5 bar or so in the system.
- (viii) O-rings used to seal port leakage in control valves can be checked by applying heavy grease around the suspected leakage area. If the noise stops, it indicates a trouble spot.
- (ix) A magnetic trap can be cleaned by holding a larger permanent magnet inside a polythene bag near it.
- (x) Bending of tubes creates pressure drop. But bending of tubes is preferable to cutting the tube and using a fitting as it will minimise the drop in pressure.
- (xi) The key to fault findings is the location of pressure and flow. Since a flow may occur even the system does not create pressure, but pressure can be generated if there is a flow. Pressure sensing is more valuable.
- (xii) For oil-handling the following cautions may be followed
    - (a) Don't store hydraulic oil outdoors unless it is absolutely needed.
    - (b) Don't store oil drums near sources of heat or contamination.
    - (c) Use drum taps for taking oil from drums. They provide a good flow for viscous oil even in cold weather.
    - (d) Use old oil first. Chances of oxidation, contamination or attack by heat or cold may increase with time at ease. If an oil is suspect because of long storage, it should be analyzed before using.

(e) Label each drum for easy and quick identification.
(f) Keep oil containers clean and covered when not in use.
(g) Store drums on their side. They should be tilted or turned so that water does accumulate around the bungs. When drums are stored upright, expanding air may escape during the day as the drums warm up.
(h) Make adequate fire safety arrangement in the oil store.

15.11 NOISE

The main sources of noise are:
(i) Pump and its prime mover, (ii) Pressure relief valve, (iii) Flow control valve, (iv) Pipes. Noise due to valves and control elements is transient and related to the degree of turbulence or cavitation produced. In general, gear and piston pumps are noisy and vane and screw pumps are quieter. Noise may increase due to:

(i) Undersized suction pipe leading to execesive flow velocity
(ii) Aeration of the oil due to in apropriate design of oil tank
(iii) Undersized filter in the suction line leading to partial starvation of the pump.

Confronted with very stringent laws and regulations from environmentalists as well as occupational health hazards in various countries, the noise behavior of the hydraulic circuit as well as the acoustic properties of hydraulic components and systems are being taken more seriously now. More compactness of hydraulic drives means higher power densities, higher cyclical forces within a very small volume and surface area bringing forth increasing system excitation. In such a situation there is an automatic rise in system vibration and escalation of noise which if not tackled properly, may be disastrous from the point of view of workers and operator's audio health. A silver lining in combatting noise related problems is that the power which gets emitted as noise is insignificant. It is found that a 10 kW hydraulic pump emits an effective sound power of hardly 1 W value of 90 dB (A) noise level. Sound levels above 80–90 dB (A) value damage hearing capability of human beings and not many individual hydraulic components produce such noise intensity. However there are a good number of elements such as pumps, hydraulic pipings etc. which are capable of generating noise above 80 dB (A) and adequate precautions have to be taken by hydraulic system designers while designing their system so that noise emission can be minimised effectively to keep the system quieter within the limits of tolerable noise. A noise may be of lower intensity, but not acceptable to the human ear. This means that noise has to be classified not only by its intensity but also in terms of comfort.

15.11.1 Sound Emission by a Hydraulic System

A hydraulic system emits structure borne sound due to vibrations in piping or casing excited by the internal operating forces of the system. This vibration induces excitation to the surrounding medium, i.e. air to make it vibrate depending on surface form, mode of vibration, wave length or atmosphere. The main source of the excitation force is the pump which produces the air-borne vibration due to pressure-pulsation caused by the variable cylindrical component of the volume displacement process of the pump. The pressure pulsation is propagated throughout the system by the transmitting oil. Moreover there is possibility of direct generation of structure-borne sound due to mechanical parts which is also propagated throughout the system. Both the pressure pulsation and mechanical excitation cause the pump, mountings, pipings and actuators to vibrate resulting in a complex oscillating pattern comprising of the fundamental pump frequency and other frequencies of the other parts as stated.

15.11.2 Noise in the Pump

A pressure pulse is created at the delivery end of the pump as soon as a pump discharges a full segment of flow of fluid under high pressure. In spite of smooth and perfectly machined circumference of the gears in a gear pump, pulsation has been found to occur in normal gear pumps amounting to about 10 to 14% of the delivery flow. However the pressure pulsation in gear pumps could be reduced by increasing the number of teeth. But this may reduce the overall efficiency of delivery performance. A better method is to off-set the two gear pairs by half a tooth width where the flows are super-imposed, thus volumetric pulsation could be decreased significantly up to 3.6%. This is shown in Fig. 15.8.

In a piston pump several individual superimposing flow delivery arrangement in one cyclical phase helps reduce the pulsation. It may be mentioned here that with reduction in the number of pistons, the flow pulsation will increase. The flow pulsation of a radial piston pump with seven pistons is around 3%.

The pressure pulsation in a pump may induce pulsation and vibration into:
(a) Pump casing
(b) Internal components of the pump like gears, shaft, bearings, etc.
(c) Piping connected to the pump
(d) Pump foundation, etc.

Moreover cyclical loading of the pump element between the pump suction and delivery also induces pressure imbalance in the system adding to the noise.

A similar phenomenon takes place in the case of piston pump also. The cyclic forces excite bearings, yoke mechanism, etc. to create vibrations and noise. However the reaction ring with which the pistons are in constant touch, absorb a portion of the cylical force and thus help reduce the effective noise.

One should however keep in mind that the natural vibration frequencies of the casing, its shape and size, the material used and its damping properties may have some lowering effect on the noise produced as each material may have different vibration characteristics of a few dB value of noise intensity.

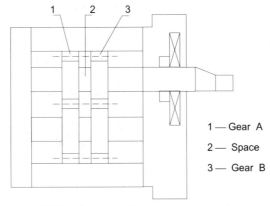

Fig. 15.8 *(a) Offset of two gears in a gear pump to reduce pump-ripple*

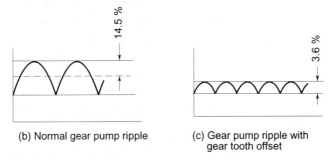

Fig. 15.8 *(b) and (c) Reduction of pump ripple in gear pump*

15.11.3 Noise in Hydraulic Pipings

It has been found that noise level of a significant value occurs in pipes and piping systems at resonance frequencies and air-borne noise with correspondingly high amplitudes is emitted. The sound energy is transmitted throughout the entire piping system and cannot remain localized. For example, the wavelength of a vibration of 1000 Hz of sound with a velocity of 3000 m/s will be 3 m which may sometimes equal the entire length of a machine-piping system. The structure-borne sound in pipes should be given due importance as the induced waves in the pipes and pipe walls creates various different type of wave patterns like flexural-wave, longitudinal wave and torsional wave. While the flexural wave creates the air-borne sound, the longitudinal wave produces the liquid-borne sound wave which gets converted to structure-borne sound at each pipe bend, each joint, each throttle point and pipe restrictions. It is also a fact that the pressurized liquid and the pipe wall continuously exchange vibration energy as both constitute a "close coupled system". Flow generated noise also occurs at the throttle and restriction points due to sudden increase in flow velocity at these points and contribute to the pipe related noise. Shock waves or water hammer as well as cavitation in piping system also contribute to the vibration excitation in piping system.

15.11.4 How to Reduce Noise in Hydraulic System

A lot of work has been done by various researchers to reduce the noise level in hydraulic system. Various internal modifications in gear and piston pumps, as discussed earlier have made it possible to reduce the pressure pulsation to a certain extent. Screw, vane and internal gear pumps are much quieter than external gear and piston pumps and hence a major noise problem in the pump is due to the pump casing, foundation, etc. One can encapsulate the pump and get noise reduction to some extent but problems like pump weight, space for installation, installation mode and heat dissipation may pose innumerable problems. To combat pump noise it is therefore more important to optimize the pump's internal design structure like selection of material, design of appropriate casing and structure so that the noise emission can be minimised at the optimum power level.

Reduction of piping related noise can be minimised to some extent by using proper clamps throughout the piping layout. Use of flexible hoses may also help reduction in structure-borne noise in pipe walls. Use of expansion vessels and viscous damping elements, geometric detuning (off resonance adjustment) for damping or declamping of noise producing subsystems are the common methods adopted in reducing vibration and noise in pipings.

15.11.5 Hose Maintenance

The following points may be noted to see that hoses do not fail prematurely. Causes must be probed and analyzed.
  (i) Ensure that the hose assembly meets the system requirement fully well.
  (ii) Maintain operating pressure and recommended temperature range.
  (iii) Ensure fluid compatibility of the hose material.
      In case of hose failure probe the following:
   1. Nature of failure one has experienced with such hoses.
   2. Find out hose part number and date of manufacturing.
   3. Nature and type of equipment where the hose is used and for how long.
   4. Location of the hose in the equipment.
   5. Maximum and minimum temperature both internal and external.
   6. Type of fluid of the hydraulic system where the hose is used.
   7. Bend radius of the hose application.
   8. The flow rate (Q) and maximum static and transient pressure in the hose.
   9. Nature of surrounding environment where the hose is used.

Exterior damage to the hose is mostly from abrasion and corrosion. In eventuality of abrasion, etc. a protective cover should be provided to the hose. For safe operation in hot condition provide fire-sleeve. Corrosion resistant cover may be used if the hose is used in a corrosive atmosphere or if it comes in contact with any corrosive element.

15.12 FAULT DIAGNOSIS OF A HYDRAULIC SYSTEM AND ITS COMPONENTS

Table 15.5 The Various Faults, their Symptoms Probable causes and also the Remedial Action that can be taken to Prevent the Damage are Enlisted Below

Sl. No.	Symptoms	Probable causes	Remedial action
1.	2.	3.	4.

1. System

(a)	(i) Excessive noise in pump	(a) Cavitation in pump	(i) Clean filters in the inlet line. (ii) Replace defective filters. (iii) Adjust drive motor to correct rpm if pump speed is too high (iv) Clean oil tank breather. (v) Check supercharge pump—replace or overhaul if defective. (vi) Check fluid temperature. If oil is too cold, warm up. (vii) Stop cock in pump inlet only partially open it fully. (viii) Suction filter too small or filter blocked to be corrected.
		(b) Air in fluid medium	(i) Check for leakage in inlet pipe to pump—joint to be tightened to stop leakage. (ii) Pump shaft seal may be damaged. To be replaced. (iii) Bleed air from system. (iv) Oil level in reservoir may be down—top up to the specified level. (v) All return lines to be checked up to see if they are below the fluid level or not. (vi) Unsuitable fluid.
		(c) Misalignment of coupling	(i) Misalignment to be corrected. (ii) Loose coupling to be tightened (iii) Pump and motor mounting to be checked for looseness. (iv) Wrong direction of rotation of motor, to be corrected. (v) Defective coupling to be replaced.
		(d) Worn out pump	Pump to be overhauled or replaced.
		(e) Pressure exceeded	Relief valve setting is too high—to reset.
		(f) Unsuitable type of pump	Replace pump.

(Contd.)

(ii)	Oil	Poor suction	(i) Fluid level too low—needs topping up
			(ii) Oil viscosity too high—use correct viscosity
			(iii) Dirty fluid—improper filtration
			(iv) Clogged filter—clean filter
			(v) Foam in oil—use proper oil
(iii)	Noisy Motor	Coupling mis-aligned	(i) Correct alignment
			(ii) Motor or coupling may be damaged and should be replaced or overhauled
			(iii) Faulty seal—replace it
(iv)	Noise in pressure relief valve	(a) Incorrect pressure setting	(i) Install pressure gauge and adjust valve setting
		(b) Valve vibrates due to worn out seat or dirt on valve seat	(i) Replace valve/clean valve
			(ii) Valve setting to be checked and corrected
		(c) Poor damping	(i) Wrong valve setting or poor valve design
(v)	Noise in suction and return line	(i) Resistance too high in inlet line to pump	(i) Reduce resistance in inbet line
		(ii) Too many bends in pipe layout	(ii) Damaged pipe line
			(iii) Improve pipe layout if possible.
		(iii) Oil level too low	(iv) Top up oil
		(iv) Return line ends above fluid level	(v) Properly install return line
		(v) Suction filter blocked.	(vi) Change/clean suction filter
(b) (i)	Excessive heat in pump	(a) Pump may generate heat due to	
		(i) cavitation	- Check inlet pipe and oil level in tank.
		(ii) Excessive air in fluid medium or improper fluid used.	- Top up oil level
		(iii) Too high setting of relief valve	- Set relief valve to desired value of pressure.
		(iv) Damaged pump	- Replace damaged pump if can't repair and take action
		(v) Excessive load on pump	- Check up load on pump
		(vi) Improper tank design	- Check up reservoir design and provide correct size and shape of tank
		(vii) Faulty estimation of reservoir capacity	Provide correct oil volume.
		(b) Heat in pressure relief valve (PRV)	

(Contd)

		PRV generates heat due to	
		(i) Incorrect valve setting	- Set the valve correctly
		(ii) Pressure override to be checked	- Causes of pressure override to be checked
		(iii) Valve seat may be worn out or dirt on valve seat	- Check up the physical condition of the valve and replace if needed
	(c) Heat due to prime mover		
		The electric motor for the pump may get heated due to	
		(i) Too high pressure relief valve setting	- Set the PRV correctly
		(ii) Excessive load	- Reduce pump load
		(iii) Damaged motor	- Replace damaged motor or repair
(iv)	Excessive heat in oil	Reservoir may get heated due to	
		(i) System operates at higher pressure for which the system is not designed.	- Check up system pressure and correct it to designed value.
		(ii) Improper selection of oil viscosity or incompatible oil used.	- Check up oil viscosity and replace oil if needed.
		(iii) Inadequate or faulty cooling system used.	- Inspect the cooling system and set right if found faulty
		(iv) Defective hydraulic components like pump, cylinder, etc.	- Replace defective components or repair
		(v) Wrong pipe lay out with lot of restrictions and bends used	- Use proper pipe layout.
		(vi) Wrong setting of pressure relief or unloading valves.	- Set the relief valve to appropriate value.
		(vii) Faulty circuit design with wrong valves used in the circuit.	- Check up the circuit and correct it if needed.
		(viii) Incorrect pump clearance	- Check up pumping system foundue leakage
		(ix) Faulty choice of oil viscosity	- Check the oil viscosity and replace oil if required
		(x) High ambient temperature	- Arrange proper protection from ambient atmosphere.
(e)	Faulty system operation	Faulty system operation may create problem in the hydraulic system due to:	- Remedial action
		(i) No flow or pressure	- Check up pump
		(ii) Inoperative sequencing device and valves	- Check limit switches and their locations.

(Contd.)

		(iii) wornout or damaged motor, cylinder, etc.	- Check up motor and replace if defective
		(iv) Defective or incorrect servo amplifier or servo valve	- Check up the servo valve and feedback system
		(v) Lack of command to servo amplifier	
		(vi) Mechanical fault like Impact, mechanical seizure shock, etc.	- Take appropriate action for remedy.
(d)	Incorrect pressure	(a) No pressure	(i) Check up pump and drive motor rotation and correct the same
			(ii) Check relief valve setting
			(iii) Check filter for clogging
		(b) Low pressure due to:	
		(i) Operating pressure set too low	- Set the desired pressure
		(ii) Pressure relief valve (PRV) seat may be dirty or damaged	- Check up and clean the PRV seat
		(iii) Mechanical damage to PRV like damaged or broken spring	- Replace PRV if damaged or replace spring of indentical specification.
		(iv) Too much internal leakage in valve.	- Inspect the system and identify the cause of leakage e.g. seal failure etc. and take corrective action
		(c) Excessive pressure due to	
		(i) Wrong setting of pressure relief or unloading valve	- Set the PRV correctly set the unloading valve correctly
		(ii) Damaged pressure control valves, i.e. PRV, unloading and pressure reducing valve.	- Inspect the valves and replace if found faulty.
		(d) Fluctuating Pressure	
		(i) Air entrained in oil	- Air to be bled off from system.
		(ii) Oil is contaminated	- Source of contamination to be detected and stopped
		(iii) Defective or damaged accumulator	- To be replaced or corrected.
		(iv) Old or damaged PRV like worn poppet	- Needs change
		(v) Piston gets sticky	- Release piston, remove burrs, check cylinder for ease of movement
(e)	Incorrect flow	(a) Excessive flow due to	
		(i) Too high setting of flow control valve.	- Correct the flow control valve setting

(*Contd.*)

	(ii)	Pump/motor rpm too high	- Check the motor rpm and correct it.
	(iii)	Wrong selection of pump flow rate and size.	- Check up pump
	(iv)	Adjusting mechanism of variable displacement pump wrongly set or damaged or inoperative.	- Check the pump
(b)		Insufficient flow	
	(i)	Inaccurate setting of flow valve.	- Set flow control valve correctly
	(ii)	Low rpm of pump/motor.	- Correct the motor rpm
	(iii)	Pressure relief and unloading valve set wrongly.	- Set the PRV correctly
	(iv)	Undesirable internal or external leakage in the system.	- Check the cause of leakage & set right to stop leakage
	(v)	Low oil viscosity	- Change oil & use correct oil
	(vi)	Excessive heat generation in the system.	- Reduce heat generation through accurate indentification
(c)		No flow	
	(i)	No flow to pump due to wrong direction of rotation of motor	- Set the motor for correct rotation
	(ii)	Drive motor damaged and pump stopped or pump damaged.	- Replace damaged pump or motor
	(iii)	Mechanical damage of coupling.	- Check the coupling and set right.
	(iv)	Direction control valve set in wrong position.	- Check the DC valve connection and correct if found wrong
	(v)	Full flow passing through PRV	- Set PRV correctly
	(vi)	Pump suction and delivery port wrongly connected.	- Connect the pump correctly
	(viii)	Excessive leakage in the system	- Check the leakage points and set right

(f) Aeration of hydraulic system

(a)	Foaming of fluid.	
(i)	Pump inlet line permits entry of air.	- Stop entry of air either through suction line or due to other reasons
(ii)	Fluid level in tank is insufficient.	- Check the fluid level and top up if level is down
(iii)	Defective shaft seal	- Change the shaft seal
(iv)	Wrong oil	- Check for correct oil and replace
(v)	Cooling system insufficient or inoperative.	- Improve cooling efficiency

(*Contd.*)

		(vi) Coolant temperature too high	- Check the coolant and take appropriate measure.
		(vii) Ambient temperature too high	- Protect the system from poor ambient condition
		(viii) Thermostat set too high.	- Set the thermoset correctly.
(g)	Slow movement	(i) Low flow	- Check up pump
		(ii) Lack of lubrication in machine slideways or linkages	– Check up lubrication
		(iii) Malfunctioning of servo amplifier	- Repair or replace
		(iv) Sticking servo valve	- Clean, check & refit
		(v) Damaged drive units.	- Repair/replace cylinder/motors
(h)	Excessive fast motion	(i) Higher flow control valve due to wrong setting	- Set it right
		(ii) Overriding of work load	
		(iii) Feed back transducer may be malfunctioning.	- May need replacement
		(iv) Servo-amplifier malfunctions	Repair or replace
(i)	Erratic movement	(i) Pressure fluctuates erratically.	Check PRV, accumulator or oil condition.
		(ii) Erratic command signal	Repair/replace signal console, feed back or wiring system
		(iii) Worn out or damaged driving unit.	Repair/replace cylinder or motor.
		(iv) Air entrains hydraulic oil	Top up oil, extend return line, bleed-off air or replace oil as per need.
		(v) Sticking servo valve	Clean, check or filter fluid
		(vi) Servo-amplifier malfunctions	- Repair/replace
(k)	No pump delivery	(i) Pump rotation may be wrong	- Should be corrected
		(ii) Oil may be of higher viscosity	- Use correct viscosity oil.
		(iii) Pressure relief valve stuck open	PRV may be cleaned and assembled.
		(iv) Oil level in hydraulic reservoir may be too low	Top up oil.
		(v) Air leak in inlet pipe	Stop leakage by tightening all connections.
		(v) Mechanical trouble like coupling disengaged or damaged	Coupling to be adjusted
(l)	Pump wear too high	(i) In case of vane pumps, vanes stuck in vane slots.	Vanes/slots may be cleaned of dirt, metal silvers or debris etc.

(Contd.)

Hydraulic System Maintenance, Repair and Reconditioning 533

		(ii) Chatter due to presence of air and cavitation	Check return line and inlet pipe for leakage
		(iii) Excessive pressure	Check relief valve setting.
		(iv) Coupling misaligned	Correct alignment
		(v) Abrasive material present in oil	Filter to be checked
		(vi) Improper oil viscosity	Oil condition to be inspected. Use correct viscosity oil and see if the system works under ideal temperature condition or not.
		(vii) Excessive pressure	Check pressure relief valve setting and sets it correct.
(m)	Pump pressure build up is low	(i) Internal leakage in pump	Identify the source of leakage and stop leakage.
		(ii) Complete loss of flow	May be PRV stuck open pipe line—inlet and return line damaged.
		(iii) Wrong setting of PRV	PRV to be reset correctly.
(n)	Poor valve response	(i) Faulty valve assembly	Check the valve and signal lines.
		(ii) Faulty valve actuation	Repair/replace actuating system.
		(iii) Excessive heat	Check up for leakage Check up for seal failure Install heat resistant seal compatible to system Check return line Remove plug if not removed

2. Components

(a)	Pressure control valve malfunctions.	(i) Sequence valve develops back pressure.	– Clean and flush valve.
		(ii) Drain line blocked	
		(iii) Valve spool sticky	– Check for oil cleanliness
		(iv) Valve spring damaged	– Remove/replace spring
		(v) Excessive leakage in valve *Pressure reducing valve malfunctions like*	– Identify source of leakage
		(i) Broken spring	– Remove/replace spring
		(ii) Sticky spool	– Clean spool/check oil
		(iii) Valve Orifice blocked	Clear block age in orifice and maintain contamination level.
		(iv) In seat type of valve, seat and valve cone does not match.	Check and replace

(Contd.)

		(v)	Pressure relief valve malunctions like pressure fluctuaters, low or no pressure, excessive noise etc.	Take appropriate action to repair PRV or replace faulty PRV.
(b)	Flow control affected	(i)	Broken differential spring	Replace
		(ii)	Throttle blocked inside valve body	Clear bore, throttle screw
		(iii)	Mismatch between valve and valve seating.	Correct it
		(iv)	Broken/damaged check valve	Replace
		(v)	Excessive internal leakage	Look for source of leakage, rise in temperature and oil condition
		(vi)	Throttle drain blocked	Clear it or correct drain line
		(vii)	Differential spool sticky	Clean spool and spool bore
(c)	Accumulator leaks		Problems mostly encountered are:	
		(i)	Leakage through bladder seam	Replace bladder. Do not fold bladder Do not keep bladder in hot and humid place Look for bladder material fault as well as manufacturing fault. Maintain oil contamination level.
		(ii)	Gas valve or oil valve leaks.	Tighten connection, replace faulty valve
		(iii)	Bladder charged, brittle or porous	High pressure ratio resulting in high temperature and failure of material
		(iv)	Cracks or holes in bladder	Material fatigue. Damaged during assembly. Charge pressure tool low.
(d)	DC valves do not provide adequate pressure oil when valve gets actuated	(i)	Valve vibrates	Solenoid may be defective
		(ii)	Defective valve	Valve worn out, correct pilot pressure and check up flow rate.
		(iii)	Fluctuation in pilot control or no oil flow in pilot line	Check for line blockage or too much drop in pilot pressure
		(iv)	Improper flow rate through valve	Incorrect port size selected or ports got blocked
		(v)	Spool sticky	Spool to be cleaned free of gummy materials and other deposits
		(vi)	Solenoid fails to responded	First check if indicating light works or not. Check then manually pushing

(*Contd.*)

Hydraulic System Maintenance, Repair and Reconditioning **535**

	(vii) Presence of excessive back pressure in the valve lines	the spool whether it moves or not. Take further appropriate action like repair/replacement if needed. Clear any restriction to the return line.
	(viii) Mechanical failure of valve	Check up if actuating spring, push button/pin, etc. are broken or damaged and replace, if found faulty
(e) Cylinder malfunctions		
(i) Cylinder produces noise	(i) Faulty alignment, loose or inadequate mounting, piping vibrations etc.	Realign, tighten mounting bolts, use adequate pipe clamps.
	(ii) Load fluctuations too much.	Check loading patterns and take corrective action.
	(iii) Internal cylinder fittings and assembly may be faulty.	Refer manufacturers' catalogue, repair or replace cylinder.
(ii) Cylinder Thrust abnormal	(i) Too much entrain of air into the system and insufficient bleeding.	Bleed cylinder correctly Ensure deaeration in the system.
	(ii) Inadequate cushioning	Check and set the cushion pin to correct position
	(iii) Incorrect flow control	Set the flow control valve to supply proper amount of flow as per speed requirement.
(iii) Cylinder produces jerky irregular and motion during movement	(i) Heavy load at slow speed, misaligned mechanical slides, wrong assembly of piston packing, air in system, etc.	(i) Take appropriate action like appropriate slide lubrication, proper alignment, proper and adequate tightening of packings etc.
(iv) Cylinder produces no thrust	(i) Low pressure, faulty piston, excessive leakage, inappropriate pressure setting, etc.	Take appropriate corrective action to see that PRV is set to correct pressure and stops all points of oil leakage.
(v) Mechanical failure of cylinder	(i) Seized piston rod, excessive rod wear broken linkages, misalignment, etc.	Cylinder bore and piston clearances may be checked Identify causes of rod wear and take appropriate action to stop wear. Replace linkages if damaged.

(*Contd.*)

(vi) Incorrect speed of cylinder	(i) Incorrect flow control setting	Reset flow control valve
	(ii) Low pump capacity	Replace with proper pump. Use correct motor rpm
(f) Pipes, hoses and fittings	(i) Faulty fittings	Check up each fitting and replace faulty one
(i) Leakage through the line fittings	(ii) Mismatch of threads	Straight male thread put into female tapered thread.
	(iii) Threads damaged due to over tightening.	Use appropriate tightening only.
	(iv) Pipe fittings loosened due to vibration.	Use adequate clamps and clamp tightening.
	(v) Damaged seals	Check for burrs, inspect roughness of sealing surface of port.
	(vi) Port spot face is small.	Enlarge spot face to allow proper fitting of sealing.
(ii) Excessive pressure drop in hose	(i) Poor bore condition or wrong size of hose	Check and replace
(iii) Hose tube cracked	(ii) Excessive heat	Protect the hose and check fluid temperature
(iv) Hose bursts	(iii) High frequency pressure surge.	Use correct hose with spiral reinforced hose.
	(iv) Hose cover might have been damaged due to mechanical abrasion, cutting, acid, chemicals, too difficult environment like heat/cold	Reconsider environmental or working condition.
(v) Fitting blowing off at the end of the hose	- Wrong fitting used Under or over crimping Hose installed without appropriate allowance to take care of shortening when pressurised.	Refit correctly Check for adequate crimping Allow at least 4% shortening during installation.
(vi) Spiral reinforced hose burst split	- Hose is too short to accommodate change in length when pressurised.	Use correct length of hose
(vii) Tube or hose badly deteriorated.	- Tube or hose material may be incompatible with oil or environment.	Check compatibility of oil with tube/hose materials.
(g) Seal failure	- Excessive heat	Check operating and surrounding temperature
(i) Extrusion of seals takes	- Excessive clearance	Check up the dimensions and replace.

(Contd.)

place	- Too high pressure	Adjust pressure setting
	- Defective or improper groove geometry	Find out correct dimensions. Consult system manufacturer.
	- Lack of support and back up for seals.	Arrange adequate back up.
(ii) Compression set	- Excessive temperature	Provide adequate lubrication and check up seal to oil compatibility
	- Excessive loading	Refer actual seal and groove dimensions, seal design and seal loading mechanism.
(iii) Seal cracked	- Age hardening of seal Seal exposed to heat Abrasion of seal	Avoid long idle periods of system. Check operating temperature Mating surfaces rough
(iv) Scoring of seals	- Poor machining of bore Axial displacement of seal undersize bore.	Avoid sharp edges and use recommended finish.
(v) Excessive swell of seal	Fluid contaminated and incompatible	Use compatible seal material
(vi) Excessive shrinkage	Seals dry out due to idle storage and age hardening.	Idle storage to be avoided.
(h) Filter		
(i) Excessive pressure drop takes place across filter	- Filter element is dirty Under size filter element used.	Filter element to be changed. Check and replace Examine appropriate filtration need of system.
(ii) Filter fails to stop down stream contamination	Filter rating may be inaccurate	Check filter and replace.

15.13 GENERAL SAFETY MEASURES FOR FLUID POWER SYSTEM

A maintenance mechanic should adhere to strict standards of safety measures. Safety consciousness amongst the work force helps to prevent occurrence of industrial accidents. Unnecessary damage and loss of property and life can be averted in many cases if strict safety procedures are practised in our industry. This is true in any field and is equally applicable in the case of fluid power systems. In case of fluid power systems both hydraulic and pneumatic—the medium of energy is in a pressurized condition and hence extra care is needed as it is more dangerous to handle if one considers the aspect of safety against injury of the individuals handling such systems. Hence it will not be wrong to say that designers and manufacturers of hydraulic and pneumatic systems and equipments have an important and responsible part to play in designing a safe system.

Most of the hydraulic fluids which are used contain hazardous chemical elements—though in a very insignificant quantity. Proper handling instruction should be given to the user to avoid injuries.

Oil is highly inflammable. Adequate protective measures must be taken against fire hazards due to accummulation of heat in the hydraulic system. This is an important point to reckon with regarding safety of fluid power system

The design of hydraulic and pneumatic equipment for machine tools should provide for a noise-free suction and discharge of oil and it should ensure that the fluid is discharged only into the concerned receiver or tank.

Adequate protective measures should be incorporated against damage due to over pressure. To ensure safe working of the hydraulic system the accumulated air in the system should be bled off.

Similarly noise level of the compressed air (exhaust) from the system should be controlled to the minimum limit. As most of the pneumatic accessories used in machine tools are either clamping or declamping operations, proper safety arrangement should be provided against self-release of the backing system. Suitable protective devices should be incorporated against undesirable drop in working pressure to prevent accidents and damage of the tool.

Leakage is a permanent problem for the system. Due to leakage the space around a hydraulic machine gets dirty and slippery. Hence the plumbing system of any hydraulic system should be so designed and manufactured to minimize this.

There are innumerable points which could be cited here to emphasize the need to follow safety rules strictly. However for a hydraulic and pneumatically operated system we can summarize the probable actions needed to ensure operational safety of the system as follows:

(i) Examine the machine, inspect the parts, starting and stopping switches, protective devices, etc.
(ii) Ensure safe pressure of the hydraulic system and pneumatic devices.
(iii) Ensure safe clamping and declamping procedures for automatic systems.
(iv) Adjust the control levers and switches against malfunctioning.
(v) Use protective guards in the case of hydraulic and pneumatic presses.
(vi) Use proper oil and have regular inspection of oil.
(vii) Use face shields for self protection against pressure actuated system and machine tools.
(viii) Never work the machine with higher than designed pressure.
(ix) For pressure containers and pressurised hydraulic and pneumatic chambers, ultimate pressure against bursting of the container should be maintained at the safe of possible limit.
(x) Equipment for protection against fire should be kept always in readiness for any eventuality.
(xi) For proper working and safe handling, hydraulic and pneumatic elements should not be hit hard by external means.

(xii) Both hydraulic and pneumatic elements should be kept in clean atmosphere in order to avoid their premature failure due to contaminants.
(xiii) Use the correct tool for dismantling parts.
(xiv) While checking for mechanical faults, disconnect the power source fully and then do the work.
(xv) Hydraulic and pneumatic hoses and pipes should not be kinked.

15.14 INSPECTION FORMAT FOR HYDRAULIC SYSTEMS

A simple format for inspection is given here.

Table 15.6 Inspection Format for Hydraulic Systems

Type of machine: Section:
Machine No. Date:

Sl. No.	Items	Condition	Inspected by	Remarks
1.	Reservoir (a) Breather (b) Inlet plug (c) Drain plug (d) Magnetic plug (e) Oil level (f) Internal condition (g) Heat exchanger (h) Oil condition (visual inspection)			
2.	Electromotor—direction of rotation			
3.	Pump (a) Motor alignment (b) Flow rate (c) Pressure (d) Noise level (e) Condition of internal mechanism			
4.	System (a) Temperature (b) Pressure (c) Vibration			
5.	Suction filter—condition			
6.	DC valve and servo valve (a) function (b) actuating element (solenoid etc.), electronics/electrical parameters (c) Spool movement (d) Seals			
7.	Pressure relief valve (a) Setting (b) Noise			

(Contd.)

(c) Spring
8. Flow control valve—
   (a) Setting and noise
   (b) Location in the circuit
   (c) Pressure drop
9. Non-return valve—
   (a) Direction
   (b) Leakage
   (c) Location
10. Hydraulic cylinder
    (a) Mounting
    (b) Piston rod condition
    (c) Friction
    (d) Parallelism of movement for rod
    (e) Jerky movement
    (f) Piston rod wiper
    (g) Rod packing/other seals for leakage
    (h) Rod end connections
    (i) Tie rod tension
11. Pipes and hoses
    (a) Joints
    (b) End fittings
    (c) Abrasion on hose
    (d) Vibration
    (e) Leakage through joints
12. Bleeder—function
13. Other parts—(if any) Pressure switch
                          accumulator
                          booster etc.

Signature of the I/C

Review Questions

1. (a) State the common faults in a hydraulic system.
   (b) "Replace all faulty hydraulic elements"—comment on this statement.
2. State the various techniques used to inspect hydraulic oils.
3. Write down the check list of fire resistant oils.
4. What is oxidation? What is the effect of oxidation on hydraulic oil and the system?
5. How does heat affect the hydraulic system? State some common sources of heat in a hydraulic system.
6. What is an electrostatic liquid purifier? State its functional feature.
7. What is condition monitoring? How does it influence the preventive maintenance of a hydraulic system?
8. What is a hose? State various synthetic materials which are used for hose manufacturing.

9. Give your comments on compatibility of synthetic materials with fire resistant oils.
10. (a) State various tests done on hydraulic hoses for their reliable functioning.
    (b) What is an impulse test? Explain the method of impulse test.

References

15.1 Sastry, MNL, "Maintenance and operational problems in hydraulic systems in plant machinery" *Fluid Power*, Dec. 1994, Bangalore, India.

15.2 Henn, AH, "Performance monitoring—A new approach to an oil problem" *Hydraulics & Pneumatics*, 614 Superior Ave., West Cleveland, Ohio, USA. Aug. 1972,

15.3 Stull, Ned R, "Fluid Power Maintenance", *Hydraulics & Pneumatics*, USA.

15.4 "Notes on Maintenance—Manual on Hydraulics", Bosch GmbH, Stuttgart, Germany, 1969.

15.5 Rao, BVA, "Diagnostic and Condition Monitoring", Notes compiled for short term course on "Industrial Hydraulics" held at IIT, Madras.

15.6 Keiper, W and Robert Bosch, "Noise and vibration in hydraulic drives", GmbH, Stuttgart, Germany, Oel Hydraulik & Pneumatik, May, 1992, Germany.

15.7 Farel Bradbury, *Hydraulic Systems and Maintenance*, Ileffe Books (Butterworth), London, UK.

15.8 Elonka, Stephen M and Johnson OH, *Standard Industrial Hydraulics Questions/Answers*, McGraw-Hill Book Co., USA.

15.9 Bhargava, M, "Refining of Hydraulic Oils", Fluid Power, FPSI Bangalore, India, Sep., 1990, p. 20.

15.10 Sasaki Akira, "Die Entwickeln von electrostatischer Fluessigskeitsrienigung," *O & P Oel Hydraulik und Pneumatik*, October, 1989. Germany, p. 816.

15.11 Leiber, W. "Hydraulische Anlagen", p. 17, Carl Hauser Verlag, Munich, Germany, 1996.

Index

Abrasive wear 23, 55
Absolute viscosity 26
Acceleration 12, 254
Accumulator 334, 341, 356, 361, 383, 534
Accumulator selection 349
Acidity 47, 50
Actuation 147, 148
Additive 43, 432
Adjustable throttle 190
Aeration 42, 341, 531
Air cooled heat exchanger 325
Air cooling 326
Air gap solenoid 229, 230, 231
Alignment 272, 273, 274, 511
Alkanity 46
Amplifier 211, 217, 218, 219
Angular speed transducer 218
Angular transducer 216
Aniline point 43, 48, 400
Axial piston motor 291
ANSI 335

Back pressure 164, 165
Bacterial prevention 43
Backup ring 413, 414, 415, 430
Baffle 341
Balanced pump 104, 105, 106
Balancing groove 158
Bar 8, 9, 10
Bernoulli 17
Bending radius,
    steel tube 468, 469
Bent axis pump 114, 116, 121, 122
Bent axis piston motor 296
Beta rating 67, 68
Biocide 43
Bio-degradable oil 4, 50
Bladder 344, 345
Blotting paper test 83
Braided hose 458, 459
Breather 337, 339, 340
Buckling load 259

Bulk modulus 22
Buna 406, 460
Burst pressure 440, 461, 462
Burt test 434, 435, 480
Butyl 404, 405, 461
Bypass filter 74, 75, 76

Cam plate 117
Cam ring 104, 105, 106, 110
Care for HWCF 42
Cartridge assembly 109, 288
Catastrophic failure 61
Causes of seal failure 429
Cavitation 21, 23, 24, 112, 320, 339,
    477, 511, 527
Centipoise 32
Center conditions, spool
    valves 154, 155
Centrifugal pump 91, 92
Centrifuge 505, 506
Centi-stoke 31
CETOP 62
Change of oil 50
Checklist, pump
    selection 141
Cleanliness class 71
Check valve 59, 172, 173, 364, 365
Classification, pumps 89
Clearance 426
Clearance, critical 87, 132
Clearance range 97
Clevis 271, 272
Closed center 154, 155, 156
Closed loop 202, 305, 306
Coefficient of heat transfer 324
Coefficient of resistance 186, 187
Cold flow 409
Commercial HWCF 44, 45
Classification of seals 393
Common faults 490, 491
Compatibility 21, 42, 70, 132,
    463, 464

Index 543

Component fitting and
 failure 500, 501, 502
Compression fitting 446, 447, 521
Compression set 399, 406, 520, 537
Condition monitoring 516, 517
Contamination 46, 48, 54, 55, 56, 72, 73,
 498, 499, 500
Contamination level 62, 63, 64
Continuity of flow 18
Cooling 41, 326, 328, 334
Corrosion 42, 43
Counterbalance valve 180, 182
Coupling and connectors 443, 472, 473
Cracking 519
Cracking pressure 177
Crankshaft 115
Crimping 470, 471
Cup seal 412
Cushioning 275
Cylinders 57, 242, 243, 535, 536
Cylinder force 256
Cylinder material 247

DC valve 59, 69, 150, 151, 534
Dead band region 233
Deaeration 319
Degradation failure 61
Degree of protection 85
Degree Engler 30
Deionization 43
Demulsification 45
Demulsibility 23
Density 10, 12, 24, 71
Depth filter 74
Derived SI units 9
Design of hydraulic
 circuit 351
Detents 167, 168
Diaphragm 343, 344
Diffuser 339
Direction control valves 59, 69, 143, 144,
 147, 150, 151, 152
Dirt holding capacity 74, 80
Dither 241
Drop in pressure 455, 456, 457
Double acting 243
Dual vane 108
Dynamic viscosity 26, 30, 31, 32
Dynamic seal 395, 422, 423

Effect of additives 432
Effects of oxidation 503
Effects of contamination 55
Efficiency 104, 111, 131, 132, 263
Elastomers 252, 253, 402, 403
Elbow 444
Electrical actuation 148, 149
Electrical motors 299
Electro-hydraulic servo 149, 207
Electromagnet 149
Electronic particle
 counter 68, 86
Electrostatic liquid
 purifier 505, 506, 507, 508, 509
Emissivity 323
Emulsifier 43
Energy loss 451
Encoder 215
Energy saving 126, 127
Engler 30, 31, 32, 33
Entrapped air 320
Entrapped oil 95
Environment 330, 331
Equivalent length 452, 453
Estimation, filter size 70
Estimation of line diameter 455, 456
Estimation of seal failure 518
Estimation of heat rise 321
Euler's formula 267, 268
Evaporation 42, 45
Exchanger 326
Extrusion 398, 406, 413, 415, 430,
 519, 536
Extrusion press 379

Factor of safety 265, 267, 440
Failure of seal 426, 430
Failure rates 491
Faraday's laws of induction 213
Fault diagnosis 527, 528, 529, 530, 531,
 532, 533, 534, 535, 536, 537
Faulty fitting of seal 434
Feed back 203, 204, 206
Ferrule 446, 447, 468
Filler cap 339
Filter 54, 66, 68, 69, 78, 80, 537
Filter, full flow 74
 lite 504, 505
 material 70

mode   70
rising   65, 69, 85
size   70
surface   73, 74
Filtration   48, 54, 65, 80, 85, 86
Fire point   22
Fire resistant   22, 42, 44, 45, 80
Fire sleeves   463
Fit, cylinder assembly   249
Fixed displacement   104
Flare fitting   446, 447
Float center   157
Flow compensation   346
Flow control valve   143, 144
Flow equalizer   366
Flow force   165
Flow gain   232, 238
Fluid   10, 60
Fluid preparation   42
   sampling   82, 83
Foaming   22, 43, 531
Foot mounting   271
Free wheeling   299, 301
Friction   19, 165, 166, 260, 319, 427, 430, 484, 485
Frictional force   257, 260
Functional diagram   383

Gain, hydraulic   206
Gasket   424, 425
Gaseous contamination   81
Gears   93, 97, 98
Gear pump   91, 93, 94, 95, 96, 97, 98, 99, 100, 101, 132
Gear motor   290
Gyrator pump   101, 102
General tips   521, 522
Glass fiber   78
Glycol   41, 45
Guidelines for seal selection   433

Hagen-Poiseuilli's law   184
Handling, tubes and pipes   521
Hardness   398, 428, 430
Hat ring   395
Head of oil   13
Heat   299, 318, 322, 323, 528, 529
Heat balance   322

Heat dissipation   22, 323, 324, 325, 338, 341
Heat exchanger   312, 313, 319, 325, 326, 327, 328, 330
Helical gear   93
Herring bone gear pump   99
High water content fluid   40, 41, 42, 43
Hi-lo pump   367, 368, 371
Hoop stress   264, 265, 266
Hose   438, 457, 473, 477, 536
Hose assembly   469, 470, 472, 486
Hose maintenance   481, 482, 526
Hose material   459
Hose reinforcement   461
Hose selection   478
Hydraulic circuits   352
Hydraulic fluids   20
Hydraulic oils   47
Hydraulic system maintenance   513, 514
Hydro-copying   373, 376, 378, 379, 380, 381
Hydrostatic drive   301, 304
Hydro steering   311
Hydraulic valves   59, 60, 143, 144, 146, 147
Hydro-transmission   305, 306, 307, 308
Hydraulic lose   166
Hysteresis   240

Idler screw   103
Impulse test   480
In-built filters   81
Intake filter   81
Influence of oil contaminants   433
Inspection of oils   493, 494
Inspection format   539, 540
Installation method, cylinder   280
Installation procedure   424
Integral reservoir   341
Internal gear pump   100, 101, 129
Intra vane   108
IS   49
ISO   4, 36, 62, 64, 65, 448, 463, 479
ISO code   64
ISO cleanliness class   69, 70
ISO viscosity grade   35, 36

JIC   448, 473
Joseph Bramah   13
Joule   9

Kelvin 6
K-factor 452
Kinetic energy 16
Kinematic
   viscosity 31, 38

Laminar flow 19, 27, 28, 34, 42, 454
Lap 239
Leakage 163, 164, 520
Leather 401, 402
Level indicator 339
Life expectancy 134
Linear Hall effect 215
Linear motion
   potentiometer 213
Lip seal 423
Lobe pump 102
Location of filter 81, 82
Loss 256, 319, 451, 452, 454
Lubrication 427
Lubricity 21, 22
LVDT 213, 214, 215

Magnetic filter 79, 85
Magnetic trapper 337, 339, 340
Maintenance 42, 99, 278, 340, 423, 481, 482, 489, 520, 521
Maintenance and performance monitoring 514, 515
Maintenance level 72
Maintaining pressure 346
Maintenance, repair and reconditioning 489, 490
Measurement of viscosity 30
Mechanical lever 15
Mechanism of oxidation 502
Media, filter 78
Meter, by pass 197
Meter-in 196, 359, 360, 361
Meter-out 196, 359, 360, 361
Microemulsion 40, 41, 43
Microprocessor control 382, 383
Micron 47, 48, 68
MIL STD 62
Misalignment of coupling 527
Modular valve 170, 171
Motor 286, 287, 288
Mounting style 269, 271, 272, 273, 274

Movement diagram 386
Multipass test rig 66, 67

NAS 47, 64, 65
Newton 4, 7, 8
Neoprene 403, 405, 406, 461
Neutralization number 46, 48
Nitrogen 345
Noise 328, 522, 523, 528
Noise level 125, 135, 136, 137
Noise in the pump 523, 524
Noise in hydraulic
   piping 525
Nominal pressure 451
Non-positive displacement 90, 91
Non-return valve 172, 173, 174
NPT 448, 472
Null bias 239
Null region 232
Null shift 239
Nylon 459

Oil compatibility 132
Oil heater 339
Oil life and temperature 496, 497, 498
Oil property 494
Oil tank 324
Oil-to-air cooler 327, 328
Oil-in-water 409
On-off valve 211
OPEC 40
Open center 154, 155
Open loop 202, 305, 306
O-ring 395, 396, 397, 413, 419, 421, 422, 424
Overall efficiency 132
Overlap 159, 160, 161, 162
Oxidation 23, 38, 41, 43, 46, 49, 57, 84

Pascal 2, 10
Pascal's law 13
Paint compatibility 42
Painting 339
Particle count 69, 86
Permanent set 400
Petroleum oil 39, 40, 41
pH 46
Phosphate ester 41, 44, 45, 46, 47, 50, 404

Pipes   438, 439
Pipe threads   448, 472, 473
Pipe wall thickness   441
Pilot controlled check valve   157
Pilot controlled PRV   177, 178, 179
Pipe anchoring   484
Pipe fittings   442
Pipe O.D. and tolerance   441
Piston motor   288, 289, 290
Piston pump   93, 113, 116, 117, 118, 119, 120, 132
Piston rod   266
Pivot   272
Poise   32
Poppet   145
Position-step diagram   181, 386
Positioning flow control valve   201
Positive displacement   90, 91, 92
Positive sealing   393
Pour point   22
Power   128, 137
Predicting wear   255
Pressure   8, 10, 12, 131
Pressure balancing   96, 97, 123
Pressure compensated vane pump   111
Pressure compensated flow control valve   192, 193, 319, 366
Pressure control valve   172, 175, 176, 177
Pressure drop   70, 71, 73, 258, 259
Pressure gain   239
Pressure reducing valve   182
Pressure transducer   216
Pressure relief   176
Pressurised reservoir   321, 335
Proof pressure   440
Problems in contamination   503, 504
Proportional flow control   197, 198
Proportional pressure control   198
Proportional valve   211
PTFE   79, 252, 400, 403, 405, 408, 409, 460
Pulsation   100, 103, 104, 114, 128, 135
PQ curves   178
Pulse speed   217
Pump   60, 318
Pump capacity   133
Pump displacement formula   129
Pump maintenance   509, 510

Pump-motor combination   302
PWM   232
Protective cover   462

Quad-ring seal   395
Quick coupling   475, 476

Radial piston motor   291, 296, 297
Radial piston pump   114, 124, 125
Reassembling tube fittings   521
Reciprocating pump   92, 93
Redwood seconds   30
Refractometer   46
Regenerative circuits   361, 362, 363
Reliability   252, 253
Reliability test   479
Relief valve   60
Reservoir   82, 324, 325, 334, 492, 493
Return line filter   81
Reynolds' number   33, 34, 453, 454, 455
Ripple   137, 138, 139, 140
Rotary pump   93
Rotary spool   145
Rotor   105, 110, 111
Rubber   402, 403, 460
Running in period   57
Rust formation   23, 47

SAE number   35, 47, 48, 62, 458, 459
Safety   344, 537, 538, 539
Sampling   48
Sayboldt universal seconds   30
Screw pump   102, 103, 104, 132
Seal   251, 392, 395, 423, 424
Seal failures   519, 520, 536, 537
Seal form   410
Seal material   397, 400, 401, 402, 403, 404, 405, 406 407, 408, 409.
Selection of pump   367, 368
Sensor   215
Sequence valve   179, 180
Servo valve   47, 48, 60, 70, 200, 211, 224, 225, 226, 235, 236
Shear   25, 26
Shock absorbing capacity   273
Shock suppression   346, 347
Shore hardness   252, 428, 430
Shrinkage   399, 422, 520
Silicone   408

Index   547

Silt index   69, 84
Single stage torque
  converter   313
SI system   5, 6
Sizing cylinder tube   263
Sizing hydraulic pump   89, 127
Sludge   48, 49, 57
Solenoid   148, 228, 229, 230, 231
Soluble oil   41
Sound emission   523
Specific gravity   23, 24, 84
Speed control   183
Speed variation in cylinder   371
Speed transducer   216
Spiral failure   431
Spiral twisting   519
Spiral wire   462
Spool valve   144, 145
Squeeze   398, 400, 421, 422, 431
Static seal   423, 424
Stoke   31
Strainer   74
Surface finish   427
Surface protection   448, 449
Surge pressure   431, 479
Swagging   471, 472
Swash plate pump   116, 117, 291, 293, 295
Swelling   399, 520
Synthetic hydraulic hose   457, 458
Synthetic oil   41

Tandem center   155, 156, 373
Tee joint   444
Teflon   460, 473, 484
Temperature   6, 49, 70, 430
Tensile load   259, 265
Testing accumulator   349, 350
Thermal expansion   339
Thermal warning device   330
Throttle form and viscosity   188, 189, 190, 191
Throttle valve   191
Throttling aperture   184, 185
Tie rod cylinder   245
Toleration limit of
  particle size   61, 62
Torque   286, 287, 293, 298
Torque converter   308, 312, 313, 314, 315

Torque motor   208, 210, 211, 221, 222, 223, 225, 226, 235
Transducer   211, 213, 214, 215, 216, 217, 218, 383
Travel-time diagram   386
T-ring   395
Trunnion   274
T-seal   416
Tubes   438, 439
Tube thickness   264
Turbulent flow   19, 34, 42, 454
Two-stage servo valve   226

U-cupring   395
Unbalanced pump   104
Underlap   159, 160, 161
Unloading valve   183
Useful formulae   129

Valves   512
Valve, position of   153
Valve size and capacity   168, 169
Vanes   104, 108, 110, 289
Vane pump   91, 104, 105, 106, 107
Vane motor   288
Variable axial piston pump   120
Variable displacement   104, 113, 119
Velocity control   358, 359
Vibration   335, 431
Vibration and signature
  analysis   517, 518
Viscosity   21, 24, 25, 26, 27, 28, 29, 30, 31, 32, 33, 35, 36, 37, 70, 83, 486
Viscosity index   21, 37, 38, 477
Viscometer   30
Visual check   495, 496
Viton   404, 405, 410
Volumetric efficiency   132
V-packing   396, 412, 413

Water   40
Water based fluid   40
Water cooled heat
  exchanger   325, 328
Watt   9
Water-in-oil   45, 410
Water glycol   45, 46, 410
Wear debris analysis   518
Wear plate   96

Wear rate   254
Weight-to-power ratio   134
Wet armature solenoid   231
Wheeled loader   308
Wiper   395, 416, 417, 418
Wire braids   458, 462
Wobble plate pump   116, 118, 119

X-ray fluorescence analysis   519

Yoke   119, 120, 123, 295

Zinc dialkyldithio-phosphate   48